Standard Grade Biology

Standard Grade
BIOLOGY

Team Co-ordinator
James Torrance

Writing Team
James Torrance
James Fullarton
Clare Marsh
James Simms
Caroline Stevenson

Diagrams by James Torrance

Hodder & Stoughton

A MEMBER OF THE HODDER HEADLINE GROUP

British Library Cataloguing in Publication Data.

Standard grade biology.
1 Organisms. – For schools
I. Torrance, James
574

ISBN 0 340 49369 0

First published in Great Britain 1989

Impression number	15	14	13	12	11	10	9
Year	1998	1997	1996	1995	1994		

Typeset in 9/10 Helvetica by Tradespools Ltd., Frome, Somerset, UK
Printed in Great Britain for Hodder & Stoughton Educational, a division of Hodder Headline Plc, 338 Euston Road, London NW1 3BH by Martin's the Printers Ltd.

Contents

Preface

This book is intended to act as a valuable resource to pupils and teachers by providing a concise set of notes which adhere closely to the syllabus for SCE Standard Grade Biology to be examined in and after 1990.

Each section of the book matches a syllabus topic. The topic 'Investigating Cells' has been moved forward to become Section 2 in the book, since the authors feel that knowledge of basic cell structure is a prerequisite to understanding the structure and function of those specialised cell types that appear in later sections. The sequence of sections is not intended, however, to be a prescriptive teaching order and many cross references have been included in the text to allow for alternative routes through the syllabus.

Each chapter corresponds to a syllabus sub-topic. The text is presented in a two-tier format with the first part suited to the needs of both General and Credit pupils and the 'More to do' sections aimed at those candidates aspiring to Credit level. The text is interspersed with Key Questions designed to continuously assess Knowledge and Understanding (Course Objective Type A). For this reason the questions are directly related to the stated learning outcomes with Extra Questions being aimed at Credit level.

Each section is followed by Problem Solving Exercises (Course Objective Type B) which reflect the stated grade related criteria and give pupils practice in handling and processing information, evaluating procedures and information, and drawing conclusions and making predictions. Again this material is differentiated with some problems pitched purely at General level, some carrying, in addition, Extra Questions at Credit level and some aimed solely at Credit level. Further problem solving exercises are available in the associated book *Problem Solving in Biology*.

The ten appendices have been included to act as reference sections and to aid the development of skills such as drawing line graphs and constructing hypotheses.

Section I The Biosphere

1 Investigating an ecosystem

An **ecosystem** is a natural biological unit which is made up of living and non living parts. Figures 1.1, 1.2 and 1.3 show some examples of ecosystems. An ecosystem contains one or more habitats. A **habitat** is the place where a living organism (e.g. plant or animal) lives. A burrow in garden soil, for example, is an earthworm's habitat. The other main part of an ecosystem is the **community** which is made up of all of the plants and animals living there.

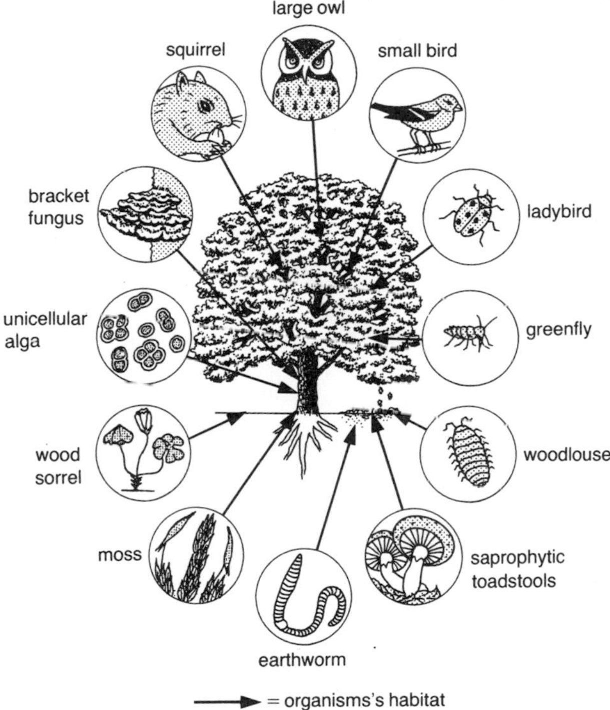

Figure 1.1 Oak tree ecosystem

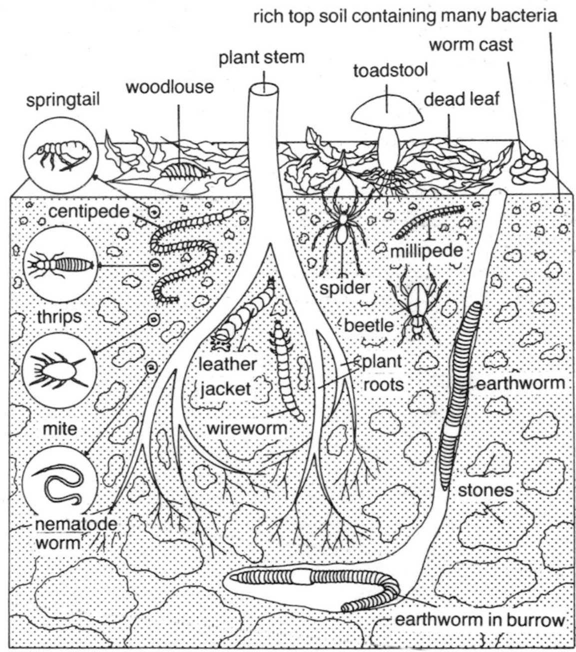

Figure 1.2 Soil ecosystem

The plants and animals form populations. A **population** is a group of living organisms of the one type (e.g. a forest of silver birch trees, a herd of deer etc.).

Thus an ecosystem is a natural biological unit which is made up of a community of living things, their own living surroundings and the factors that affect the lives of all the members of the community.

Biotic and abiotic factors

Factors related to living things such as amount of available food, number of predators, incidence of disease and competition for the necessities of life are called **biotic** factors.

Non living factors such as temperature, rainfall, light intensity and pH are called **abiotic** factors (also see page 9).

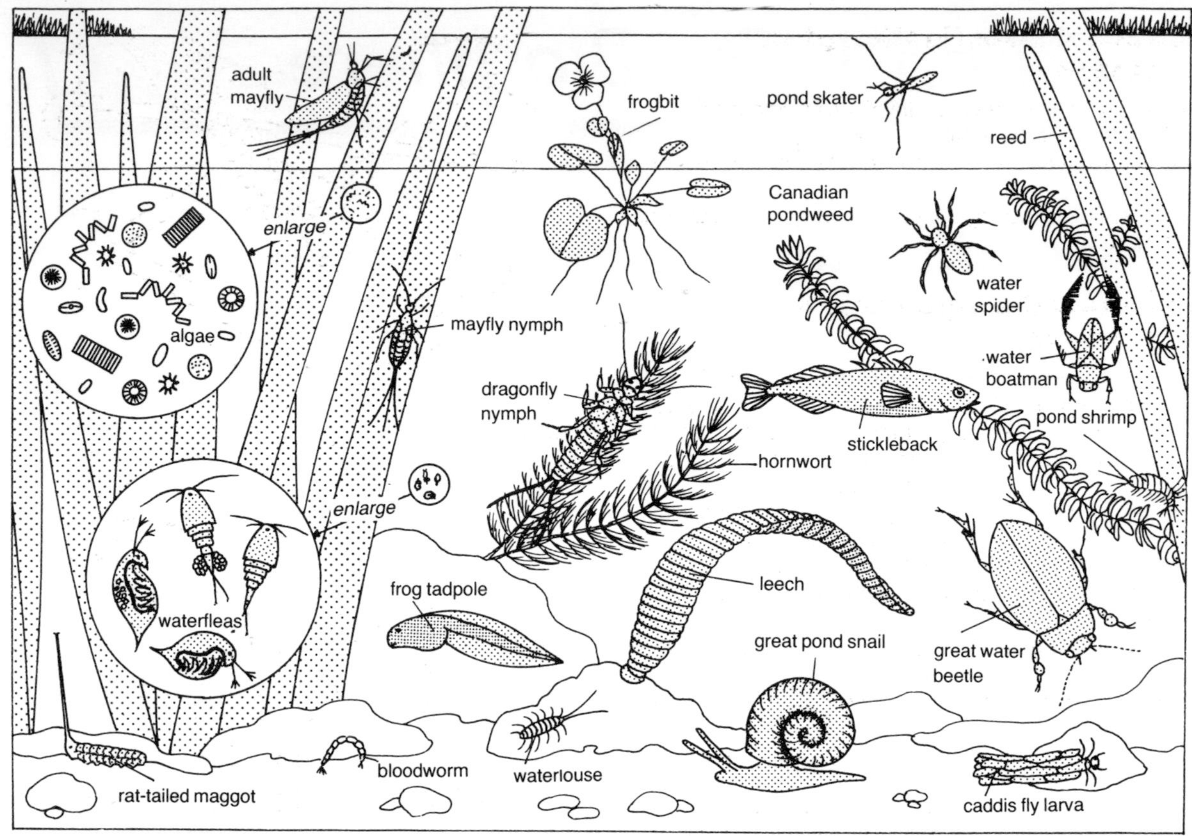

Figure 1.3 Fresh water ecosystem

ecosystem	sampling technique	possible sources of error	ways in which errors may be minimised
tree	tree beating using **beating tray** and **stick** (figure 1.4) A walking stick is used to give the branch of the tree a few sharp taps. Small animals drop onto the tray held beneath the branch.	(a) The numbers and types of animals may not be representative of the ecosystem as a whole. (This source of error is true of all sampling techniques.) (b) Some organisms may miss the tray or drop off the edges or fly away. (c) Some animals may not be dislodged by beating.	(a) Take several samples by beating several different branches on the tree and pool the results. (b) Use a large tray with raised edges. Quickly insert the tray into a large plastic bag. Once back in the laboratory use a pooter (figure 1.5) to trap animals before they escape. (c) Examine branch after beating and use blunt instrument to collect remaining animals in a jar with a screw-top lid.

Table 1.1 Sampling by tree beating

Practical investigation of an ecosystem

The study of an ecosystem involves finding out:
which plants and animals live there;
how abundant (rare, common, etc.) these organisms are;
and then investigating the reasons why the organisms live there.

Sampling the organisms in an ecosystem

It is rarely possible to count all of the plants and animals in an ecosystem because this would take too long and would probably cause permanent damage to the ecosystem. Instead, small samples which represent the whole ecosystem are taken. To make the procedure fair each sampling unit must be equal in size and be chosen at random (so that the results are not biased). From the samples an estimate of an organism's abundance for the whole ecosystem can be calculated.

Many different techniques (see tables 1.1 to 1.3) are used to sample organisms, each suited to the particular ecosystem being studied.

Quadrats

A further method of sampling uses quadrats. A **quadrat** is a rectangular-shaped sampling unit of known area. Its frame is normally made of wood, metal or string. This form of sampling is often used to estimate the numbers of plants in an ecosystem (or slow-moving animals such as shellfish on a rocky shore).

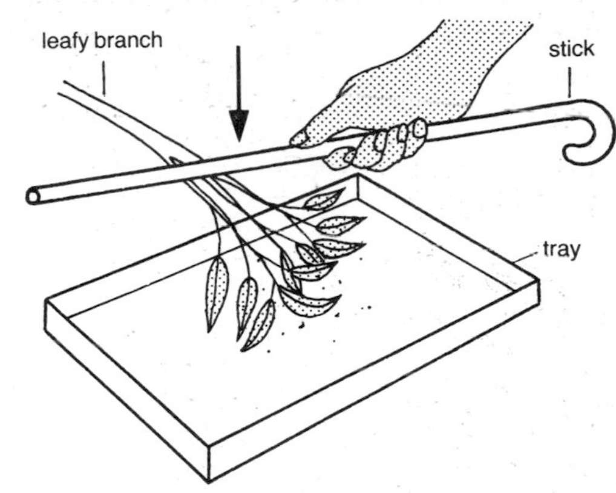

Figure 1.4 Tree beating

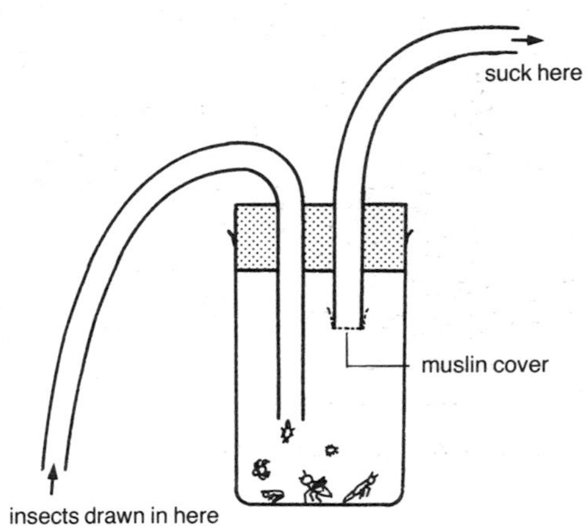

Figure 1.5 Pooter

ecosystem	sampling technique	possible sources of error	ways in which errors may be minimised
soil	trapping using **pitfall trap** (figure 1.6) Animals that are active on the soil surface and amongst leaf litter fall into the trap and are unable to climb out again.	(a) As before (b) Birds may eat trapped animals. (c) Some animals may eat others.	(a) Set up several traps. (b) Disguise the opening with a lid (e.g. leaf or stone) supported on sticks (figure 1.7). (c) Check traps regularly or put preservative liquid (e.g. 50% ethanol) in the bottom of the traps.
	trapping using **Tullgren funnel** (figure 1.8) Tiny animals that live in the air spaces in the soil move down and away from the hot, dry, bright conditions created by the light bulb and fall through the sieve.	(a) As before. (b) Soil sample may be too thick and may still contain many organisms at the end of the experiment. (c) Sieve mesh may be too fine and some animals fail to fall through.	(a) Set up several funnels. (b) Make layer of soil on sieve thin or keep trap set up for a longer period. (c) Use mesh with larger size of holes.

Table 1.2 Sampling by trapping

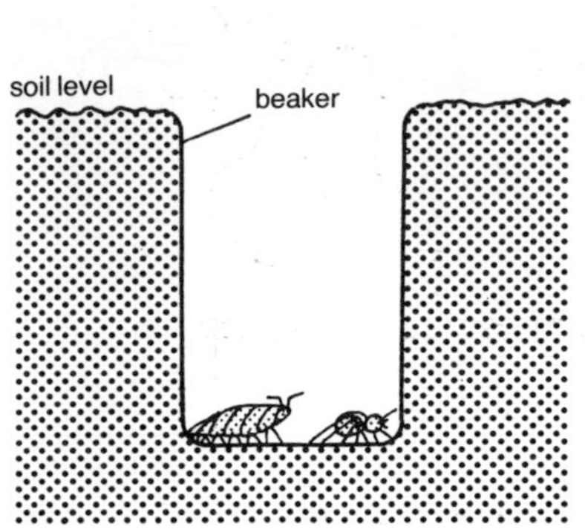

Figure 1.6 Pitfall trap

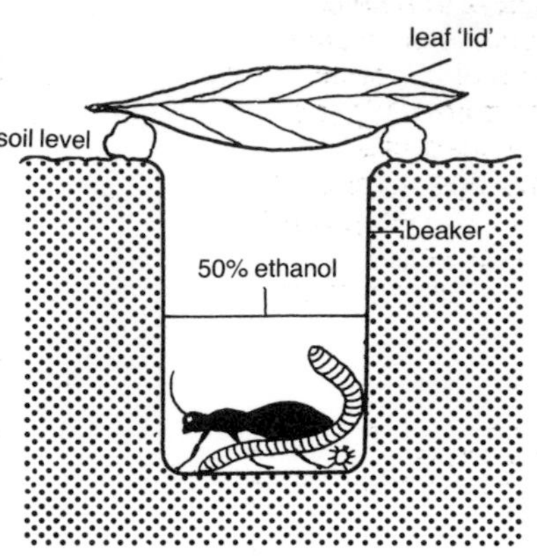

Figure 1.7 Improved pitfall trap

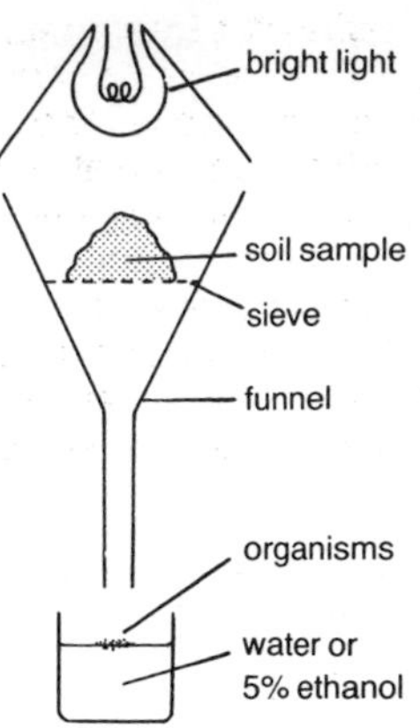

Figure 1.8 Tullgren funnel

ecosystem	sampling technique	possible sources of error	ways in which errors may be minimised
fresh water pond	netting using **water net** (figure 1.9) The net is moved rapidly through the water, catching animals which are quickly transferred to screw-top jars containing pond water.	(a) As before. (b) Small animals may escape through the holes in the mesh. (c) When investigating the bottom of the pond some animals in the net may have been caught from the water near the surface as the net passed through.	(a) Repeat the procedure many times. (b) Choose net with finer mesh. (c) Rotate handle of the net to close the opening on the way down and back up (figure 1.10).

Table 1.3 Sampling by netting

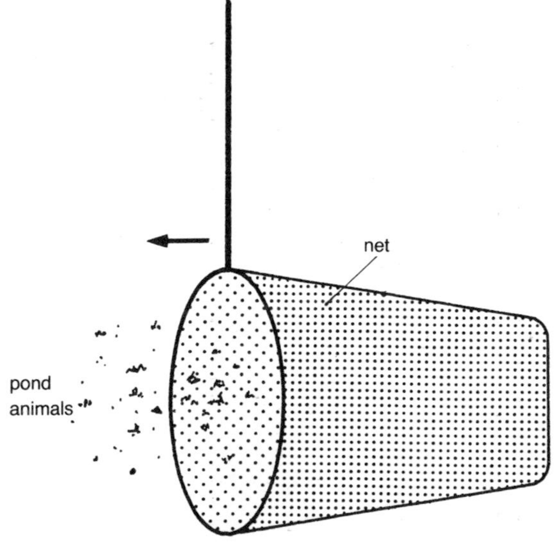

Figure 1.9 Net

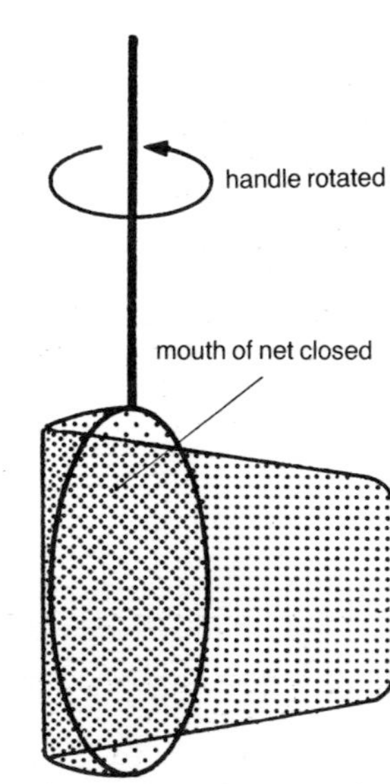

Figure 1.10 Careful use of net

Estimating the total number of thistles in a field

Figure 1.11 shows a quadrat which encloses an area of one square metre. The sites for several quadrats (say ten) are chosen at random (see figure 1.12) and the number of thistles present in each quadrat is counted. From this information the average number of thistles per square metre is calculated. The length and breadth of the field are measured and its area calculated. An estimate of the total number of thistles in the field is then worked out.

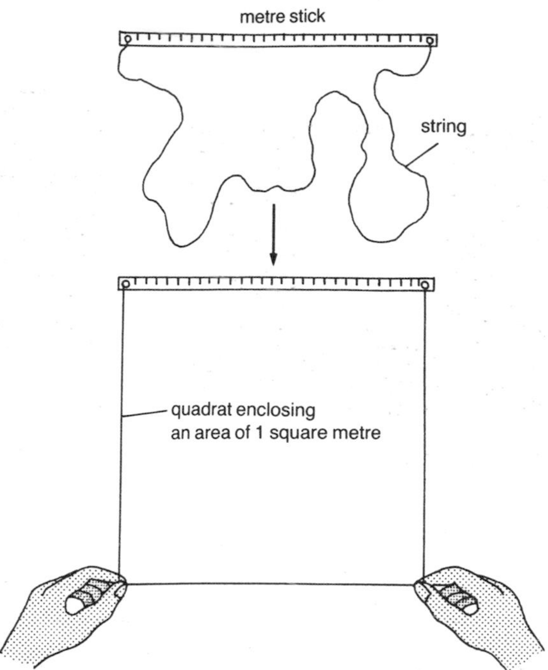

Figure 1.11 Quadrat

quadrat	number of thistles
1	11
2	0
3	5
4	15
5	1
6	7
7	3
8	7
9	9
10	4

Table 1.4 Quadrat results

Example calculation

Table 1.4 shows a typical set of results.
Total number of thistles in ten quadrats = 62
Average number of thistles per quadrat
(i.e per m^2) = 62/10 = 6.2
Total area of field (length × breadth)
= 50 m × 20 m = 1000 m^2
Estimate of total number of thistles in field
= 6.2 × 1000 = 6200 thistles

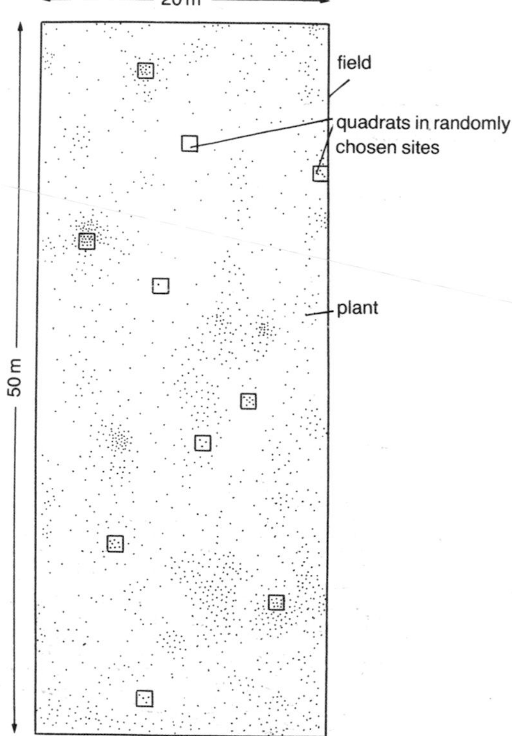

Figure 1.12 Sampling a field

Possible sources of error

a) Ten quadrats may be too small a number of samples to give a fair representation of the number of plants present in the ecosystem. This is especially likely if the plant being investigated tends to grow in clusters. Most or all of the ten quadrats might just happen to land in positions where none of the plants under investigation grow (see figure 1.13).

b) Some of the type of plant being considered might be located partly inside and partly outside a quadrat. Should they be included in the count or not?

Ways in which errors may be minimised

a) A much larger number of quadrats could be studied by the class working as, say, ten groups with each group doing ten quadrats and then the whole class pooling their results.

b) A basic rule could be established and followed by everyone involved (see figure 1.14). Any plant partly in and partly out falling on the bottom or left hand side of the quadrat counts as in, whereas any plant on the top or right hand side counts as out.

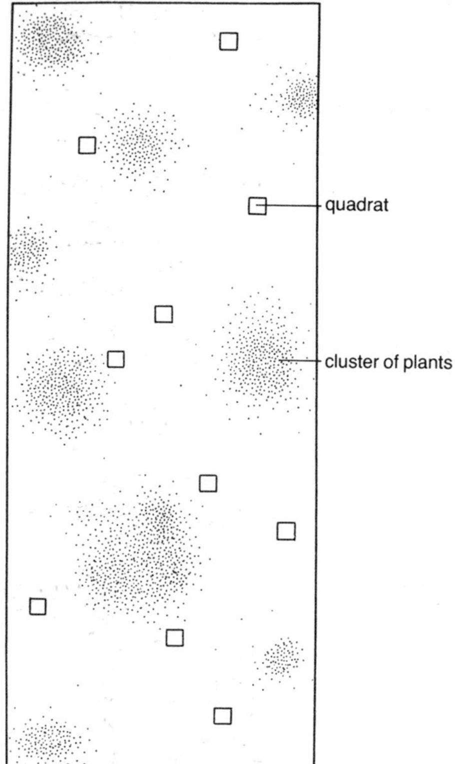

Figure 1.13 Too few quadrats

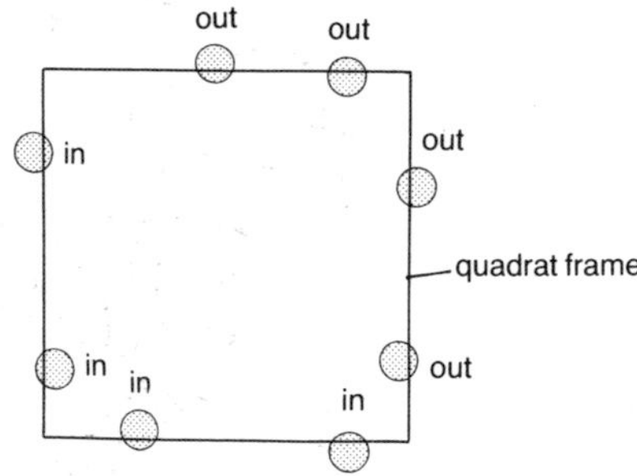

Figure 1.14 In or out?

Identifying the organisms collected

Each organism can be identified by comparing it to pictures in a suitable book. However this method can take a very long time. A much better method is to use a **key**. A key may be branched or in the form of paired statements. Figures 1.15, 1.16 and 1.17 each give a key to the animals commonly found in the ecosystems illustrated at the start of this chapter. (After identification organisms should be returned to their natural habitats so that as little disturbance as possible is caused to the ecosystem.)

KEY QUESTIONS

1 Identify the main parts of an ecosystem.

2 a) Name a technique that could be successfully employed to sample the organisms living in each of the following ecosystems; (i) tree, (ii) soil, (iii) fresh water pond.
b) For each technique state the apparatus that you would use and describe how you would use it.
c) Identify ONE possible source of error for each sampling technique.

3 Why is using a key a better method of identifying an organism than comparing it to pictures in a book?

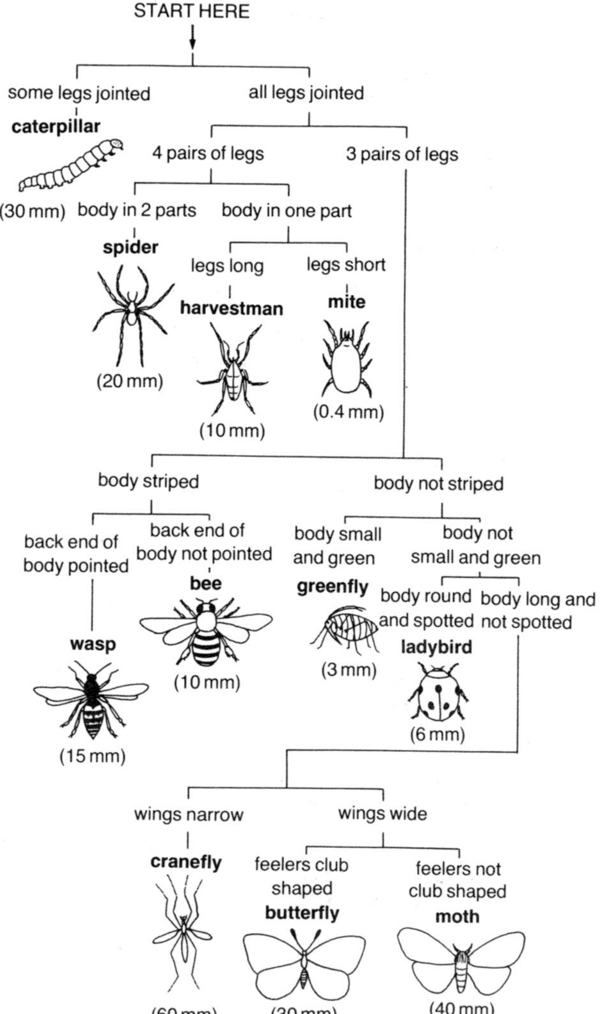

Figure 1.15 Key to tree animals

1	body has no legs	go to 2
	body has legs	go to 6
2	body not divided into sections (segments)	go to 3
	body divided into segments	go to 5
3	body worm-like	**nematode worm** (a)
	body not worm-like	go to 4
4	shell present	**snail** (b)
	no shell present	**slug** (c)
5	no more than 13 segments present	**fly maggot** (d)
	more than 13 segments present	**earthworm** (e)
6	6 jointed legs present	go to 7
	more than 6 jointed legs present	go to 11
7	grub-like insect	go to 8
	adult insect	go to 9
8	non jointed legs present on abdomen	**caterpillar** (f)
	non jointed legs absent from abdomen	**beetle larva** (g)
9	thin waist between thorax and abdomen	**ant** (h)
	no thin waist between thorax and abdomen	go to 10
10	spring attached to abdomen	**springtail** (i)
	no spring attached to abdomen	**beetle** (j)
11	8 legs present	go to 12
	more than 8 legs present	go to 13
12	body divided into 2 parts	**spider** (k)
	body not divided into 2 parts	**mite** (l)
13	14 legs present	**woodlouse** (m)
	more than 14 legs present	go to 14
14	each body segment has 1 pair of legs	**centipede** (n)
	each body segment has 2 pairs of legs	**millipede** (o)

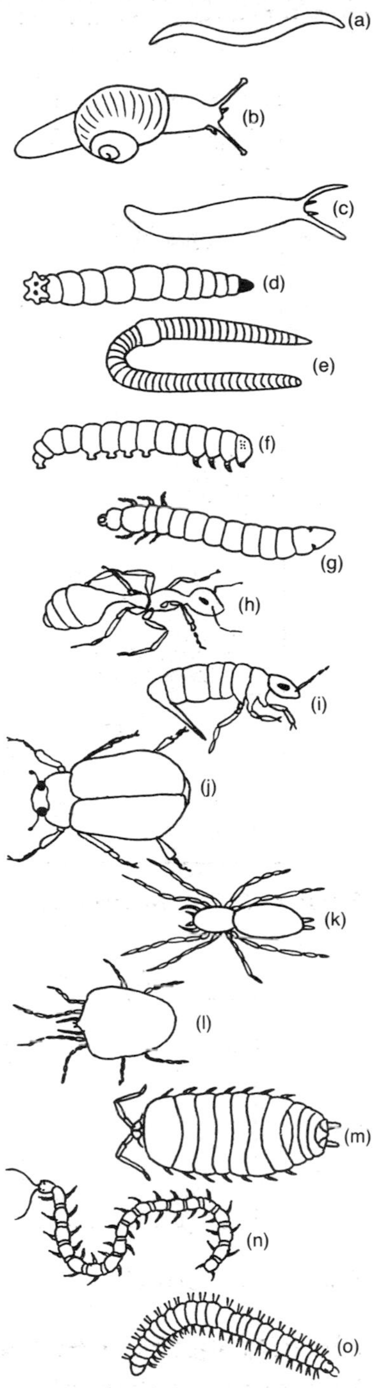

(a) (b) (c) (d) (e) (f) (g) (h) (i) (j) (k) (l) (m) (n) (o)

(not drawn to scale)

Figure 1.16 Key to soil animals

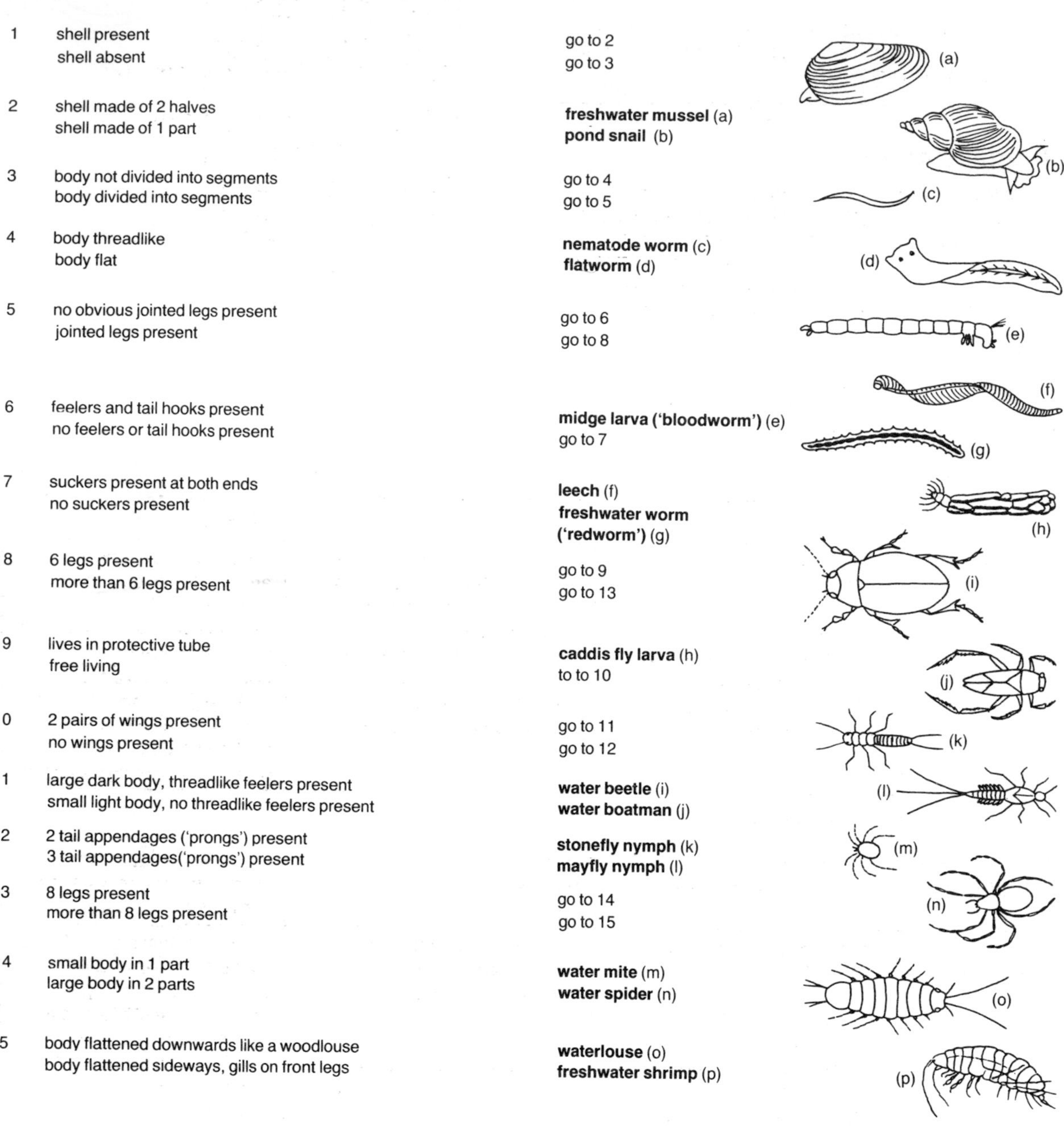

1	shell present	go to 2
	shell absent	go to 3
2	shell made of 2 halves	**freshwater mussel** (a)
	shell made of 1 part	**pond snail** (b)
3	body not divided into segments	go to 4
	body divided into segments	go to 5
4	body threadlike	**nematode worm** (c)
	body flat	**flatworm** (d)
5	no obvious jointed legs present	go to 6
	jointed legs present	go to 8
6	feelers and tail hooks present	**midge larva ('bloodworm')** (e)
	no feelers or tail hooks present	go to 7
7	suckers present at both ends	**leech** (f)
	no suckers present	**freshwater worm ('redworm')** (g)
8	6 legs present	go to 9
	more than 6 legs present	go to 13
9	lives in protective tube	**caddis fly larva** (h)
	free living	to to 10
10	2 pairs of wings present	go to 11
	no wings present	go to 12
11	large dark body, threadlike feelers present	**water beetle** (i)
	small light body, no threadlike feelers present	**water boatman** (j)
12	2 tail appendages ('prongs') present	**stonefly nymph** (k)
	3 tail appendages('prongs') present	**mayfly nymph** (l)
13	8 legs present	go to 14
	more than 8 legs present	go to 15
14	small body in 1 part	**water mite** (m)
	large body in 2 parts	**water spider** (n)
15	body flattened downwards like a woodlouse	**waterlouse** (o)
	body flattened sideways, gills on front legs	**freshwater shrimp** (p)

Figure 1.17 Key to pond animals

Extra Question

4 For each of the sources of error that you gave as your answer to question 2(c) explain how the error might be minimised.

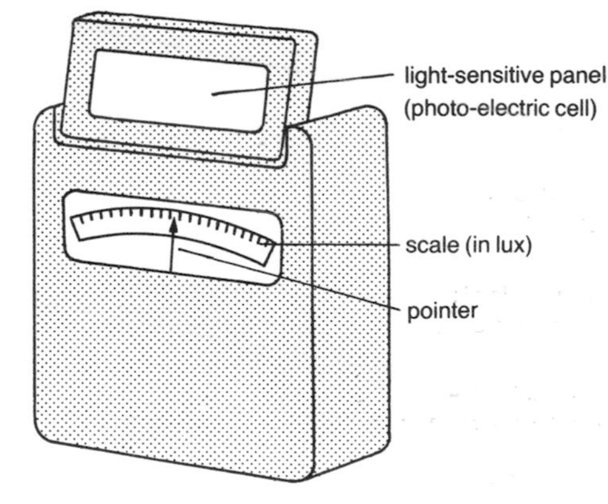

Figure 1.18 Light meter

Abiotic factors

Abiotic factors are non living factors present in an ecosystem such as light intensity, temperature, air humidity (dampness), pH (see Appendix 7), wind speed, oxygen content, etc. A living organism is only able to survive in a certain habitat and play its part in an ecosystem if a combination of abiotic factors suited to its needs is present there.

Measuring abiotic factors

The techniques used to measure four different abiotic factors are given in tables 1.5 to 1.8.

abiotic factor	technique for measurement	possible sources of error	ways in which errors may be minimised
light intensity	using a **light meter** (figure 1.18) The light meter is held so that the light sensitive panel is held at the same angle to the light as the surface whose light intensity is being measured.	(a) The observer may stand in the way of the light and cast a shadow on the light meter. (b) Light intensity may change from one moment to the next (because of cloud cover) making comparison between different locations invalid.	(a) Ensure that all observers are standing to one side of light meter. (b) Take all measurements as near the same time as possible during a period of similar light intensity or, if that is impossible, take several readings at each site and calculate an average for each.

Table 1.5 Measuring light intensity

abiotic factor	technique for measurement	possible sources of error	ways in which errors may be minimised
air humidity (dampness)	using an **evaporimeter** (figure 1.19) Water is introduced along the long arm of the capillary tubing using a rubber teat. The disc of blotting paper is stuck onto the flooded surface to give an air-tight seal. Evaporation from the disc is measured by following the progress of the meniscus along the scale. The more rapid its movement the less humid the surrounding air.	(a) Air may be trapped under the blotting paper thus reducing flow of water. (b) Heat from hand may affect rate of movement of meniscus.	(a) Ensure that rubber disc is flooded with water when blotting paper is being pressed against it to give an air-tight seal. (b) Take care not to touch glass tubing during experiment.

Table 1.6 Measuring air humidity

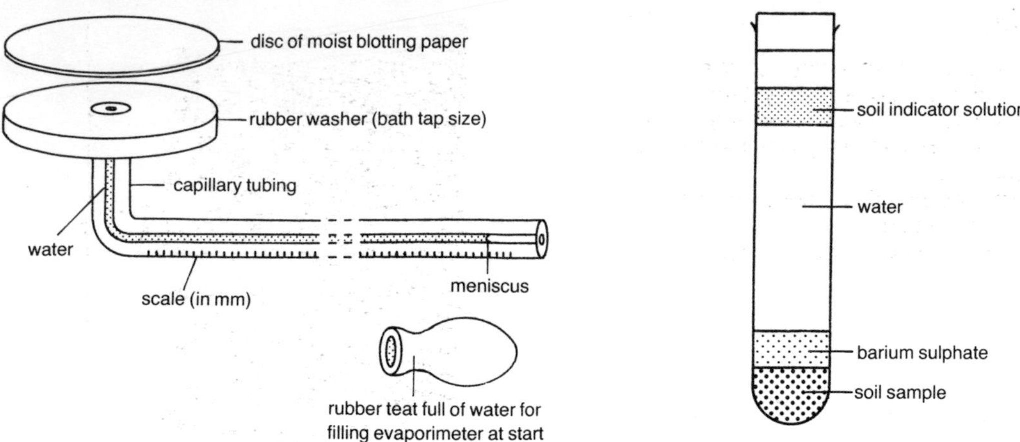

Figure 1.19 Evaporimeter

Figure 1.20 Measuring pH of soil

abiotic factor	technique for measurement	possible source of error	way in which error may be minimised
water content of soil	drying soil sample in **oven** 100 g of fresh soil is dried in an oven at 95°C for two days. Its new mass is then found and its loss in mass calculated. This is then converted to a percentage as follows: % water content = $\frac{\text{loss in mass}}{\text{original mass}} \times \frac{100}{1}$	It is possible that after two days a little of the original soil water is still present in the 'dry' soil.	Repeat the drying and weighing process until the soil sample loses no further mass and has therefore reached constant mass.

Table 1.7 Measuring water content of soil

abiotic factor	technique for measurement	possible source of error	way in which error may be minimised
pH of soil	using soil **indicator solution** (figure 1.20) Indicator solution, barium sulphate and water are added to a soil sample and shaken up. (Barium sulphate makes clay particles stick together and settle out, leaving the coloured solution clear.) The resulting colour is compared with a pH chart and the soil's pH read off.	The water being used may affect the final pH reading.	Ensure that water being used is neutral (pH 7)

Table 1.8 Measuring pH of soil

KEY QUESTIONS

1 Name TWO abiotic factors that play a part in determining the character of an ecosystem.
2 a) Name the technique that you would employ to measure each of the factors that you named in your answer to question 1.
b) Describe how you would go about measuring each of these named factors.
c) For each factor identify a possible source of error.

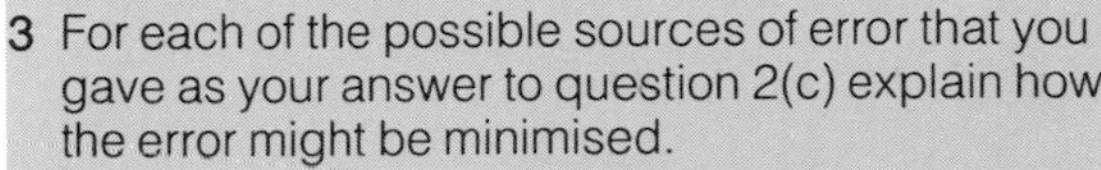

3 For each of the possible sources of error that you gave as your answer to question 2(c) explain how the error might be minimised.

Effect of light intensity on distribution of *Pleurococcus*

Pleurococcus is a simple plant (with a green powdery appearance) which grows on the trunks of trees (see figure 1.21). To measure its distribution on a tree, a string is first tied round the tree at, say, one metre from ground level and then tabs indicating the eight points of the compass are hung from it as shown in figure 1.22.

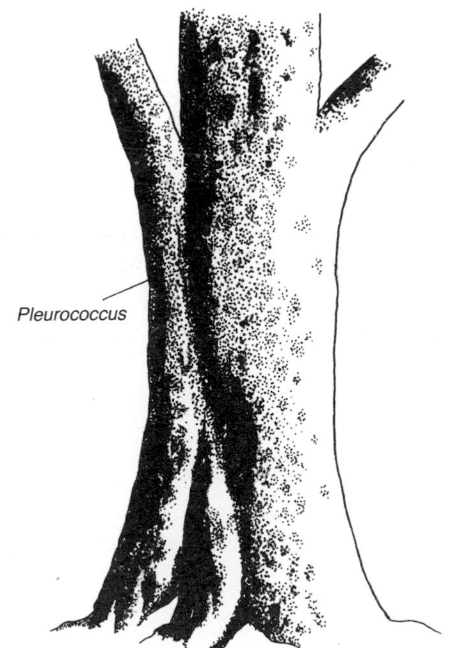

Figure 1.21 Pleurococcus

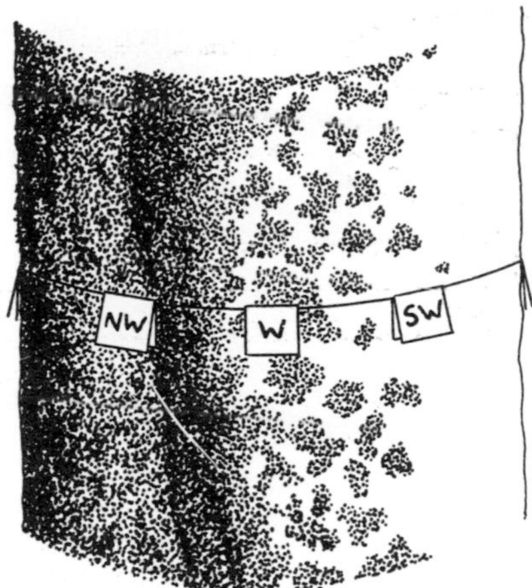

Figure 1.22 Compass points

One way to estimate the percentage of bark covered with *Pleurococcus* at each compass point is to hold a small quadrat against the bark (see figure 1.23) and then decide what percentage of *Pleurococcus* is present inside it. This is called a subjective method. This means that the results depend on the opinion of the investigator. However, such an opinion may be inaccurate and/or variable from quadrat to quadrat and person to person.

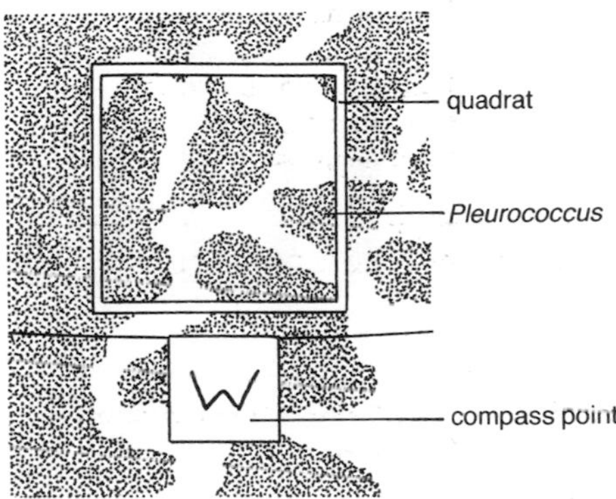

Figure 1.23 Subjective method

It is better where possible to use an objective method. This means obtaining results that are independent of personal opinion. One way of doing this is to use the quadrat shown in figure 1.24 which is designed so that the wires cross over at a hundred points. These are the sample points. The percentage cover is estimated by counting how many of these points have *Pleurococcus* growing directly beneath them. Table 1.9 gives a typical set of results. Figure 1.25 shows this data displayed as a bar graph.

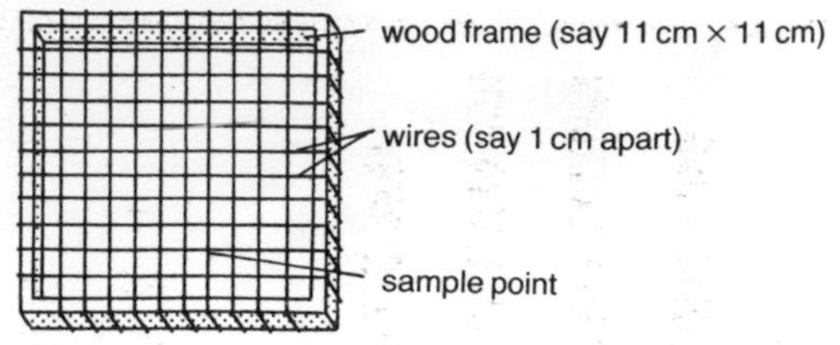

Figure 1.24 Quadrat for objective method

compass point	% *Pleurococcus*
W	56
NW	91
N	100
NE	100
E	87
SE	32
S	10
SW	24

Table 1.9 % Pleurococcus *cover*

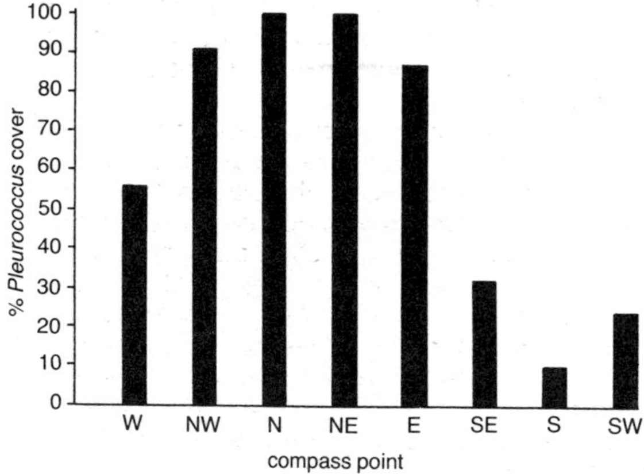

Figure 1.25 Graph of % Pleurococcus *cover*

Light intensity
The light intensity falling on the tree trunk at each of the compass points is also measured, using a light meter. The results shown in table 1.10 were obtained on a bright autumn day. Figure 1.26 shows this data displayed as a bar graph. Since the sun rises in the east, shines from a southerly direction during the day and sets in the west, the northern sides of the tree receive least light.

When the two graphs are compared, it can be seen that most *Pleurococcus* is present on those surfaces that receive light of lowest intensity and that least *Pleurococcus* is present where the light is most intense. It is possible therefore that light intensity in some way affects the distribution of *Pleurococcus*.

compass point	light intensity (in lux)
W	1000
NW	750
N	250
NE	250
E	250
SE	500
S	2000
SW	1500

Table 1.10 Light intensities

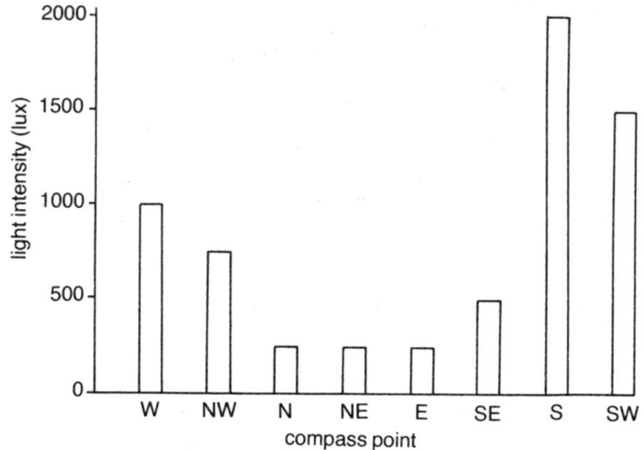

Figure 1.26 Graph of light intensities

More to do

Since green plants need light for photosynthesis, it seems strange that *Pleurococcus* grows best in conditions of low light intensity. How can this be explained? It is possible that intense sunlight acting on a surface of a tree causes much of the surface water to evaporate, creating dry conditions. Perhaps *Pleurococcus* only grows well in moist conditions.

Effect of humidity (dampness) on distribution of *Pleurococcus*

An **evaporimeter** (see figure 1.27) is used at each of the eight compass points round the tree to measure water loss during a period of, say, thirty minutes. The evaporimeter's design is adapted so that its evaporating surface lies parallel to the vertical surface of the tree trunk.

The distance travelled by the water meniscus along the evaporimeter is an indication of air humidity. The further it travels, the drier the air surrounding the blotting paper (and nearby *Pleurococcus*) and vice versa. Table 1.11 shows a typical set of results. These are presented as a bar graph in figure 1.28.

When the information in figure 1.28 is combined with the original bar graph of percentage *Pleurococcus* cover shown in figure 1.25, the two are seen to be related (see figure 1.29). *Pleurococcus* is more abundant in damper conditions (e.g. north) and less abundant where the air is drier (e.g. south).

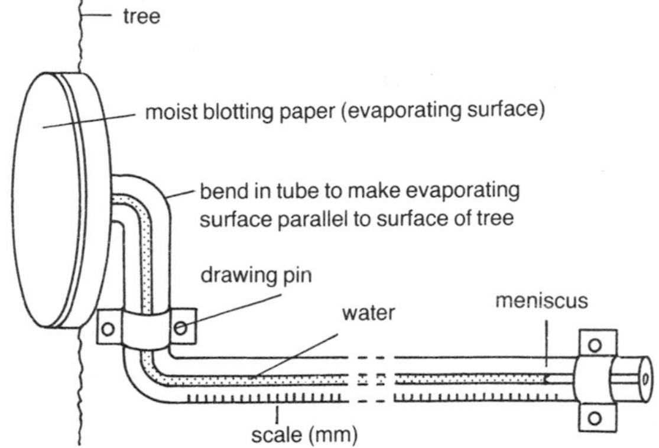

Figure 1.27 Using an evaporimeter

compass point	distance travelled by meniscus along scale (in mm)
W	55
NW	35
N	20
NE	20
E	25
SE	50
S	75
SW	65

Table 1.11 Evaporimeter results

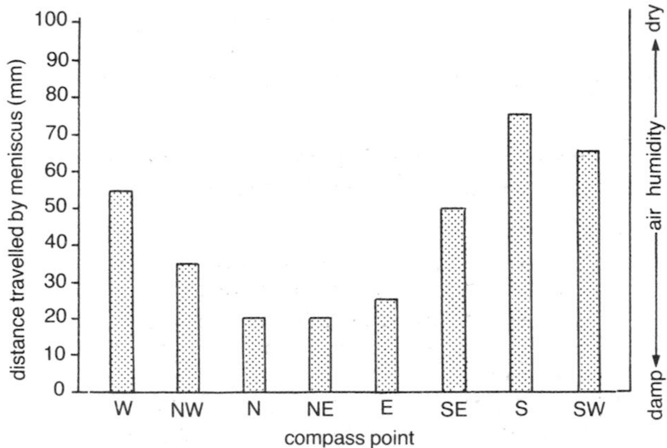

Figure 1.28 Graph of evaporimeter results

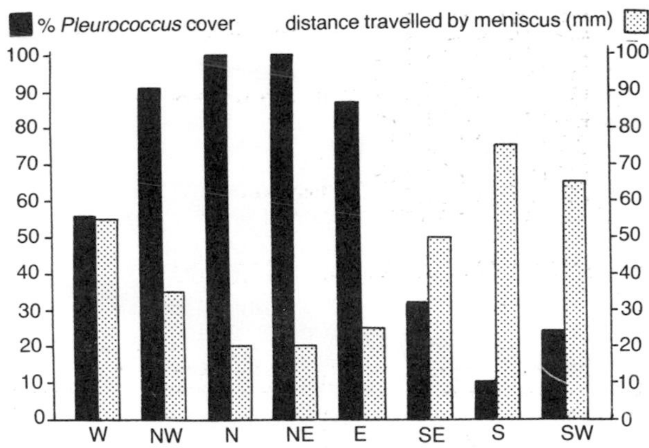

Figure 1.29 Graph of combined results

Mechanisms determining distribution of *Pleurococcus*

It is possible that *Pleurococcus*'s distribution is determined mainly by the moisture content present in its habitat. Perhaps it becomes desiccated (dried out) and dies in dry places. However it is a fact that *Pleurococcus* fails to grow on moist tree trunks that are very heavily shaded, showing that its distribution would seem to be affected to a certain extent by light intensity. Being a green plant *Pleurococcus* clearly needs a certain minimum amount of light for photosynthesis.

It must be kept in mind that there are many other factors that may also influence the distribution of *Pleurococcus*. For example, this plant is dispersed through the ecosystem by wind which may not deposit an equal number of cells on all surfaces of a tree.

From this investigation it is only possible, therefore, to suggest mechanisms by which abiotic factors influence the distribution of *Pleurococcus*. Solving the problem of exactly which factors are responsible for an organism's distribution, and the reasons why, is normally a complex one involving a long series of investigations.

Main features of an ecosystem

Look again at the ecosystems shown in figures 1.1, 1.2 and 1.3. An ecosystem consists of a living community of plants, animals and micro-organisms which interact with one another and their non living environment to form a balanced biological unit. The parts of an ecosystem are said therefore to be **interrelated**.

More to do

Interrelationships

Figure 1.30 represents an ecosystem (e.g. a woodland) as a unit consisting of living and non living parts through which there is a continuous flow of **energy** and **raw materials** such as mineral salts, oxygen, carbon dioxide and water.

All living things need raw materials and energy to survive. Green plants obtain raw materials such as carbon dioxide from the air and mineral salts and water from the soil. Green plants get their energy from sunlight. Animals obtain oxygen from the air and get other raw materials and all of their energy by eating plants or other animals.

Within an ecosystem a **balance** exists. The carbon dioxide given out by living organisms is taken in during photosynthesis by green plants. These in turn give out oxygen which is used by living organisms during respiration.

Plants depend on decomposers (such as micro-organisms in the soil) to break down waste materials and dead organisms. This process releases essential mineral salts back into the ecosystem to be used again by the plants.

In addition to these interrelationships, many others may also exist in an ecosystem. Plants may be **competing** with one another for light, space, water and minerals. Animals may depend on plants for **shelter** from extremes of abiotic factors (such as temperature and light intensity). Plants may also **camouflage** animals from their enemies. Plants often depend in turn on animals for **pollination** and **seed dispersal**.

Thus an ecosystem is a delicately balanced unit of living things interacting and interrelating with one another and the non living environment.

KEY QUESTIONS

1 Describe how the percentage distribution of *Pleurococcus* on the sides of a tree can be estimated.

2 a) Name ONE abiotic factor that may affect the distribution of *Pleurococcus*.
b) What apparent effect does this factor have on the distribution of *Pleurococcus?*

3 Rewrite the following sentence choosing only the correct word(s) from the brackets in each case.

The living and non living parts of

{a community / an ecosystem} are {abiotic / interrelated}.

Extra Questions

4 a) Name a second factor that may affect the distribution of *Pleurococcus*.
b) What apparent effect does this factor have on the distribution of *Pleurococcus?*
c) Give an explanation of the possible mechanism by which the named factors might influence the distribution of *Pleurococcus*.

5 Name TWO living parts of the ecosystem shown in figure 1.30 and explain how they are interrelated.

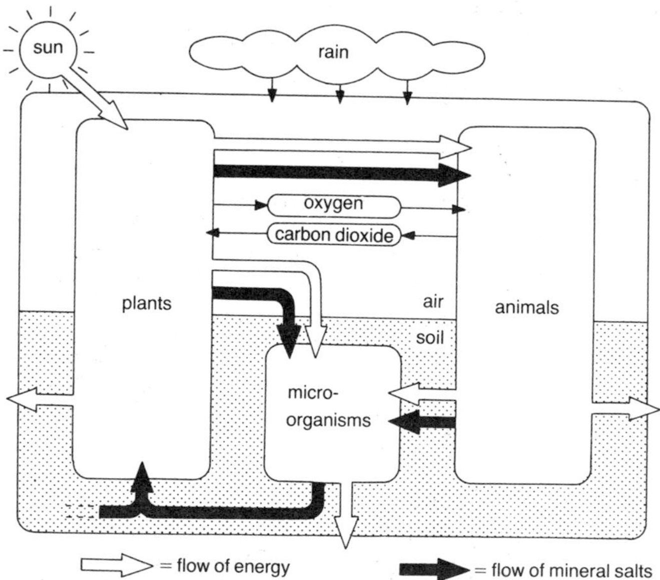

Figure 1.30 Interrelationships in an ecosystem

2 How it works

Producers and consumers

All the energy needed by the living things in an ecosystem comes from the **sun**. Green plants are called **producers** because they are able to produce their own food by converting the sun's light energy into chemical energy (contained in food). This process is called photosynthesis (see page 87).

Animals (and non green plants such as fungi) cannot produce their own food from sunlight. They are called **consumers** because they must consume plants or other animals in order to obtain the energy they need to stay alive and grow.

Amongst animals there are different types of consumers. A **herbivore** (e.g. sheep) eats only plant material. A **carnivore** (e.g. dog) eats only animal material. An **omnivore** (e.g. man) eats a mixture of plant and animal material.

Predators and prey

An animal which hunts another animal for its food is called a **predator**. The hunted animal is called the **prey**.

Energy flow

When a plant is eaten by an animal, energy is transferred from the plant to the animal (the primary consumer). When the primary consumer is eaten by a second animal (the secondary consumer) energy is again transferred and so on through a series of organisms as shown in figure 2.1.

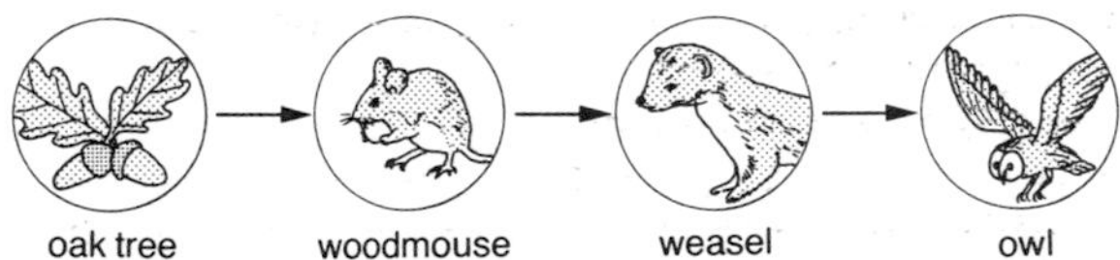

Figure 2.1 Food chain

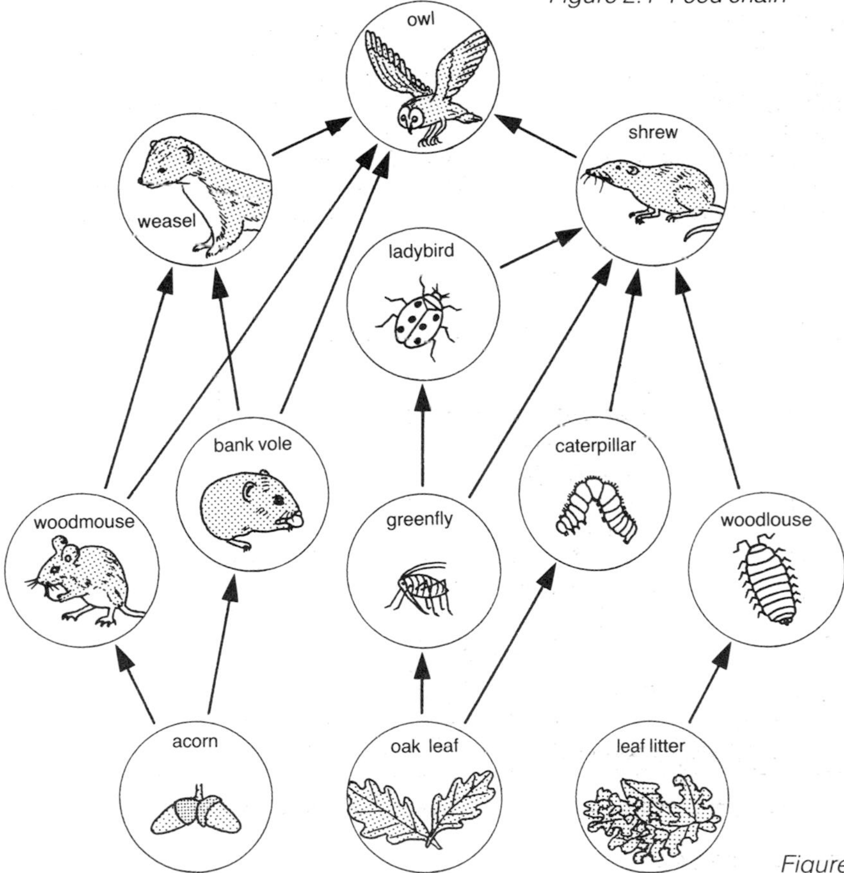

Figure 2.2 Oak tree food web

Food chain
A relationship where one organism feeds on the previous one and in turn provides food for the next one in the series is called a **food chain**. A food chain always begins with a green plant. Each arrow indicates the direction of energy flow.

Food web
A food chain rarely occurs in isolation in nature because the producer is normally eaten by several animals which are in turn preyed upon by several different predators. Various parts of an oak tree, for example, are eaten by woodmice, bank voles, greenfly, woodlice and many other animals. Greenfly are eaten by ladybirds. Greenfly and ladybirds are eaten by shrews. Shrews and woodmice are eaten by owls, and so on.

Under natural conditions an ecosystem really contains many inter-connecting food chains. This more complex relationship is called a **food web**. Two examples are shown in figures 2.2 and 2.3 where each arrow again indicates direction of energy flow.

KEY QUESTIONS

1 What is the source of the energy entering a food chain or web?
2 Describe the meaning of the terms **producer** and **consumer**.
3 Present each of the following groups of organisms as a food chain:
   a) lion, grass, zebra
   b) weasel, fieldmouse, owl, wheat
   c) greenfly, oak leaf, thrush, ladybird
   d) herring, animal plankton, man, plant plankton
   e) frog, hawk, grass, snake, grasshopper
4 Construct a food chain in which you are the final consumer.
5 Convert the following three food chains into one food web:

   oak (acorn) → wood-pigeon → fox

   oak (acorn) → woodmouse → fox

   oak (acorn) → woodmouse → weasel
6 Copy and complete the following marine food web using at least four additional arrows.
7 What does each arrow in a food chain or web mean?

Figure 2.3 Pond food web (organisms not drawn to scale)

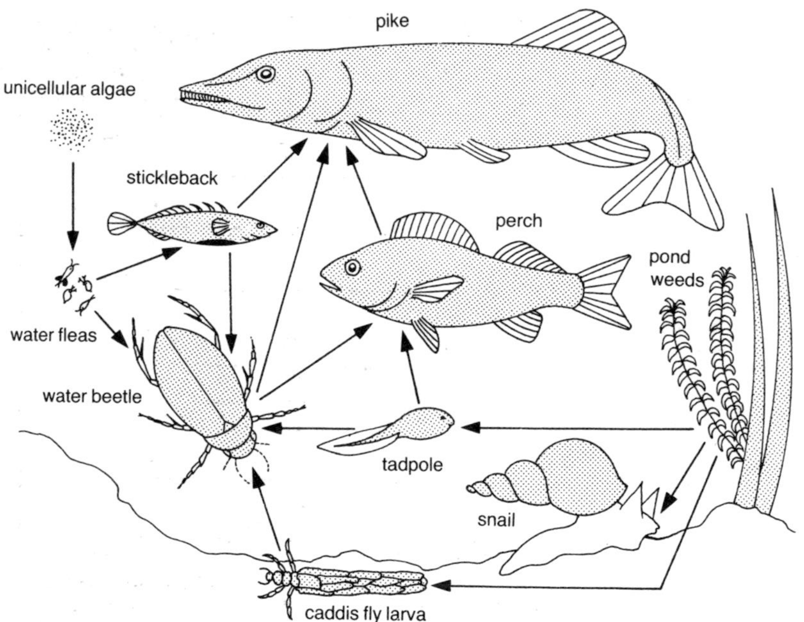

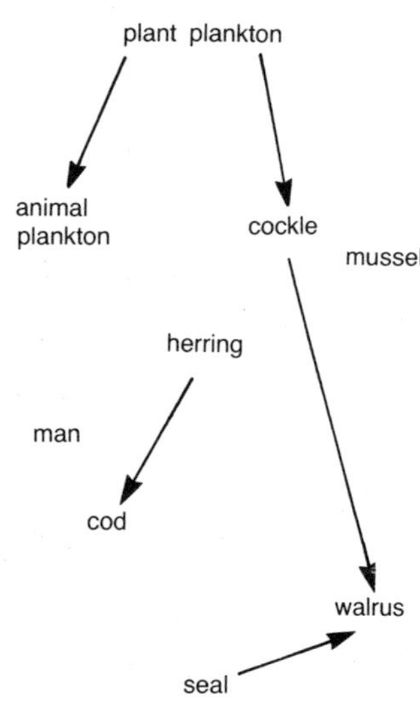

Figure 2.4 See question 6

Energy loss

As energy flows through a food chain or web, a progressive loss occurs for two reasons. Firstly, an organism uses up energy to build its body. However, this may include parts such as cellulose cell walls, or bone or skin or horns which when eaten by the next consumer may turn out to have little nutritional value.

These parts tend to be left uneaten or to be expelled, undigested, as faeces. As a result, energy is lost from the food chain though some of it may be gained by the ecosystem's decomposers (see page 181).

Secondly, most of the energy gained by a consumer in its food is used for moving about and, in warm-blooded animals, keeping warm.

A lot of energy is therefore lost as heat and only about 10 per cent of the energy taken in by an organism is incorporated into its body tissues. Figure 2.5 shows the fate of energy as it is transferred along a marine food chain. Figure 2.6 shows an alternative method of presenting this energy transfer and loss information.

KEY
= energy built into organism's tissues
= energy lost
eaten by krill
eaten by whale
100 tons **plankton** (producer)
10 tons **krill** (herbivore)
1 ton **blue whale** (carnivore)

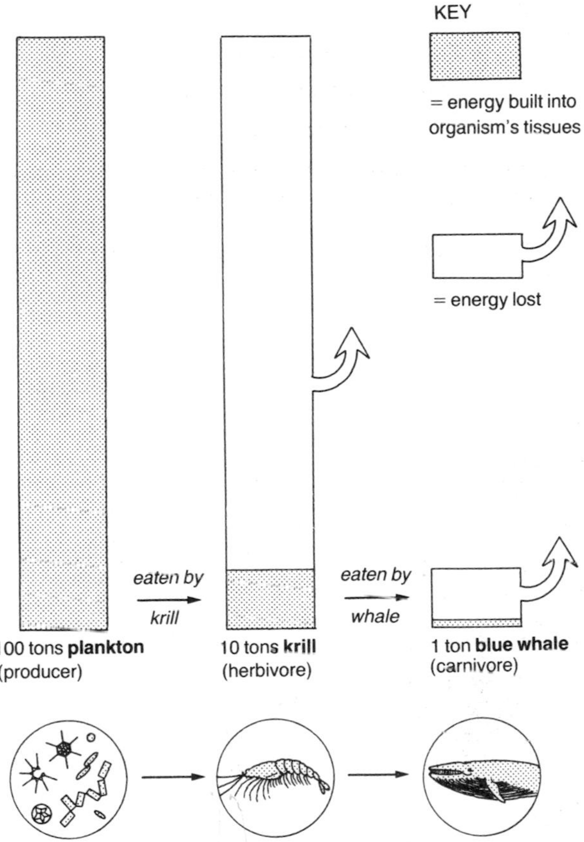

Figure 2.5 Energy loss in a food chain

Length of a food chain

More efficient use is made of food plants by humans consuming them directly rather than first converting them into animal products, since this cuts out at least one of the energy-losing stages in the food chain.

sunlight
small amount of sun's energy captured by plant plankton
energy lost by plant cells during respiration
energy built into new plant plankton cells
energy in plant plankton eaten by krill
energy lost by krill cells during respiration, e.g. as heat and movement
energy lost in krill faeces (e.g. in undigested plant material)
energy built into new krill tissue
energy in krill tissue eaten by blue whale
energy lost by whale cells during respiration, e.g. as heat and movement
energy lost in whale faeces (e.g. in undigested parts of krill)
energy built into new whale tissue
energy used by decomposers

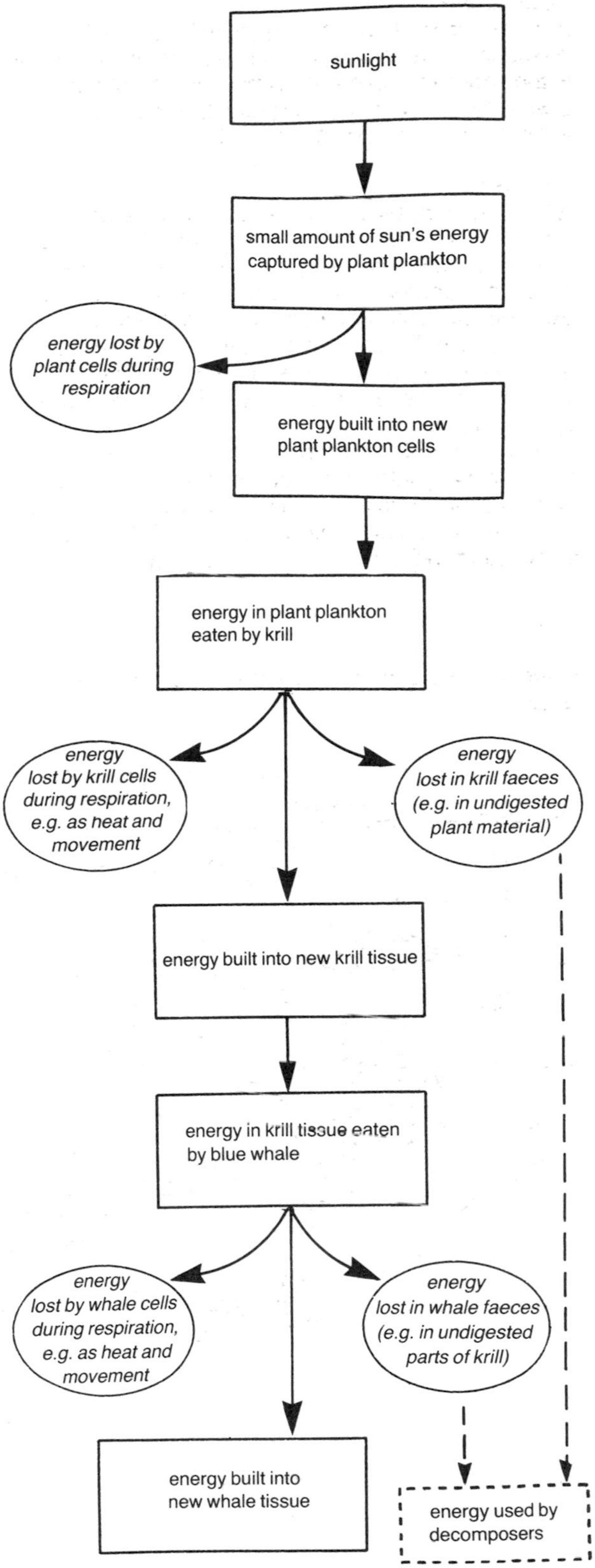

Figure 2.6 Energy transfer and loss

More to do

Pyramid of numbers

Consider the following food chain:

alga → waterflea → stickleback → pike
(simple
green plant)

In terms of numbers, the producers are found to be the most numerous, followed by the primary consumer, and so on along the chain, with the final consumer being the least numerous. This numerical relationship is called a **pyramid of numbers** and is often illustrated in the form shown in figure 2.7.

This relationship takes the form of a pyramid because (a) the energy loss at each link in the food chain limits the amount of living matter that can be supported at the next level and (b) the final consumer tends to be larger in body size than the one below it and so on.

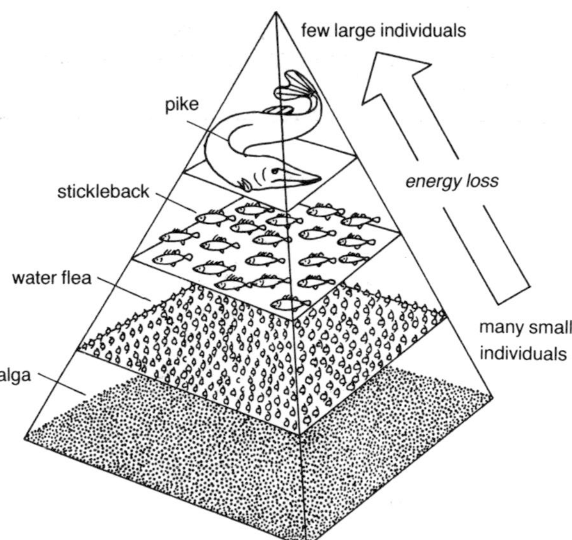

Figure 2.7 Pyramid of numbers

Pyramid of biomass

The **biomass** of a population is its total mass of living matter. In a food chain the biomass of the producer is greater than that of the primary consumer which in turn is greater than that of the secondary consumer and so on along the chain. Since biomass decreases at each level, it can also be represented as a pyramid (see figure 2.8).

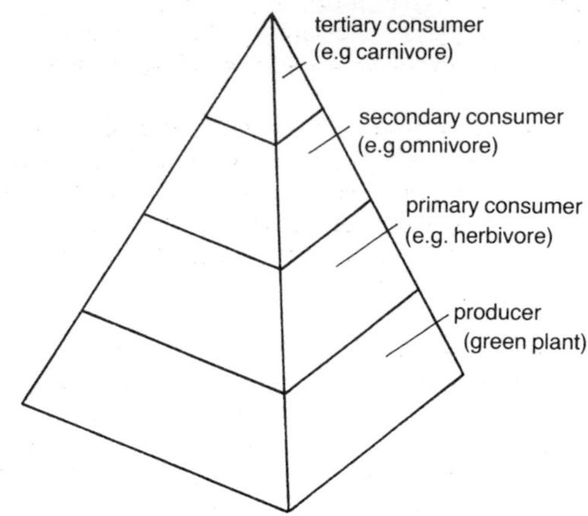

Figure 2.8 Pyramid of biomass

Disturbing a food chain or web

In a balanced food chain (or web) a certain amount of green plant material grows continuously and supports a fairly constant and large number of primary consumers which are in turn consumed by a fairly

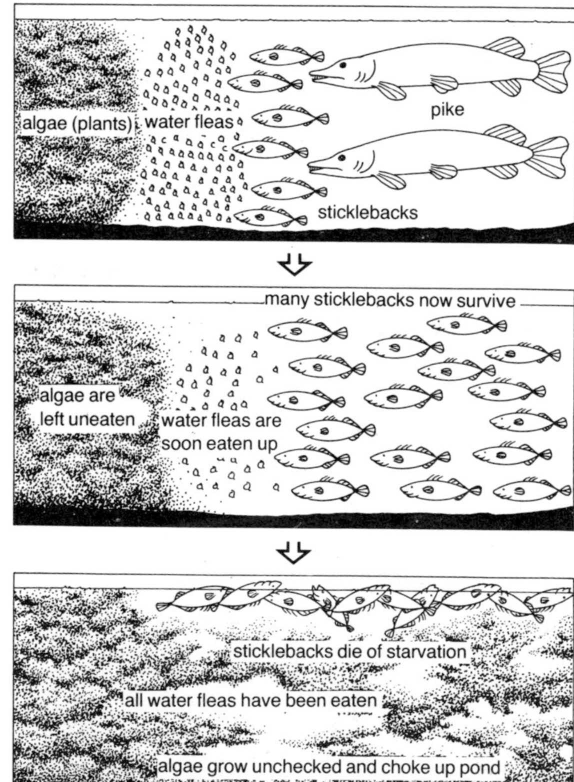

Figure 2.9 Disturbing a food chain

constant but smaller number of secondary consumers, and so on. However, this balance is disturbed if one of the species is removed from the ecosystem.

Consider the sequence of events shown in figure 2.9. Removal of pike by fishermen initially leads to a population explosion of sticklebacks. However these soon eat all the waterfleas and die of starvation, leaving the algae to grow unchecked and choke up the pond.

Number of links

If a food web has only a few links then the effect of removing one species on the remaining organisms can be severe. Figure 2.10 shows a food web where rabbits are the main source of food for foxes and birds of prey. However, in 1954/5 a disease called myxomatosis wiped out almost the entire population of rabbits. As a result, more tree seedlings and grass grew but many more lambs than normal were attacked by foxes and birds of prey.

If a food web has many links, then removal of one species may not have such a drastic effect. For example, removal of the limpets from the food web shown in figure 2.11 would leave more large seaweeds and small algae for other consumers but would not seriously alter the ecosystem.

KEY QUESTIONS

1 a) State TWO ways in which energy may be lost from a food chain or web.
b) Explain how the energy is lost in each case.

Extra Questions

2 a) Copy the pyramid of numbers shown in figure 2.12 and complete it using the following organisms: waterflea, pike, alga, stickleback.
b) Which of these organisms is the secondary consumer?
c) Which population of organisms in this pyramid contains the most energy?
d) Compared to the other organisms, what rule applies to the individual body size of the organisms occupying the top position in a food pyramid?
e) Why do the numbers decrease towards the top of a food pyramid?

3 Explain what is meant by the term **pyramid of biomass**.

4 Predict the effect on the plants of removing the thrushes from the food web in figure 2.13. Explain your answer.

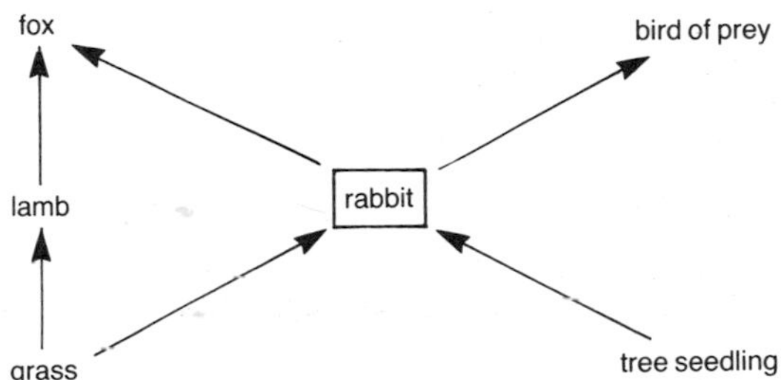

Figure 2.10 Disturbing a food web with few links

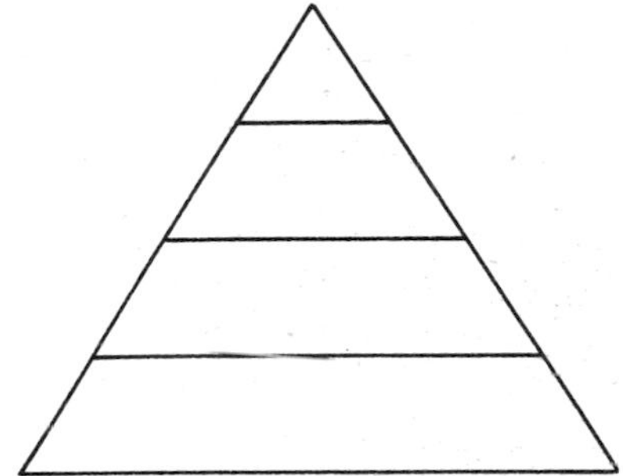

Figure 2.12 See question 2

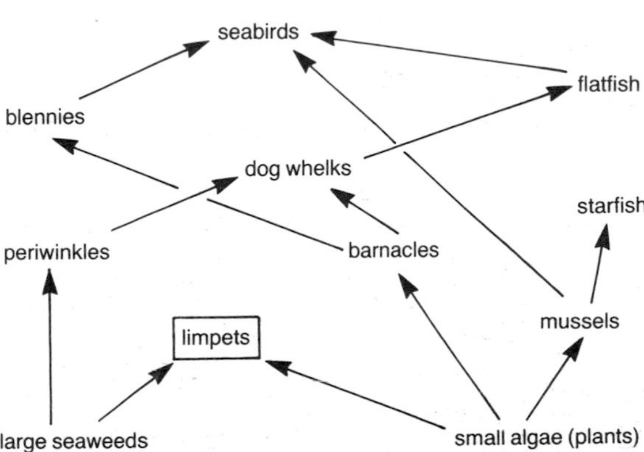

Figure 2.11 Disturbing a food web with many links

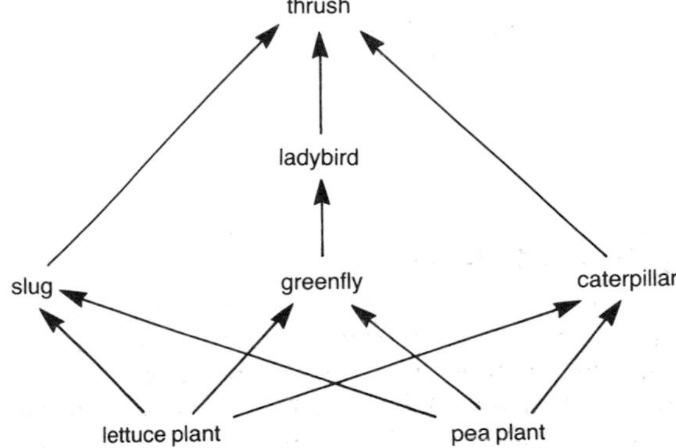

Figure 2.13 See question 4

Birth and death rate

The number of organisms of the same species present in an ecosystem is called a **population**.

The **birth rate** of a population is measured by counting the number of live births that occur within that population during a certain interval of time (e.g. number of human births in one year per 1000 of population).

The **death rate** of a population is measured by counting the number of deaths that occur within that population during a certain interval of time (e.g. number of human deaths in one year per 1000 of population).

The **growth rate** of a population depends on both the birth and death rates. For a population to grow, the birth rate must be greater than the death rate. The population then increases by the **difference** between the two rates as indicated by the shaded areas in the graphs shown in figure 2.14.

In country A (which is still developing) the death rate is decreasing more rapidly than the birth rate. The difference between the two is therefore increasing with time and population growth is speeding up.

In country B (which is already developed) the birth rate is decreasing more rapidly than the death rate. The difference between the two is therefore decreasing with time and population growth is slowing down.

Uncontrolled population growth

If a population consisting of a few individuals enters an unoccupied area where the conditions are ideal (no shortage of food, no predators, no disease, etc.) then reproduction occurs and the number of individuals increases.

If the population grows by, say, doubling itself for each unit of time, and none of its members die, then the population shows uncontrolled growth. This is illustrated in figure 2.15 which gives the growth curve

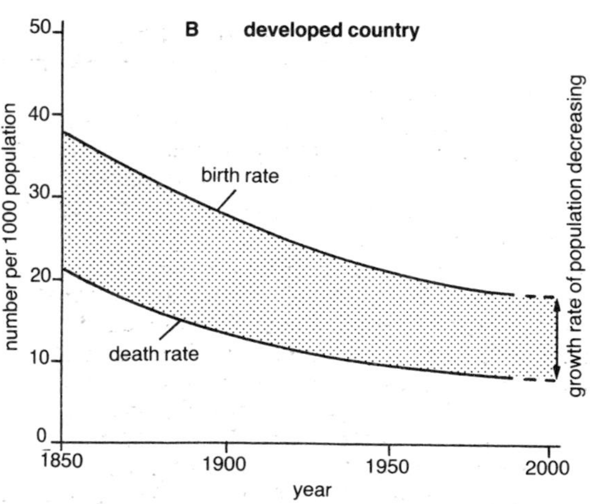

Figure 2.14 Birth and death rates

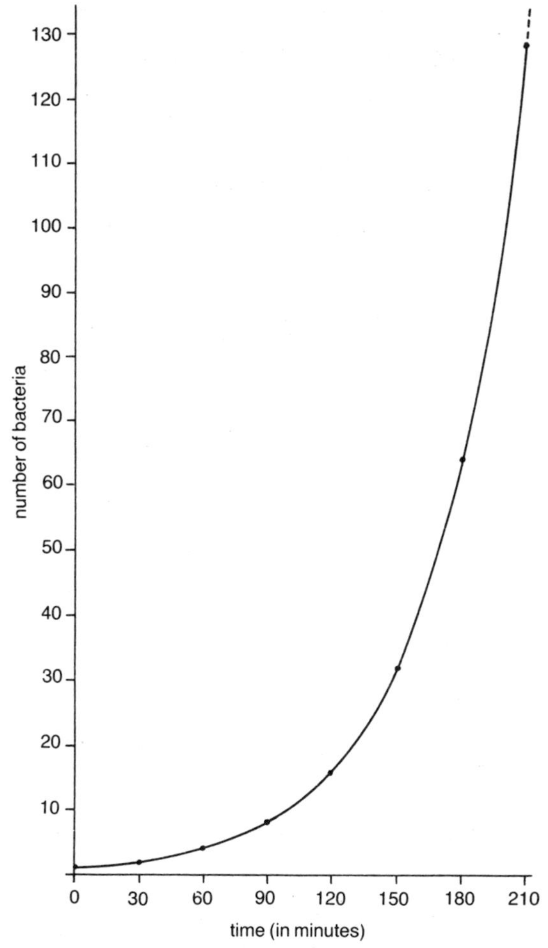

Figure 2.15 Graph of uncontrolled population growth

for a population of bacteria growing in ideal conditions and doubling its number every thirty minutes as shown in table 2.1. If the ideal conditions continue then the population curve will rise indefinitely.

time (in min)	number of bacteria
0	1
30	2
60	4
90	8
120	16
150	32
180	64
210	128

Table 2.1 Uncontrolled growth of bacteria

More to do

Explanation of growth curve

The growth curve of a population in ideal conditions continues to rise indefinitely because the birth rate is high and the death rate is low (or non existent) and therefore the population continues to increase and realise its full reproductive potential without being controlled by factors such as predators, lack of food, disease, etc.

Human population explosion
In man a population explosion has occurred during the last century (see figure 2.16). Survival of so many people is due mainly to our exceptional brainpower. This has enabled man to improve public health and develop vast areas of land for food production. As a result, the world's human birth rate in 1970 was 33 per 1000 but the death rate only 13 per 1000 of population. If this annual increase of 20 people per 1000 of population continues unaltered then the world's population will double itself by the year 2005.

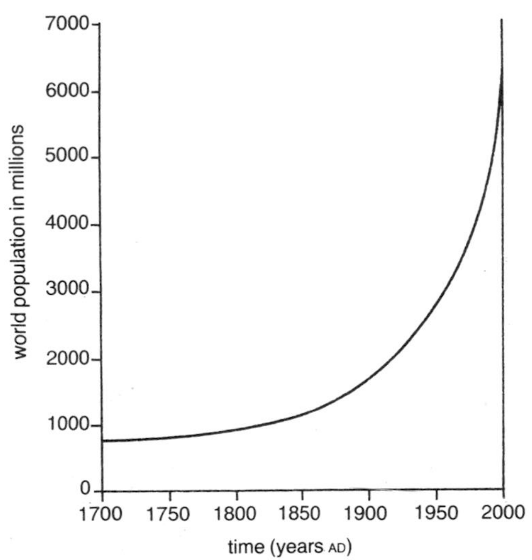

Figure 2.16 Increase in world population

Factors limiting population growth

Uncontrolled growth of a population does not usually occur for long in nature because ideal conditions do not continue to exist for an indefinite period of time. Sooner or later the size of a population is limited by one or more of the following factors.

Shortage of food and/or water
If only a limited supply of one or both of these essentials is present in an ecosystem then only a certain size of population can be supported.

Lack of space
If only a limited amount of space is available in the ecosystem then overcrowding occurs as the population size increases and soon the weaker members die.

Predators
If a species is preyed upon by predators then its population size is prevented from increasing indefinitely (also see page 22).

Toxic wastes
The build-up of toxic wastes made by the members of the population and passed out into the ecosystem creates poisonous conditions which prevent further growth of the population.

Disease
A disease-causing organism arrives in the ecosystem and many members of the population die, thus limiting the size of the population.

All of the above factors increase the death rate within a population and in reality, therefore, the growth curve of a population is normally found to take the form shown in figure 2.17, which represents the growth of a population of yeast cells.

The broken line in the graph shows the direction that the curve would have taken if uncontrolled growth had occurred. However the real growth curve (solid line) does not continue to rise indefinitely. Instead, after about 25 hours, growth begins to slow down and eventually at about 50 hours the population stops growing and the curve levels off. This could be because the food supply has run out or poisonous wastes have built up. If the nutrient solution containing the yeast cells is not changed, the cells will start to die and the population will decrease in number.

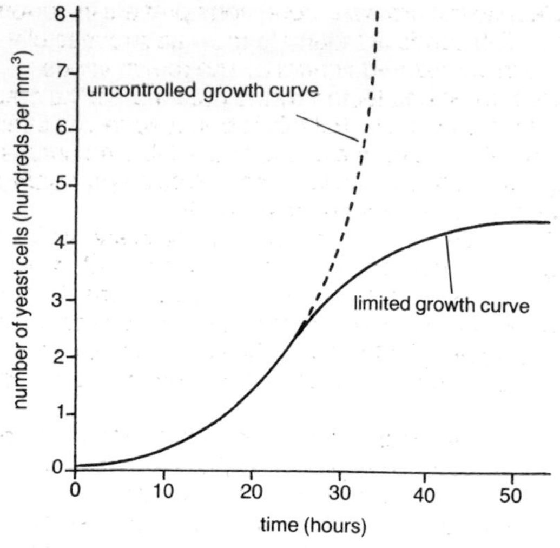

Figure 2.17 Limited growth curve

As the population of *Paramecium* increases, the yeast population decreases since they are being eaten at a faster rate than their rate of reproduction. However, at about day seven *Paramecium* cannot find enough yeast to eat and its numbers now begin to decline. This is followed by the yeast recovering and so on, with the *Paramecium* population exactly mirroring but slightly lagging behind the variations shown by the yeast population.

This pattern is also found to occur in predator-prey relationships. Since the predator is always at a higher level in the pyramid of numbers than the prey, the number of predators present in an ecosystem is smaller than the number of prey.

KEY QUESTIONS

1 Rewrite the following sentence and complete the blanks. The growth rate of a population depends on the ________ and ________ rates.
2 How is the growth rate of a population measured?
3 Describe the form that the growth curve of a population under ideal conditions takes.
4 a) Name FOUR factors which limit the growth of a population.
b) For each of these factors explain the effect that it has on the population's growth.

More to do

Predator and prey relationships

Figure 2.18 shows the effect of adding *Paramecium* (a tiny animal) at day three in the growth of a population of yeast cells living in ideal conditions. Uncontrolled growth of the yeast cells stops at day five since *Paramecium* is eating many of the yeast cells.

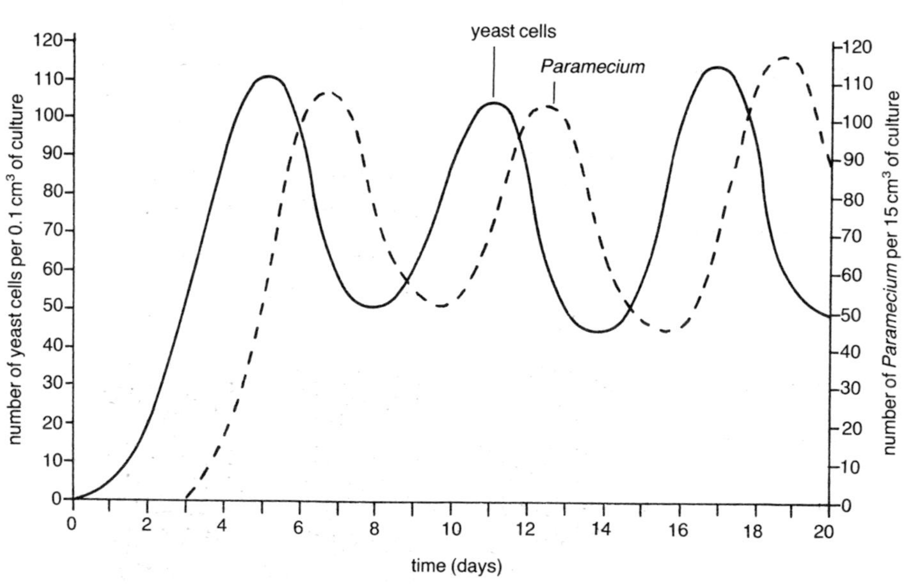

Figure 2.18 Predator and prey relationship

Extra Questions

5 Explain why the growth curve of a population under ideal conditions takes the form you described in your answer to question 3.
6 Explain why the growth curve of a population of predators always mirrors but lags behind that of its prey.

Investigating competition

The experiment shown in figure 2.19 is set up to investigate the effect of **competition** between radish and cress seeds. In carton A, 100 radish and 100 cress seeds are placed on damp cotton wool and allowed to grow. Control carton B receives 100 radish seeds and control carton C 100 cress seeds. Table 2.2 gives a typical set of results, where 'successful growth' means that the seed produces a shoot bearing two green leaves.

carton	% number of seeds showing successful growth	
	radish	cress
A	96	67
B	94	absent
C	absent	89

Table 2.2 Plant competition results

Despite the crowded conditions present in carton A, the radish seeds are found to grow as successfully as those in uncrowded control B. The radish seeds appear to be unaffected by the presence of the cress seeds. It is possible that this is because radish seeds germinate quickly, make use of available resources (e.g. water) and establish themselves by producing leafy shoots faster than cress seeds.

The cress seeds in crowded carton A are not found to grow as successfully as those in uncrowded control C. Cress seeds seem to be affected in some way by competition. Perhaps this is because cress seeds are slower to germinate and by the time their leaves emerge, many of them are shaded by the wider radish leaves.

A struggle for existence follows as the two different species of plant compete for light (which is only shining from above). Some cress seedlings die of starvation since they are unable to get enough light to make food by photosynthesis. Among those that do survive, many have a sickly appearance with curved shoots and yellow leaves. However, a few cress seedlings do manage to grow as well as radish.

It is concluded from this experiment that radish seedlings have competed more successfully than cress for the **limited resource** (in this case light).

Competition in nature

In a natural ecosystem, competition also occurs whenever two or more members of the community need a particular resource which is in short supply. Green plants compete for light, water, soil nutrients, etc. Animals compete for water, food, shelter, nesting sites and mates.

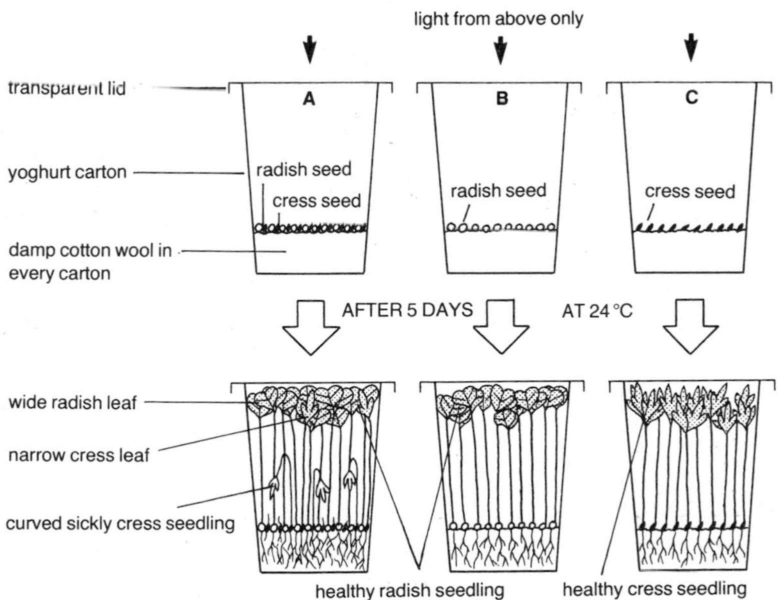

Figure 2.19 Plant competition

If the competing individuals are members of the same species, their needs will be almost identical and competition will be even more intense than between members of different species with differing needs.

Effects of competition

Whenever competition occurs, some of the competitors grow more slowly than others. These 'weaker' individuals lose out in the struggle for existence and often die before reaching reproductive age. The 'stronger', more successful competitors survive and become the parents of the next generation.

KEY QUESTIONS

1 Under what conditions does competition between organisms occur?
2 Name TWO resources for which (a) plants and (b) animals could be competing.
3 Describe two effects of competition.
4 Explain why competition between members of the same species is likely to be more intense than that between members of two different species.

Nutrient cycles

Nutrients are chemical substances such as nitrogen, oxygen, mineral salts, carbon dioxide and water which are needed by living things to stay alive and grow.

When an organism dies, the chemicals present in its dead body (and in wastes produced during its lifetime) are released back into the ecosystem by the action of **decomposers** (bacteria and fungi). These chemical nutrients then become available for use by other organisms in the ecosystem.

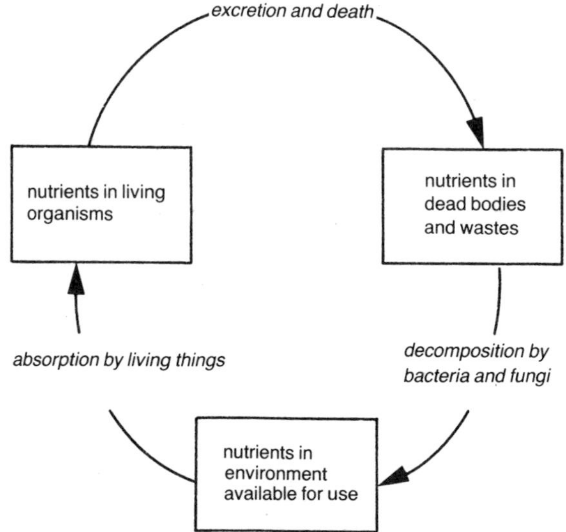

Figure 2.20 Cycling of nutrients

The flow of each chemical nutrient in an ecosystem takes the form of a **cycle** as summarised in figure 2.20. This constant recycling of nutrients is important because the supply of many chemical substances in an ecosystem is limited. If the chemicals remained 'locked up' in dead bodies and wastes, supplies would soon run out and certain nutrients would become unavailable for use by new members of the community.

More to do

Nitrogen cycle

About 80 per cent of air consists of nitrogen gas. All living things need nitrogen to make protein. However, plants and animals cannot make use of nitrogen gas directly. Plants absorb nitrogen from the soil in the form of **nitrate**. Animals must eat plant or animal **protein** to obtain their supply of nitrogen.

Nitrogen is a chemical element which is cycled round an ecosystem. Figure 2.21 shows the sequence of the main processes which occur during the **nitrogen cycle**. Further details of the various roles played by bacteria in this cycle are described on page 183.

KEY QUESTIONS

1 Give TWO examples of chemical nutrients needed by living organisms.
2 a) In general what form does the flow of such nutrients in an ecosystem take?
b) Explain why this sequence of events is important to the living organisms in an ecosystem.

Extra Question

3 Starting and finishing at 'nitrogen in nitrates' (figure 2.21), describe one possible route that could be taken by an atom of nitrogen as it passes round the nitrogen cycle.

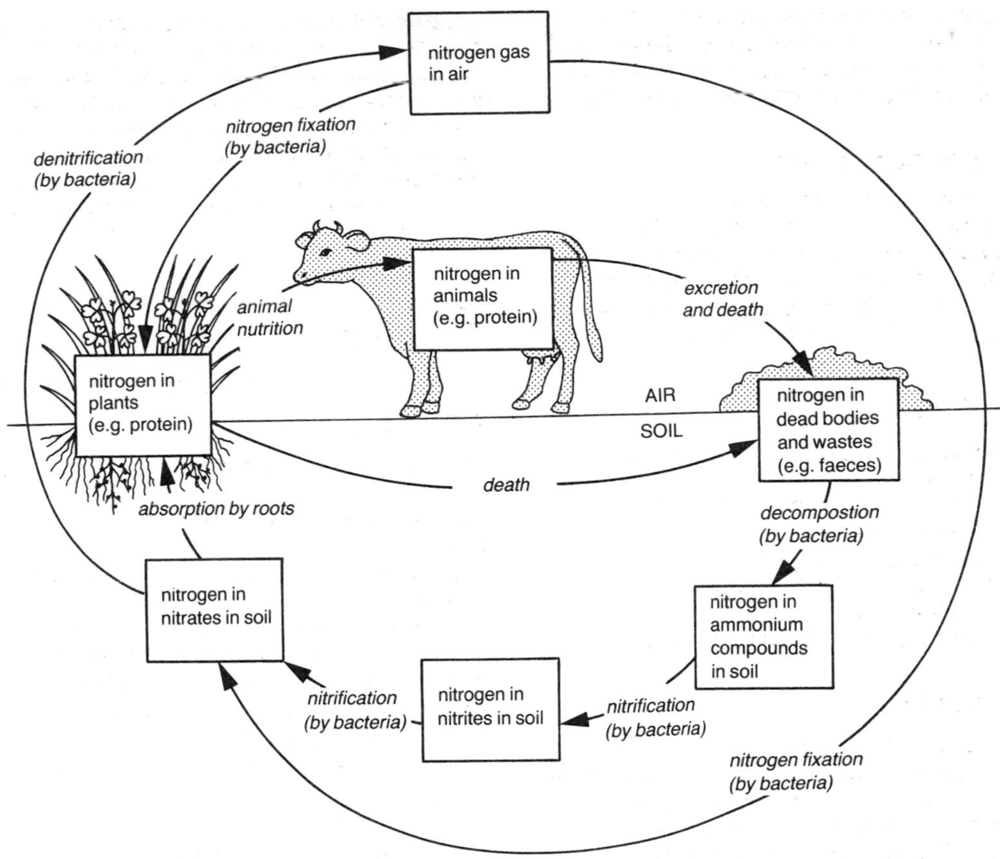

Figure 2.21 Nitrogen cycle

3 Control and management

Pollution

Environmental pollution is the contamination of our surroundings by substances which harm living things, often causing disease and, in extreme cases, death.

Test for sulphur dioxide

When sulphur burns in air, a gas called **sulphur dioxide** (SO_2) is produced. When this gas is passed through purple potassium permanganate solution, the purple colour disappears. This is the test for sulphur dioxide.

Testing coal smoke for sulphur dioxide

Look at the experiment shown in figure 3.1. Since the smoke from the burning coal turns the purple potassium permanganate solution colourless, it is concluded that coal smoke contains sulphur dioxide.

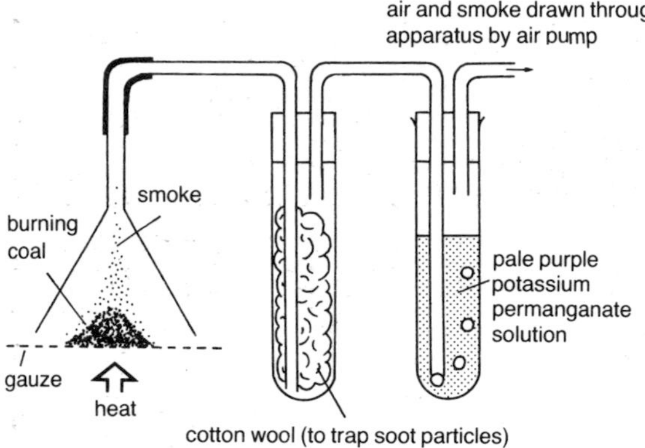

Figure 3.1 Testing coal smoke for SO_2

Effect of sulphur dioxide on cress seedlings

Look at the experiment shown in figure 3.2. From the results it is concluded that sulphur dioxide harms growing cress seedlings. Sulphur dioxide is a poisonous gas.

Importance of control

A **control** is a copy of the experiment in which all factors are kept exactly the same except the one being investigated in the original experiment. When the results are compared any difference found between the two must be due to that one factor.

If bell jar B (the control) had not been set up in the cress seedlings experiment, it would be valid to suggest that the plants in bell jar A had not been poisoned to death by sulphur dioxide but had died for some other reason (e.g. being enclosed in an airtight chamber).

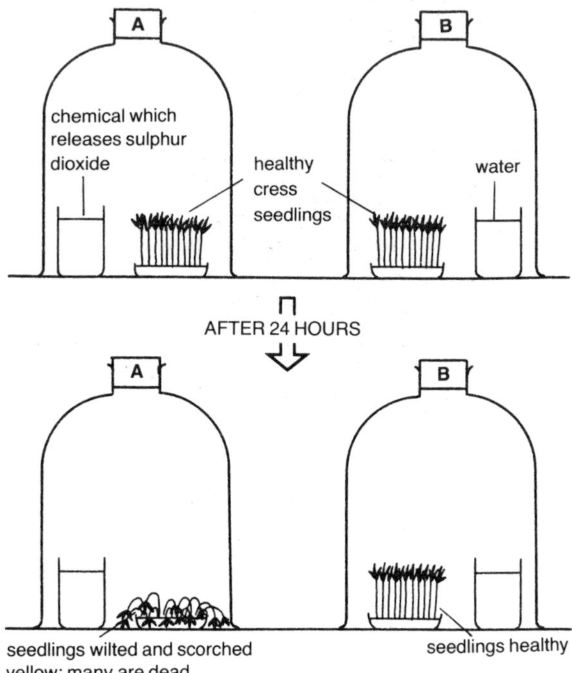

Figure 3.2 Effect of SO_2 on cress seedlings

Further effects of SO_2 on living things

When fossil fuels such as coal are burned in power stations to generate electricity, sulphur dioxide and other poisonous gases are released into the air. Even in low concentrations, SO_2 aggravates human respiratory ailments and causes leaf damage to plants. **Lichens** are simple plants composed of a mixture of fungus and alga (see figure 3.3) which are often found growing on the trunks and branches of trees. Lichens are especially sensitive to SO_2. Their almost total absence from many cities (see figure 3.4) indicates atmospheric pollution by this harmful gas.

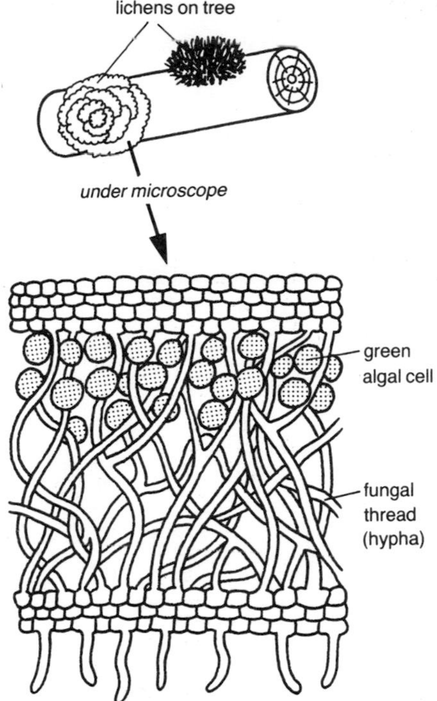

Figure 3.3 Lichens

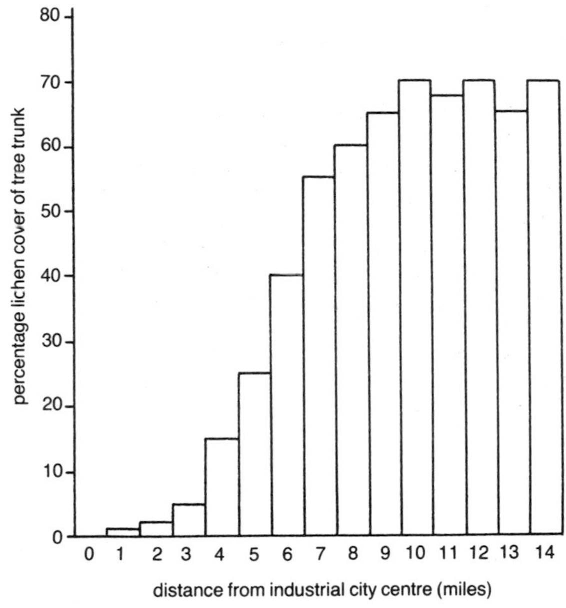

Figure 3.4 Lichen bar graph

Sources of pollution

Study figure 3.5 carefully. The four diagrams illustrate many ways in which pollution can occur. These are summarised in table 3.1.

Figure 3.5 Sources of pollution

part of environment polluted	main source of pollution	example of pollutant	possible control measure
air	industry and coal-fired power stations	smoke containing SO_2 and other poisonous gases	scrubbing smoke fumes before release into air
	car exhausts	fumes containing poisonous carbon monoxide gas and lead	fitting filters to car exhausts and using lead-free petrol
land	homes and motor vehicles	domestic rubbish and old car parts	recycling, burying and burning
fresh water	industry	inorganic chemical waste, e.g. mercury;	recycling and alternative method of disposal
		organic waste e.g. paper fibres	decomposition by micro-organisms (see page 184)
	homes and farms	raw sewage, e.g. faeces	
	agriculture	excess fertiliser and pesticides	use of minimum amounts, especially on farmland near waterways
sea	oil tankers washing out tanks at sea or becoming damaged	oil	decomposition by bacteria, sinking and burning, collection and removal
	nuclear power stations	radioactive waste	sealing waste in lead containers and dumping on ocean bed, seeking alternative energy sources, e.g. harnessing energy of sun, wind, waves and tides

Table 3.1 Sources of pollution

More to do

Nuclear power

Nuclear power is used instead of fossil fuel in some power stations to generate electricity. Many of the radioactive wastes produced are especially dangerous since they continue to give out harmful radiation for many years. If these wastes are not carefully disposed of, then they pollute the environment and have an adverse effect on living things. In addition, an accident at a nuclear power station may occur, releasing harmful radiation over a wide area and causing a higher than normal incidence of radiation-related diseases such as leukaemia and other forms of cancer.

Fossil fuels

The burning of fossil fuels can have an adverse effect on the environment because the sulphur dioxide and other poisonous gases released pollute the atmosphere and may even be converted into **acid rain** as shown in figure 3.6.

Controlling acid rain

The problem of acid rain can be tackled at source by scrubbing the fumes produced before releasing them. A scrubber contains a slurry (muddy suspension) of limestone which removes SO_2 from the flue gases. In addition, the fuel burners can be redesigned to operate at lower temperatures and produce less of the poisonous gas, nitrogen oxide.

Acid rain can also be tackled by 'treating the disease symptoms'. This involves adding lime to the affected water (and even surrounding land) as shown in figure 3.7, in order to raise the pH back to a level that living things such as fish can tolerate.

KEY QUESTIONS

1 a) Name FOUR parts of the environment that can be severely affected by pollution.
b) For each of these, name a main source of pollution, the pollutant involved and a possible control measure.

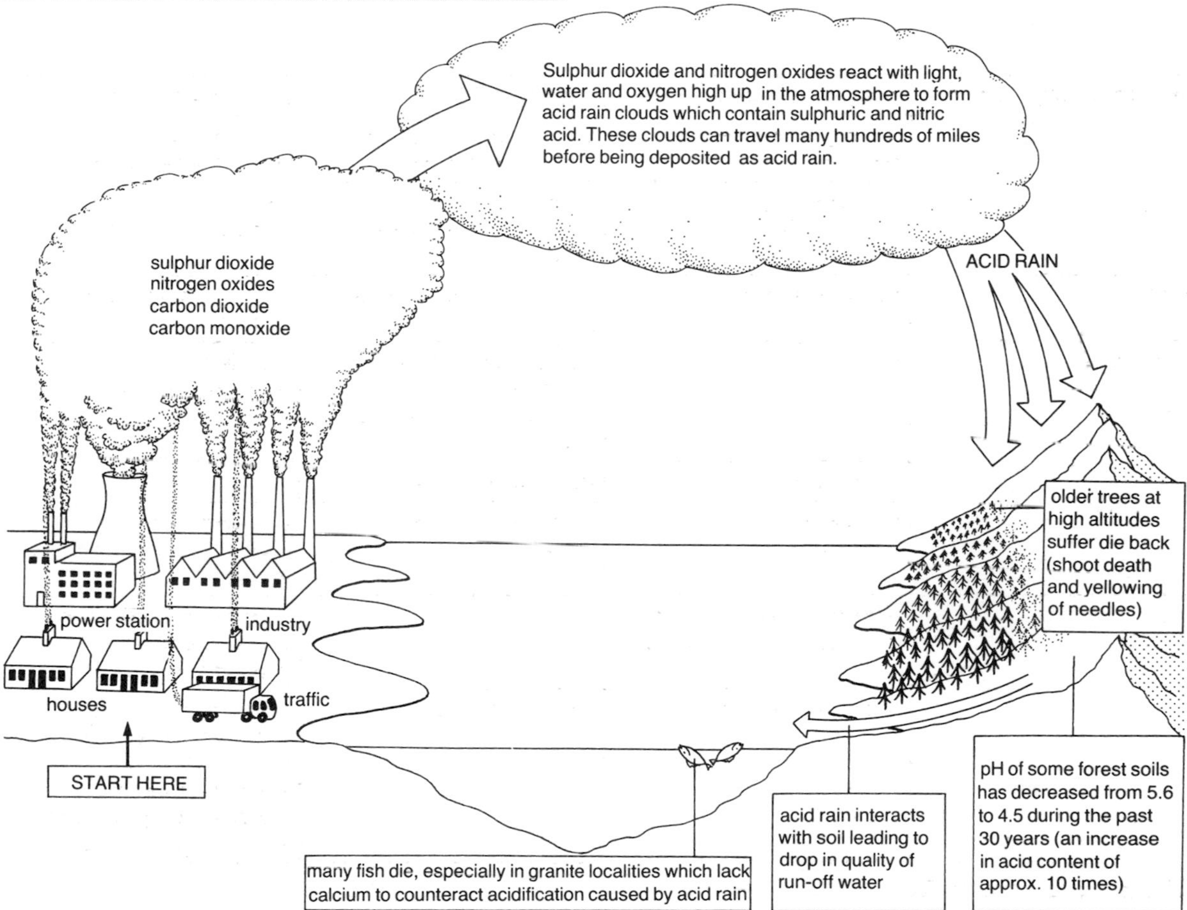

Figure 3.6 Acid rain

Figure 3.7 Adding lime

Extra Questions

2 In what way can the use of nuclear power as an energy source adversely affect man?
3 a) Name a harmful gas released from coal during burning.
b) Explain how the use of fossil fuels can have an adverse effect on the environment.

Organic waste

The unwanted products of living things, such as raw (untreated) sewage, dead leaves and stale food, are examples of **organic waste**. Organic waste is an ideal food source for the decomposer micro-organisms (bacteria and fungi) that bring about decay. In Britain and many other countries, raw sewage is biologically treated at a sewage works by being fed to these decay microbes which break it down into harmless substances (see page 181).

River pollution

In some densely populated areas the sewage works may become **overloaded**. As a result the liquid being discharged into the local river is rich in untreated sewage. In some areas, paper mills discharge untreated organic waste (paper fibres) into the river. Both of these organic materials pollute the river because they provide food for bacteria which rapidly multiply and use up the river water's supply of dissolved **oxygen**.

Comparing oxygen content
Ferrous sulphate (see figure 3.8) is a chemical which reacts with the oxygen in water and only decolourises the red dye when the oxygen supply becomes exhausted. It is found in this experiment that less ferrous sulphate is needed to decolourise the dye in polluted water than in tap water, showing that polluted water contains less oxygen.

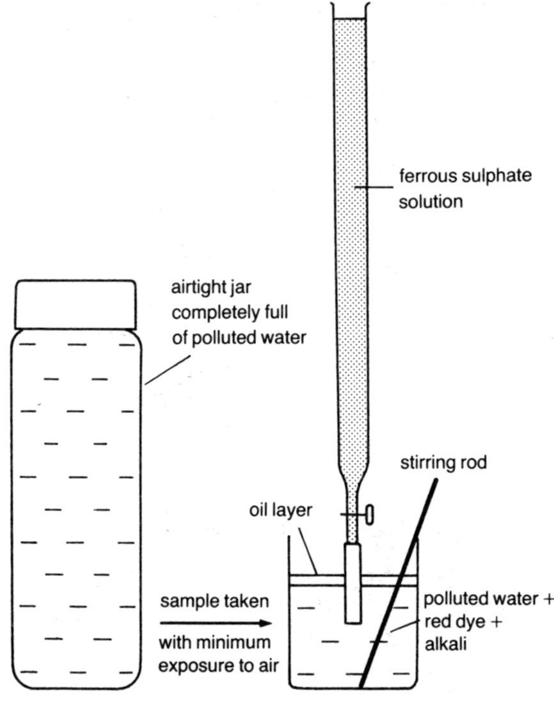

Figure 3.8 Estimating oxygen content

Effect of oxygen shortage

Research into the effects of low oxygen concentration on aquatic animals shows that species differ in their ability to tolerate the varying oxygen levels found in a river polluted with organic waste (see figure 3.9).

Some animals (e.g. **mayfly** and **stonefly** nymphs) are commonly found in clean water rich in oxygen. However, they cannot survive when the river water's oxygen content is greatly reduced by the presence of a huge number of micro-organisms.

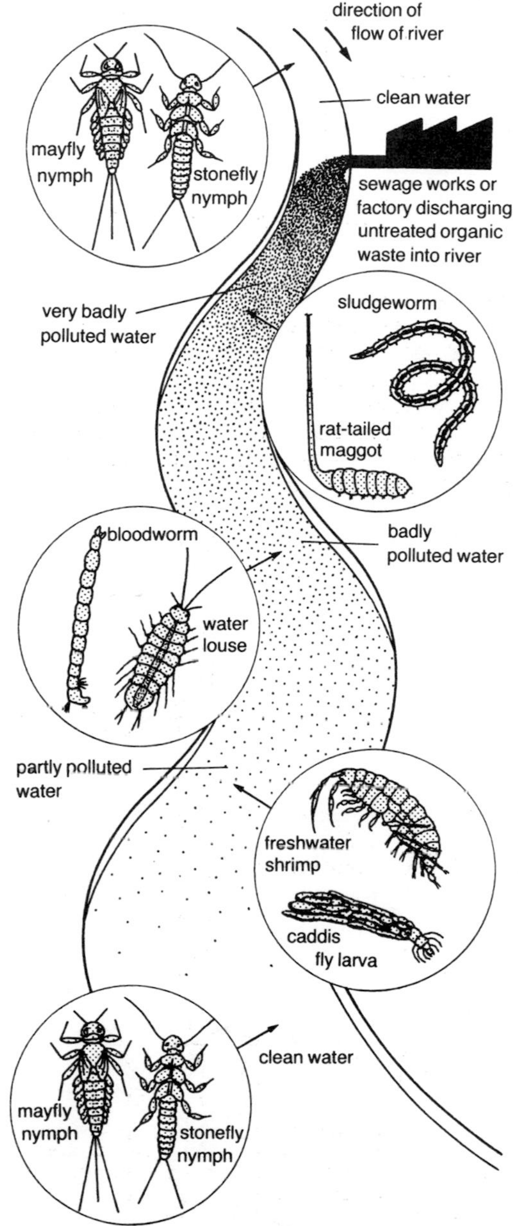

Figure 3.9 Polluted river

If the water becomes extremely polluted and contains almost no oxygen, then the only animals that survive are **rat-tailed maggots** and **sludgeworms**.

As the water gradually becomes a little less polluted, animals such as **midge larva** (**'bloodworm'**) and **waterlouse** which do not require very much oxygen are commonly found.

When pollution eventually decreases to a fairly low level, water plants can survive and produce oxygen during photosynthesis. The oxygen content of the water begins to rise and **freshwater shrimps** and **caddis fly larvae** are now found to become numerous since they can tolerate low levels of pollution.

Mayfly and stonefly nymphs are only found again when the water is very clean and contains plenty of oxygen.

Indicator species

A species that only thrives well under certain environmental conditions and whose presence shows that these conditions are present in an environment is called an **indicator species**. Table 3.2 gives a summary of the species that indicate the varying oxygen concentrations found in a river polluted with organic waste.

level of pollution	oxygen concentration	indicator species
absent or very low	high	mayfly nymph stonefly nymph
low/medium	↑	freshwater shrimp caddis fly larva
high	↓	'bloodworm' waterlouse
very high	low	rat-tailed maggot sludgeworm
extreme	zero	no animals present

Table 3.2 Indicator species

Different species of vertebrate animals such as fish also differ in their ability to tolerate varying levels of oxygen. Trout, for example, need oxygen-rich water whereas roach are more tolerant of lower oxygen concentrations. However, the invertebrate species are more easily caught and are therefore more commonly used as indicators of pollution.

More to do

Effect of pollution on number of micro-organisms

The dramatic rise in the number of micro-organisms (e.g. bacteria) that occurs in water polluted with organic waste is accompanied by a sharp drop in the water's oxygen concentration as shown in figure 3.10. This occurs because the bacteria use up the oxygen during respiration.

Effect of oxygen shortage on numbers of species

A clean water ecosystem normally contains a large number of different species of animals (and plants). Although the clean water species such as mayfly and stonefly nymphs are the most numerous within this community, a few representatives of those species that can tolerate (or even thrive well in) polluted water are also present but only in very small numbers owing to the intense competition for food, shelter, etc.

This natural balance between the living organisms in the ecosystem is upset by a very high level of pollution since it completely eliminates all the clean water species. The variety of species within the community is now reduced and only a few pollution-tolerant species (e.g. sludgeworm) survive. These animals increase dramatically in number owing to lack of competition.

Further downstream where the water is not quite so dirty, species such as waterlouse that can tolerate fairly high levels of pollution are found to rise in number and become dominant in the ecosystem.

This trend continues (see figure 3.11) until eventually the river recovers its original state and clean water animals return to their dominant position. Table 3.3 summarises the changes in thc numbers of different species.

KEY QUESTIONS

1 Give TWO examples of materials that could be described as organic waste.
2 Name a group of living things that use organic waste as a food source and in doing so bring about the process of decay.
3 a) Under what circumstances might a river become polluted with organic waste?
b) In what way does the number of micro-organisms present in such a river change?
c) What effect does this have on the amount of oxygen available to the other organisms in the water?
4 a) What is meant by the term **indicator species**?
b) What does the presence of a large population of (i) 'bloodworm' and (ii) mayfly nymph indicate about the oxygen concentration of a river's water?

Extra Questions

5 In a river polluted with organic waste, changes occur to the number of micro-organisms present and to the oxygen concentration of the water. Explain why.
6 In what way does the pollution of a river by organic waste affect the numbers of species present? Explain why.

	state of water				
type of animal species	clean		very badly polluted	partly polluted	clean
clean water species	many large populations	organic effluent added	none	few small populations	many large populations
semi-tolerant species	few small populations		none	many large populations	few small populations
dirty water species	few small populations		many large populations	few small populations	few small populations
overall variety within ecosystem	large number of different species		very small number of different species	small number of different species	large number of different species

Table 3.3 Effect on number and variety of species

untreated organic waste added
oxygen
bacteria
oxygen
bacteria
relative amounts of bacteria and oxygen
distance downstream

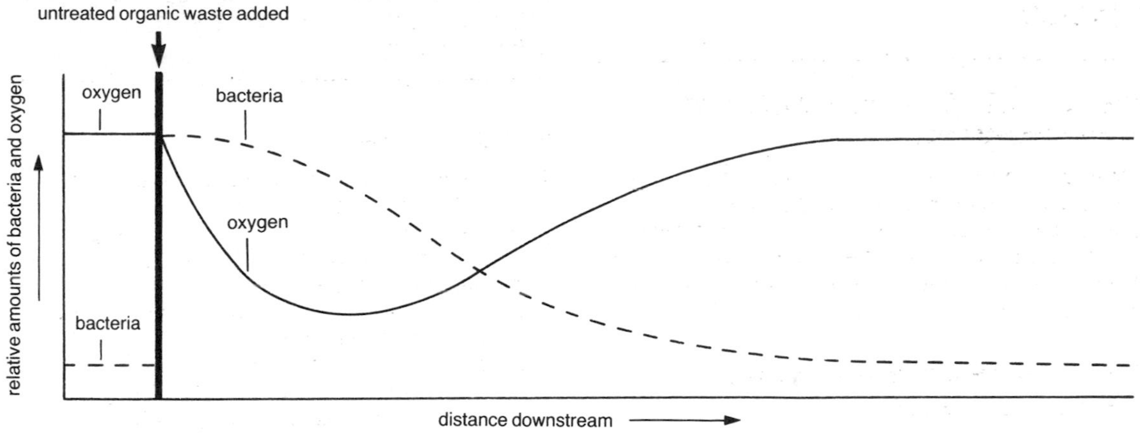

Figure 3.10 Effect on number of micro-organisms (bacteria)

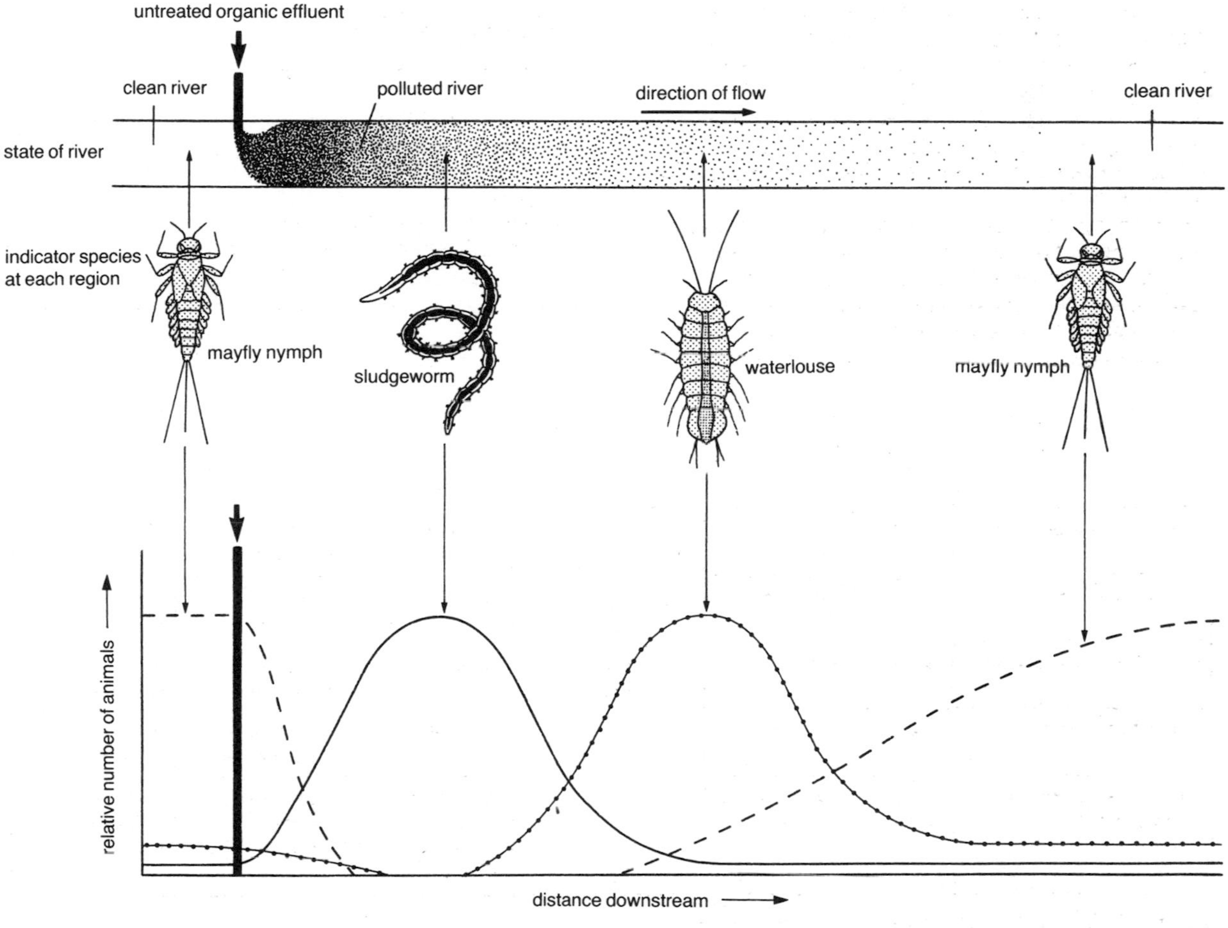

untreated organic effluent
clean river
polluted river
direction of flow
clean river
state of river
indicator species at each region
mayfly nymph
sludgeworm
waterlouse
mayfly nymph
relative number of animals
distance downstream

Figure 3.11 Effect on number of species

Natural resources

Continued human existence on Earth is dependent upon natural resources such as the air that we breathe, the water that we drink, the soil that we stand on and the plants and animals with which we share our planet.

About one quarter of the world's surface is land. Only about 11 per cent of this is suitable for cultivation. This fertile land is one of our primary natural resources.

Managing the land for agriculture

Humans have been growing crops for at least ten thousand years. In recent times, with the world's population increasing more rapidly than ever before, people demand more and more from the land. Sometimes this has involved poor management practices which in turn have led to problems as shown in table 3.4.

More to do

Control of components of an ecosystem

By employing agricultural practices, people deliberately interfere with an ecosystem in order to

poor management practice	reason for practice (background information)	effects of bad management	possible solutions or improvements
overuse of fertilisers and chemical pesticides	harvesting of crops leaves soil short of nitrate and mineral salts. Cultivation of one species of plant presents ideal conditions for attack by parasites.	excess fertilisers are washed by rain into rivers and lakes, causing pollution. Pesticides which fail to decompose naturally enter food chains and accumulate, causing harm especially to final consumer (see also question 5 on page 37).	use minimum fertiliser, especially on land near waterways. Develop a variety of biodegradable pesticides. Find methods of controlling parasites biologically (e.g. using another animal to eat them, disturbing their cycle of reproduction).
use of vast quantities of grain to feed domesticated animals	people in developed countries have become accustomed to consuming a diet containing much animal protein (e.g. meat and milk products).	vast amounts of energy wasted to produce meat-rich diet for developed countries at the expense of the developing ones. Many people in rich countries suffer malnutrition in form of obesity, heart disease, etc., while those in poorer countries suffer malnutrition in form of starvation due to lack of protein.	persuade people in developed countries to eat less meat. Feed less grain to animals and more to starving people.
growth of cash crops by developing countries instead of food crops	many poor countries (lacking mineral wealth, fossil fuels, etc.) use land to grow cash crops such as coffee for export to wealthy countries in order to buy tools and technology for development.	poor developing country unable to feed its populace and forced to import food and go into debt.	supply tools and technology to developing countries without economic strings. Persuade poor countries to grow fewer cash crops and more food crops to feed local people. Supply economic aid for more agriculture and irrigation projects.
overgrazing of grassland	livestock regarded as main source of wealth in poor developing countries.	overgrazed plants die and no longer retain water in soil. Wind-blown sand buries other healthy vegetation, leading to increase of desert area. People become poorer and more prone to malnutrition and disease.	provide economic aid to plant shelter beds of trees and resilient grass cover. Develop irrigation schemes.

Table 3.4 Poor management practices

produce vast numbers of a useful species (e.g. crop of cereal plants). During this process they attempt to control certain components of the ecosystem such as nutrient supply in the soil and habitat of the plant.

Nutrient supply
This is controlled by adding **fertilisers** to ensure that a rich supply of nitrate and mineral salts is present in the soil and available for use by the growing plant (e.g. wheat crop). Natural fertiliser (e.g. manure) is ploughed in during autumn to allow time for micro-organisms to release its minerals before the spring when the seeds are planted. This cheap method also maintains the soil's humus content.

Humus is dark brown organic material formed during the decomposition of plant and animal remains. In addition to providing nutrients for plant growth, humus increases the soil's water-holding capacity. Its sticky nature improves soil texture by holding the soil particles together as **crumbs** (see figure 3.12). This crumbly texture contains air spaces which guarantee an oxygen supply for respiring roots.

Farms specialising in crop plants and therefore lacking large supplies of animal manure are often forced to use artificial fertilisers such as ammonium sulphate or sodium nitrate. (Look back to figure 2.21 on page 25 and see how these would play an important part in helping the nitrogen cycle to turn.)

Artificial fertilisers are added to the soil in spring so that the chemicals will not be washed out by rain before the plants start to grow. Although such chemicals are quick-acting, they are expensive and fail to add humus. As a result the soil's crumb structure gradually breaks down, the soil tends to lack air and the growth of the crop plants may not be greatly improved.

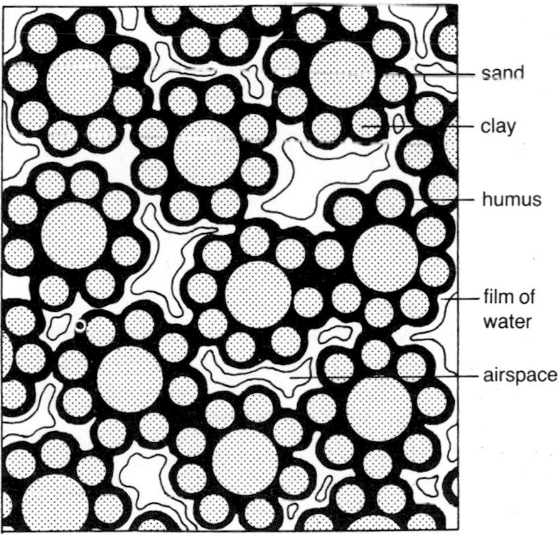

Figure 3.12 Soil crumbs

Habitat of crop plant
The farmer attempts further control of the crop plant's habitat (the soil) by clearing the land of all other possible competitors (e.g. weeds, using weedkillers). He/she cultivates the land, adds fertiliser, controls the soil PH (using lime if necessary) and maintains the water content using sprinklers and irrigation. He/she erects fences around the habitat to keep out herbivorous consumers (e.g. cattle), puts up scarecrows to keep hungry birds away from newly planted seeds and sprays the crop with pesticides to prevent growth of parasites (e.g. fungi).

By these means people produce an altered (and artificial) ecosystem in order to promote optimum growth of the desired plant. In a natural ecosystem, the plant would be unlikely to grow so well since it would have to compete for mineral salts, space, etc., and to survive attacks by hungry animals and parasites.

KEY QUESTIONS

1 a) Give TWO examples of poor management of a named natural resource.
b) Explain how each of these examples can produce problems.
c) Suggest a possible solution or improvement in each case.

2 a) Name TWO components of an ecosystem that a farmer tries to control.
b) Describe how each of these is controlled.

PROBLEM SOLVING

1 The following information was collected by studying eight different species of snail present in six Scottish lochs.
(+ = snails present) (ppm = parts per million)

Loch	Species of snail								Concentration of calcium (ppm) in loch water
	A	B	C	D	E	F	G	H	
1	+	+		+	+	+	+	+	19.9
2		+	+		+	+	+	+	17.8
3			+	+	+		+	+	15.3
4				+		+	+	+	9.5
5							+	+	7.1
6								+	5.2

a) What concentration of calcium was present in loch 5?
b) Which loch contained 9.5 ppm of calcium in its water?
c) Which species of snail was found in every loch?
d) Which loch possesses only five different species of snail?
e) How many different species of snail are found in loch 2?
f) (i) What relationship seems to exist between the number of species of snail present and the concentration of calcium in the water?
(ii) Give a possible explanation for this relationship based on the fact that one of the components of a snail's shell is calcium.

2 The following diagram shows the results from an investigation into the numbers of two closely related species of spider, X and Y, living on a climbing ivy plant. The size of circle at each sample site indicates the relative size of the total population of spiders found there. The shaded portion indicates species X, the unshaded portion species Y.

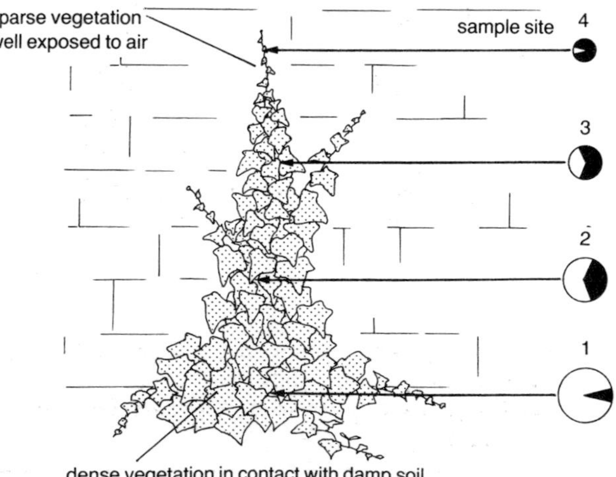

a) What relationship exists between the total number of spiders and the distance from the ground?
b) Which species is more numerous at ground level?

Extra Questions

c) Describe the trend in relative numbers of the two species of spider that occurs from site 1 to site 4.
d) Predict what will happen when equal numbers of the two species of spider are placed in the following plastic tube. Explain your answer.

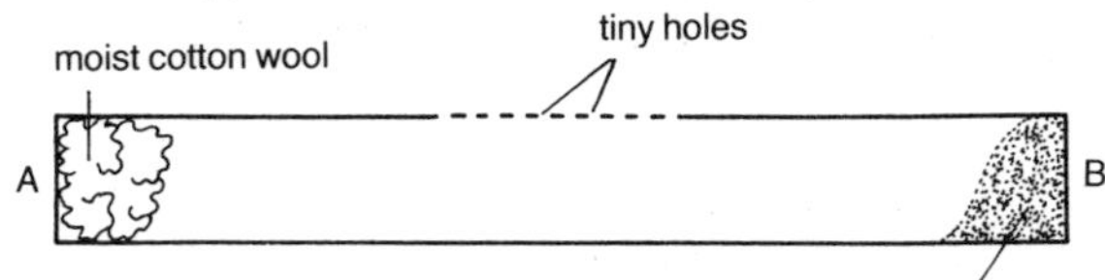

3 Like other owls, barn owls are unable to digest the fur and bones of their prey. Once swallowed, these materials are bound into dark-coloured pellets in the bird's gut. The owl then gets rid of the pellet by regurgitation ('throwing it up'). An analysis of the pellets littering the ground near a barn owl's roost allows scientists to identify the bird's prey as shown in the following table.

prey animal	number present in large sample of owl pellets
field mouse	32
shrew	16
vole	12
small bird	4

a) Why do owls make pellets?
b) Present the information in the table as a pie chart.
c) Construct a food chain which includes a barn owl.
d) Some farmers shoot barn owls on sight, claiming that they are a threat to their poultry. Does the information given in the table support this view? Explain your answer.
e) Other farmers welcome the presence of barn owls on their land. Suggest why.

4 The diagram below shows a form of atmospheric pollution.

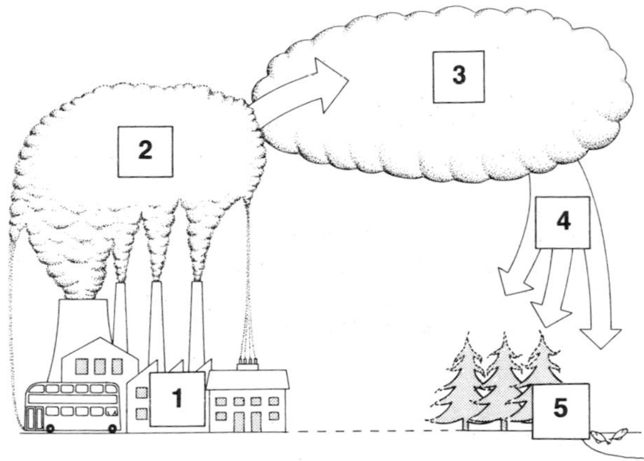

Match the numbered boxes in the diagram with the following lettered descriptions to show the correct sequence of events.

A acid rain falls on land and water environments
B fossil fuels burned by industry, power stations and motor vehicles
C acid gases form acid rain clouds
D trees suffer dieback and fish die
E acid gases released into the atmosphere

5 DDT is a poisonous chemical that used to be sprayed on crops to kill insect pests. The molecules of DDT are non-biodegradable (i.e. are not broken down by decay microbes into harmless substances). They therefore gathered in the soil and were washed by rain into lakes near farmland.

The following diagram shows the concentration of DDT in parts per million (ppm) found in the cells of the members of a lake community.

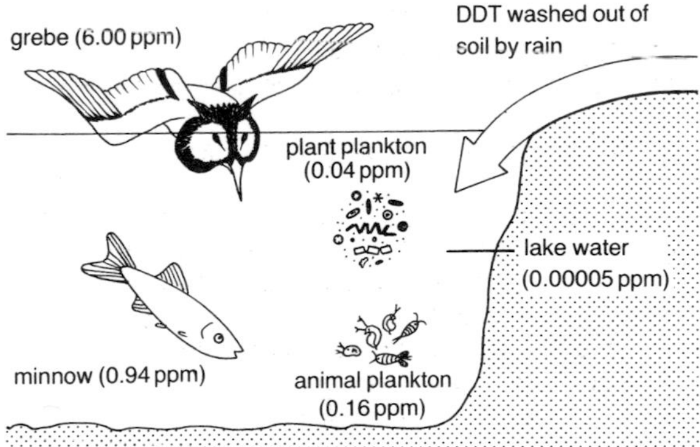

a) Construct a food chain which includes the four types of organism shown in the diagram.
b) Draw up a table to show the concentration of DDT found in the cells of each type of organism and arrange the information in increasing order.
c) (i) What relationship exists between the food chain and the concentration of DDT in the cells of its members?
(ii) Which animal do you think was found to suffer most and often die as a result of DDT poisoning? Explain your choice of answer.
d) Suggest why use of DDT has now been banned in Britain.

Extra Questions

e) Explain the relationship that you gave as your answer to question c)(i) using the terms non-biodegradable, accumulation and pyramid of numbers in your answer.
f) The accompanying bar graph shows the concentrations of pesticide residue found in the muscle tissues of three different water birds.

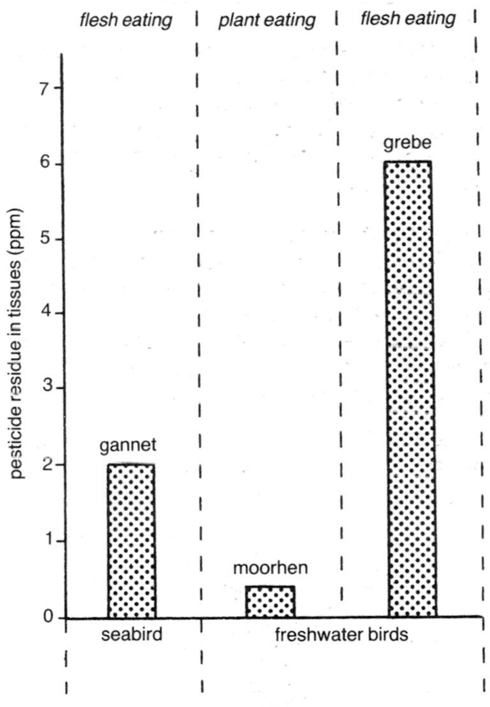

(i) Account for the difference in concentration found between grebe and moorhen.
(ii) Suggest why gannets are less severely affected than grebes.

Section 2 Investigating Cells

4 Investigating living cells

Examining a living tissue under the microscope

Look at figure 4.1 which shows one leaf from a pond weed called *Elodea* being mounted in water on a microscope slide. When living tissue is observed under the microscope (see also Appendix 5), it is found to be made up of many smaller units called **cells** as shown in figure 4.2. Cells are the basic units of all living things. Every living organism is made of one or more cells.

Staining

Some of a plant cell's parts can be clearly seen when the cell is mounted in water. An *Elodea* leaf cell, for example, is seen to possess a cell wall and several green chloroplasts. Other cell structures which are not so obvious can often be shown up more clearly by the addition of dyes called **stains**. Iodine solution is used to stain nuclei as shown in figure 4.3.

Examining animal cells

Figures 4.4, 4.5 and 4.6 show examples of animal cells viewed under the microscope. The unicellular animals are mounted in a drop of pond water, which is their natural environment. The human cheek cells have been stained with iodine solution to show their nuclei. The blood cells have had Leishman's stain added to show up the white blood cells.

Comparing plant and animal cells

On comparison, a typical plant cell and a typical animal cell are found to have structural similarities and differences as shown in figure 4.7.

Parts of a cell

The **nucleus** controls the cell's activities.

The **cytoplasm** (transparent jelly-like material) is the site of the chemical reactions essential for life. (Nucleus + cytoplasm = protoplasm.)

The thin, flexible cell **membrane** enclosing the cell contents controls which substances may enter and leave the cell.

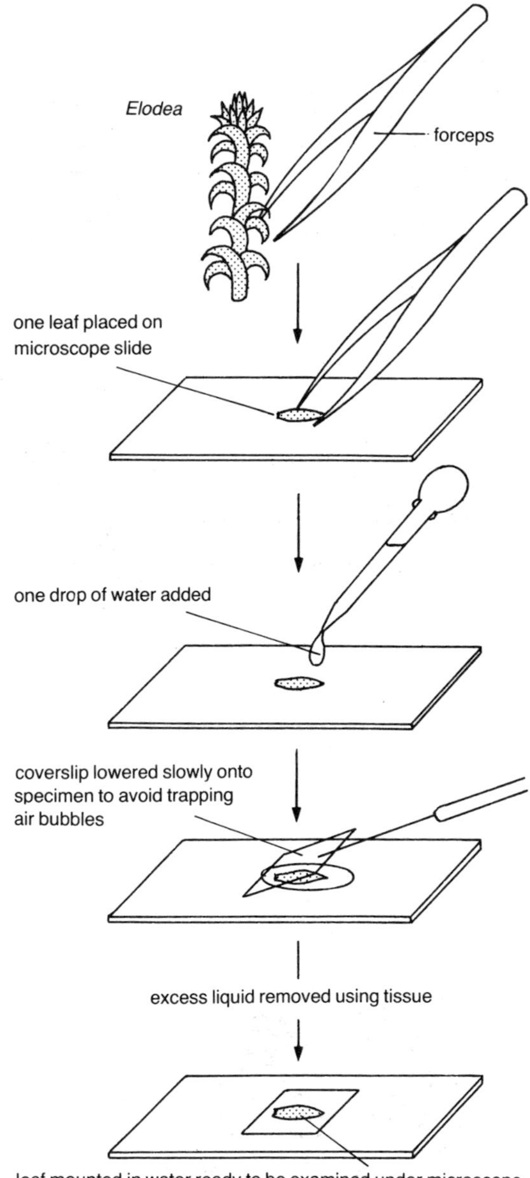

Figure 4.1 Preparing a slide as a wet mount

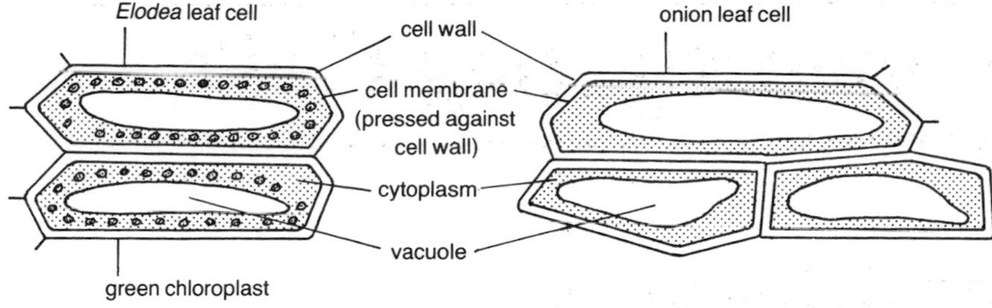

Figure 4.2 Plant cells in water

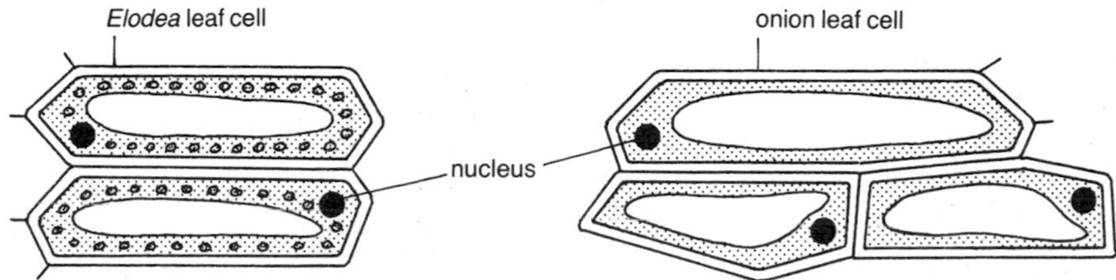

Figure 4.3 Plant cells in iodine solution

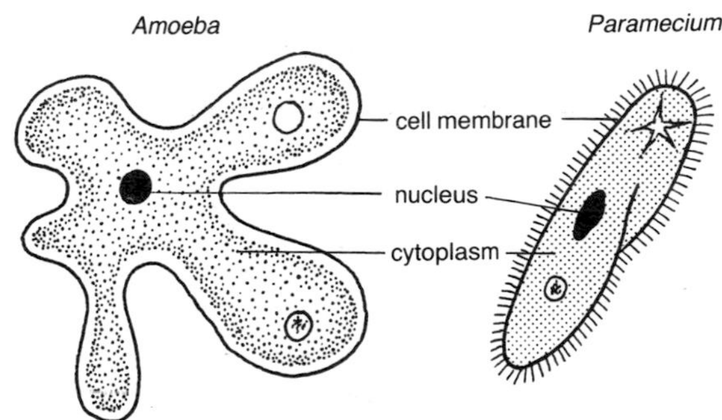

Figure 4.4 Live unicellular animals

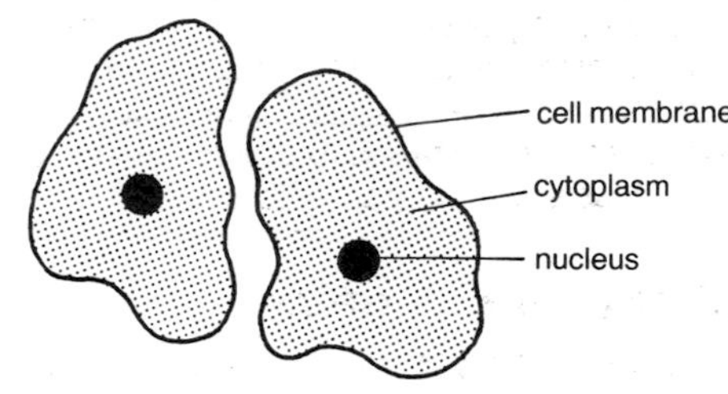

Figure 4.5 Human cheek cells

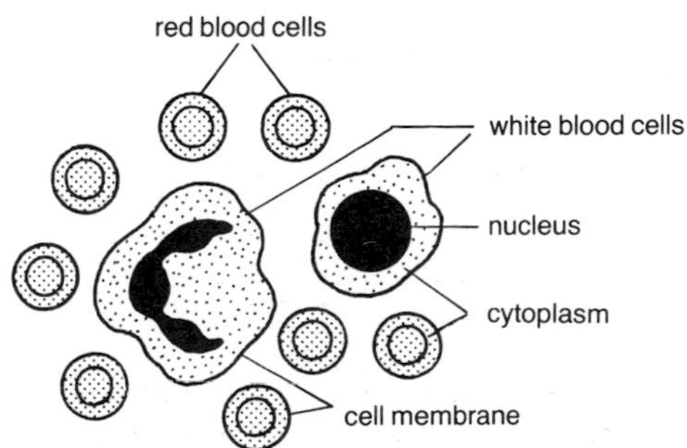

Figure 4.6 Human blood cells

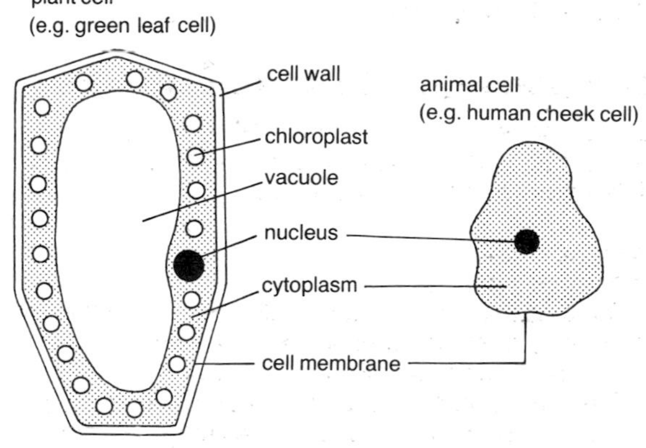

Figure 4.7 Comparison of typical plant and animal cells

The **cell wall**, which bounds plant cells only, is made of cellulose (see page 96). Although it is able to stretch slightly when water is absorbed by the cell, it is fairly rigid and therefore gives the plant cell definite shape.

A large permanent **vacuole** occurs only in plant cells. It often occupies most of the cell and always contains cell **sap** (a solution of sugars and salts). Some animal cells do, however, have small temporary vacuoles (see also page 45).

Chloroplasts containing green **chlorophyll** are found in some plant cells where they carry out photosythesis.

More to do

Size of cells

When a small piece of graph paper bearing several pinholes, 1 mm apart, is viewed under the low power lens of a microscope, the pinholes are easily spotted since light passes through them. Thus the diameter of the microscope's field of view can be estimated. Look at figure 4.8 (a), for example, where the diameter is 2 mm. Since one millimetre (mm) = 1000 micrometres (μm), the diameter of the field of view for this microscope = 2000 μm.

Next a specimen of cells is viewed and the average number of cells lying along the diameter of the field is found. Look at figure 4.8 (b) where this number is ten for rhubarb epidermal cells. Since the length of ten cells = the diameter of the field of view = 2000 μm, the average length of one rhubarb cell = $\frac{2000}{10}$ = 200 μm.

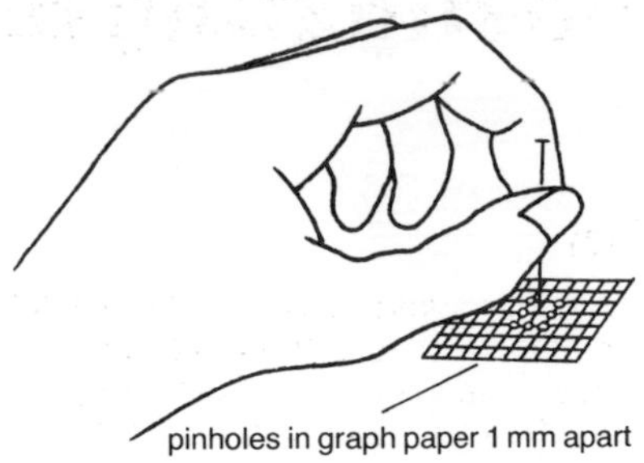

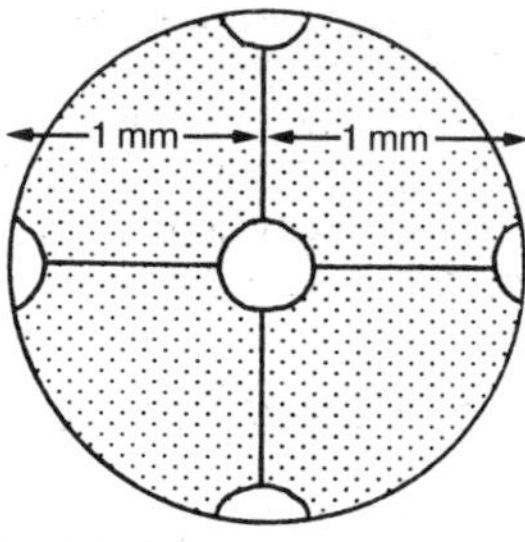

(a) pinholes under microscope

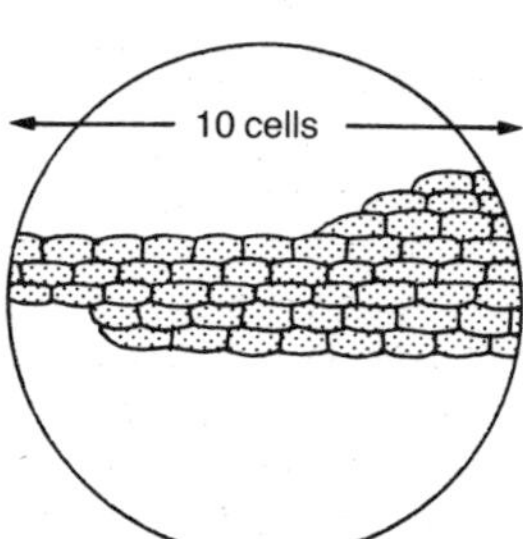

(b) cells under microscope

Figure 4.8 Measuring cell size

KEY QUESTIONS

1 What name is given to the basic units of which all living things are composed?
2 Name THREE structural features common to both a typical plant and a typical animal cell.
3 Give THREE ways in which the leaf cells shown in figure 4.3 differ in structure from a unicellular animal.
4 What is the purpose of staining a cell?

5 Investigating diffusion

Diffusion

The molecules of a liquid (and a gas) move about freely all the time. In the experiment shown in figure 5.1, the purple molecules move from a region of high concentration (the dissolving crystal) to a region of low concentration (the surrounding water) until the concentration of purple molecules (and water molecules) is uniform throughout.

Diffusion is the name given to this movement of the molecules of a substance from a region of high concentration of that substance to a region of low concentration until the concentration becomes equal.

Importance of diffusion to the cell

In a living cell (e.g. the unicellular animal *Amoeba* shown in figure 5.2) oxygen is constantly being used up by the cell contents during respiration. This results in the concentration of oxygen molecules inside the cell being lower than in the surrounding water. The cell membrane is freely permeable to the tiny oxygen molecules. Oxygen therefore diffuses into the cell from a high concentration to a low concentration.

At the same time the living cell contents are constantly making carbon dioxide (CO_2) by respiration. Thus inside the cell is a region of higher CO_2 concentration than outside. The cell membrane is also freely permeable to tiny CO_2 molecules and these diffuse out as shown in figure 5.3.

Diffusion is important to a cell since it is the means by which useful substances such as oxygen enter and waste materials such as carbon dioxide leave.

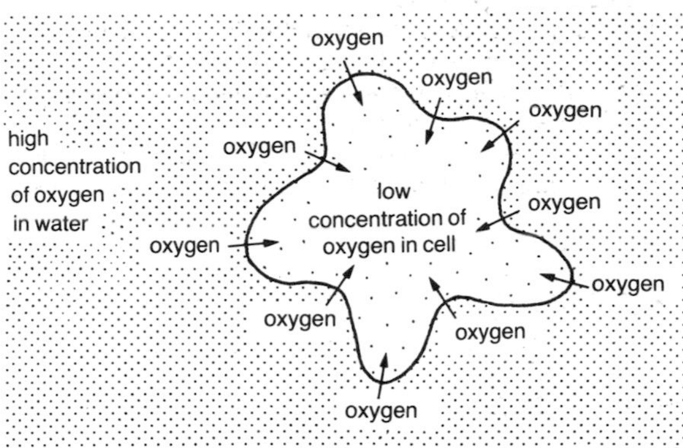

Figure 5.2 Diffusion into a cell

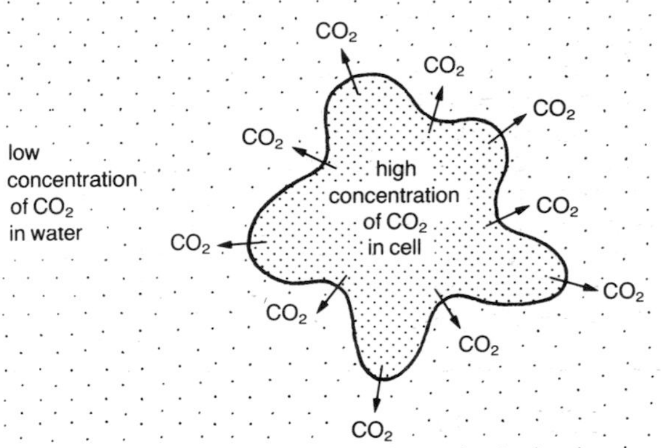

Figure 5.3 Diffusion out of a cell

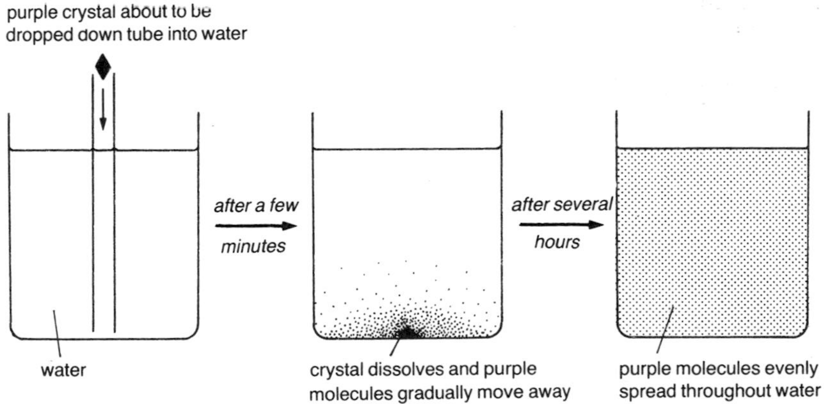

Figure 5.1 Diffusion

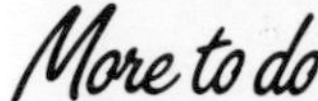

Importance of diffusion to the organism

In multicellular organisms such as man, diffusion also plays an important role. For example, blood returning to the lungs from living cells (see figure 5.4) contains a higher concentration of carbon dioxide and a lower concentration of oxygen than the air in the air sac. CO_2 therefore diffuses out of the blood and oxygen diffuses in. When the blood reaches living body cells, the reverse process occurs and the cells gain oxygen from the blood and lose CO_2 by diffusion.

Diffusion in a cell model

In the experiment shown in figure 5.5, the visking tubing is used to act as a cell membrane. After an hour, the water surrounding the visking tubing is found to give a positive result with Benedict's solution (the test for simple reducing sugar – see Appendix 6) but a negative result with iodine solution (the test for starch).

It is therefore concluded that glucose molecules are small enough to diffuse out through the visking tubing but that the starch molecules are too large to pass through.

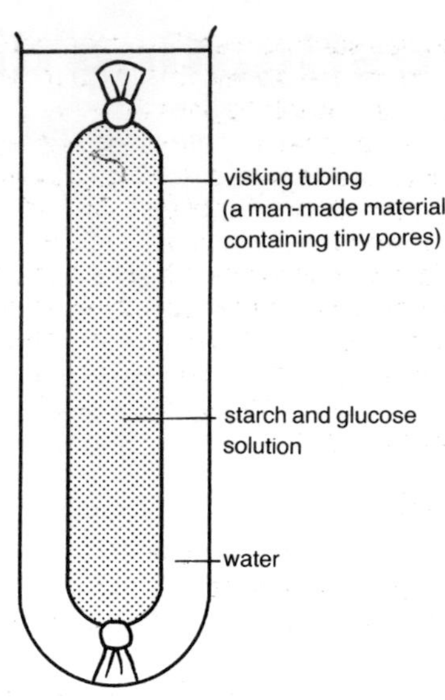

Figure 5.5 Diffusion in a cell model

Role of cell membrane

Although the membrane of a cell is freely permeable to small molecules such as oxygen, carbon dioxide and water, it is not equally permeable to all substances.

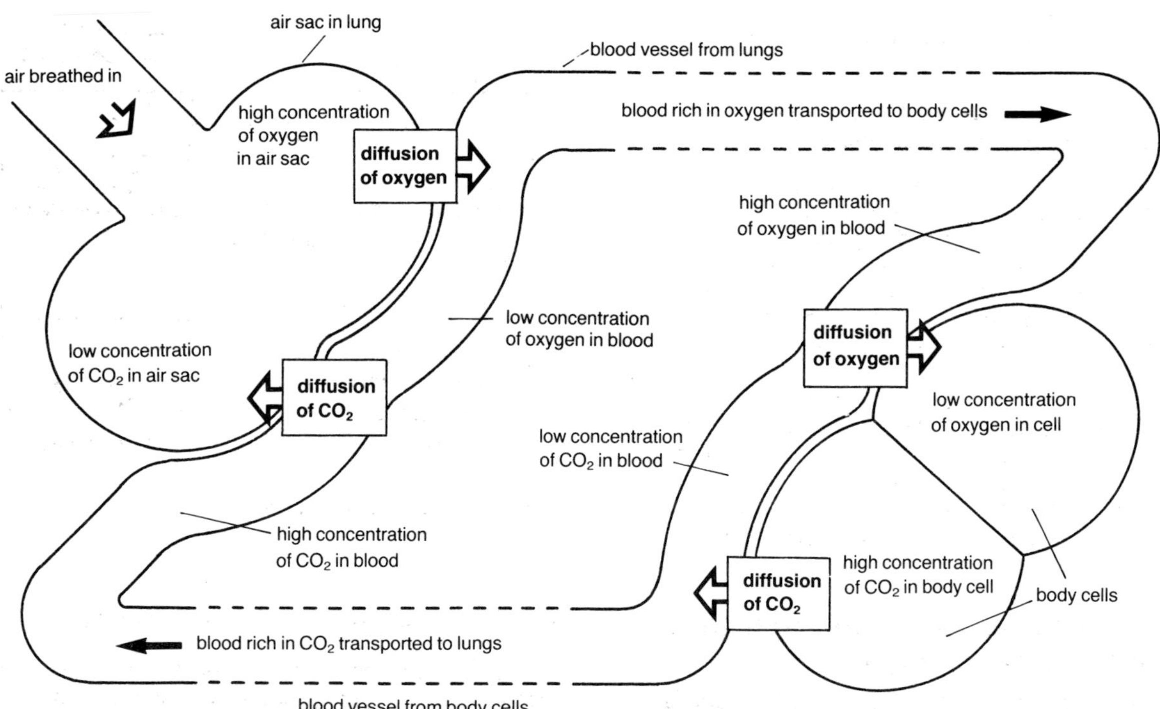

Figure 5.4 Diffusion in man

Larger molecules such as dissolved food can pass through the membrane slowly. Even larger molecules such as starch are unable to pass through.

Thus the cell membrane controls the passage of substances into and out of a cell. The exact means by which it exerts this control is not yet fully understood. It is thought that some molecules pass into and out of the cell by dissolving in the membrane. It is also known that most cell membranes possess tiny pores. It is possible that some molecules enter or leave by these pores, whereas other molecules are kept inside or outside the cell by the fact that they are too big to pass through the pores.

KEY QUESTIONS

1 Give the meaning of the term *diffusion*.
2 Why is diffusion important to a living cell?
3 a) Name an essential substance that enters an animal cell by diffusion.
b) Name a waste material that diffuses out of an animal cell.
c) What structure controls the passage of substances into and out of a cell?

4 'Diffusion is an important process that occurs inside the human body.' Justify this statement by describing an example.

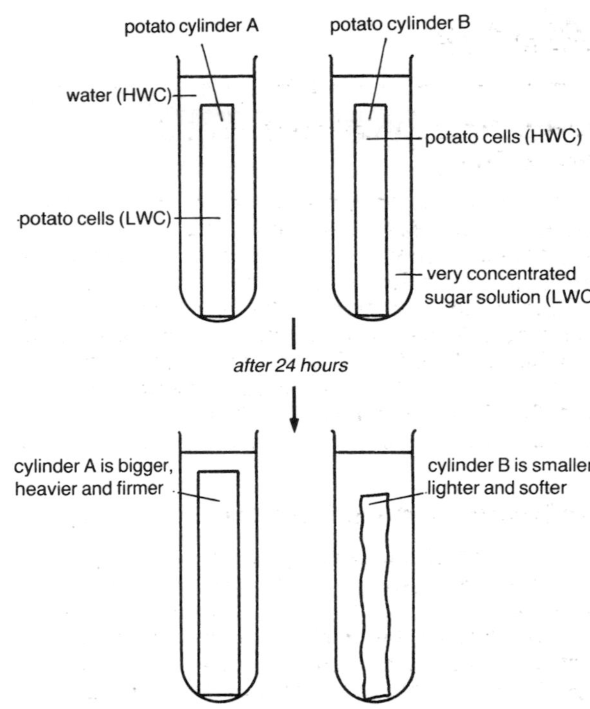

Figure 5.6 Osmosis in potato cylinders

Effect of water and concentrated sugar solution on potato cylinders

In the experiment shown in figure 5.6, the water outside potato cylinder A has a higher water concentration (HWC) than the contents of the potato cells which have a lower water concentration (LWC).

The contents of the potato cells in cylinder B have a higher water concentration (HWC) than the surrounding sugar solution which has a lower water concentration (LWC).

After 24 hours, cylinder A is found to have increased in volume and weight and to have become firmer in texture. Cylinder B, on the other hand, has decreased in both volume and weight and has become softer.

It is therefore concluded that water molecules have diffused into cylinder A and out of cylinder B through the cell membranes.

Effect of water and concentrated salt solution on eggs

In the experiment shown in figure 5.7, each egg is surrounded by its membrane only. Its shell has been removed by acid treatment prior to the experiment.

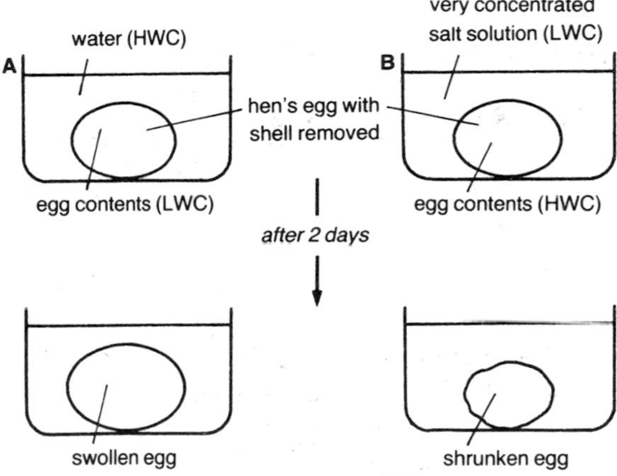

Figure 5.7 Osmosis in eggs

After two days, egg A swells up whereas egg B shrinks.

It is therefore concluded that water has diffused into egg A (from a higher water concentration to a lower water concentration) and out of egg B (from a higher water concentration to a lower water concentration).

Osmosis

In both the potato cylinder and the egg experiments, water always passes from a region of higher water concentration (HWC) to a region of lower water concentration (LWC) through a membrane. This movement of water molecules only through a membrane is a special case of diffusion called **osmosis**.

More to do

Osmosis continued

A **selectively permeable** membrane contains tiny pores which allow the rapid passage through it of small water molecules. Larger molecules such as sugar (sucrose) can only pass through slowly. Even larger insoluble molecules (e.g. starch) cannot pass through at all.

In the experiment shown in figure 5.8, the level in the tube is found to rise after a few minutes. It is therefore concluded that water molecules have passed easily and rapidly through the selectively permeable membrane.

Concentration gradient

The difference in concentration that exists between two regions before diffusion occurs is called the **concentration gradient**. During diffusion molecules always move along a concentration gradient from high to low concentration.

At the start of the experiment in figure 5.8, a water concentration gradient exists between the two sides of the membrane. Osmosis is the movement of water along this concentration gradient from high to low water concentration through a selectively permeable membrane. (Strictly speaking, water molecules move across the membrane in both directions, but mainly from HWC to LWC.)

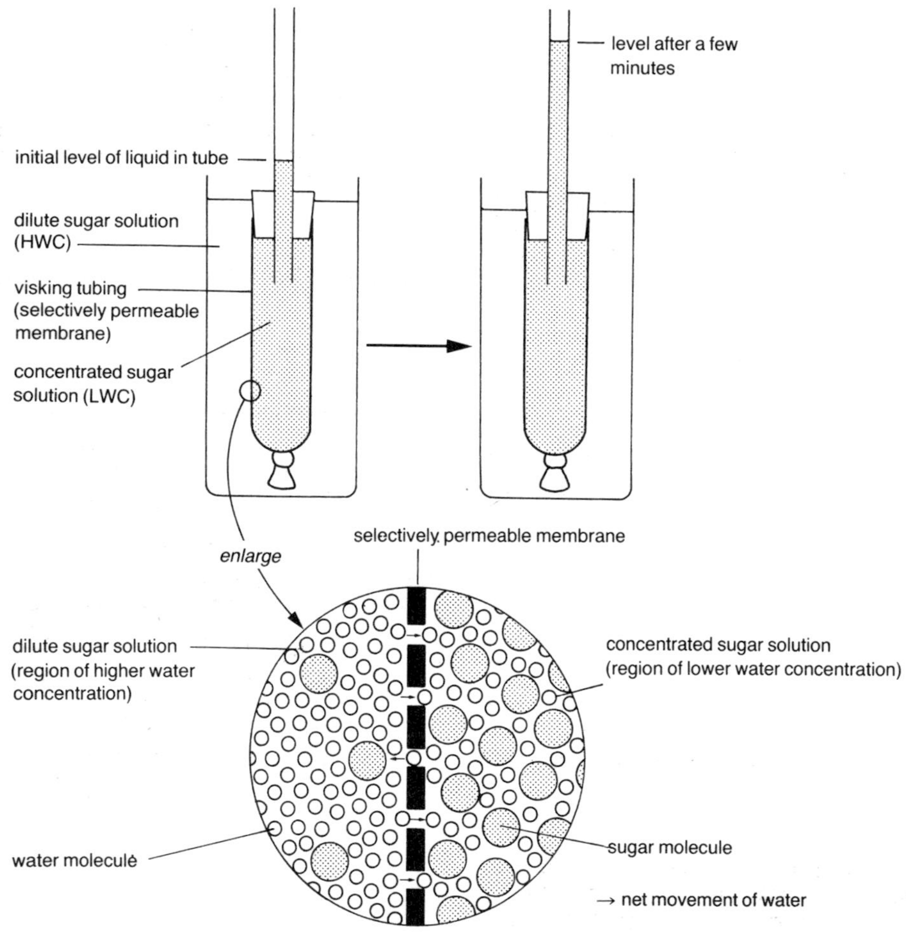

Figure 5.8 Osmosis

Water relations of cells

Since a cell membrane is selectively permeable, water passes into and out of living cells by osmosis. The direction depends on the water concentration of the liquid in which the cell is immersed compared with that of the cell contents.

Red blood cells

Since pure water has a higher water concentration than the contents of red blood cells, water enters by osmosis until the cells burst (see figure 5.9). Since 0.85 per cent salt solution has the same water concentration as the cell contents, there is no net flow of water into or out of the cell by osmosis. Since 1.7 per cent salt solution has a lower water concentration than the cells, water passes out and the cells shrink.

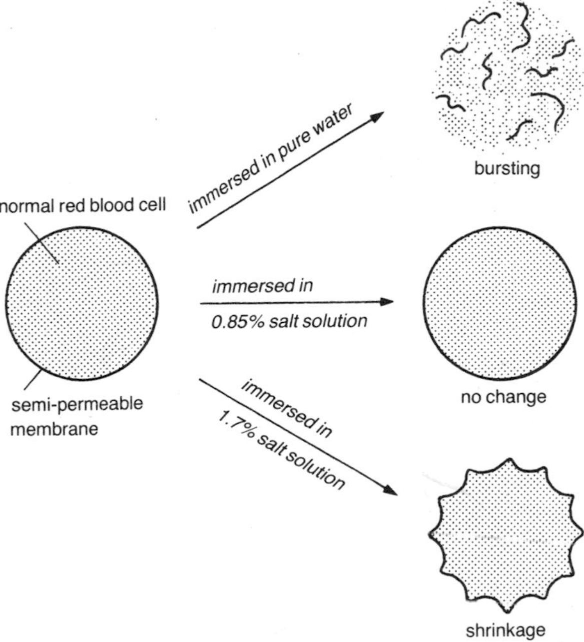

Figure 5.9 Osmosis in red blood cells

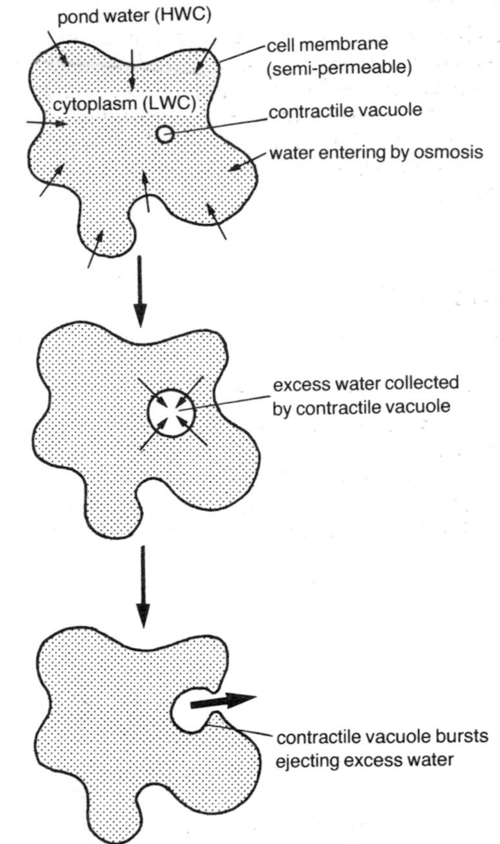

Figure 5.10 Role of contractile vacuole in Amoeba

Amoeba

Unicellular animals that live in fresh water take in water continuously by osmosis. Bursting is prevented by the **contractile vacuole** (see figure 5.10) removing excess water.

Plant cells

Since pure water has a higher water concentration than the contents of a normal plant cell (figure 5.11), water enters the cell by osmosis. The vacuole swells up and presses the cytoplasm against the cell wall which stretches slightly and presses back, preventing the cell from bursting. Cells in this swollen condition

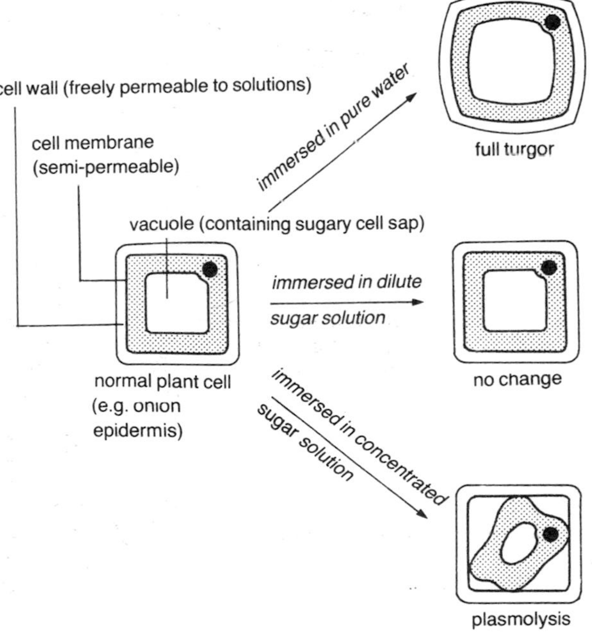

Figure 5.11 Osmosis in plant cells

are said to be **turgid**. A young plant depends on the turgidity of its cells for support.

Since the dilute sucrose (sugar) solution (figure 5.11) has the same water concentration as the cell contents, no net flow of water occurs.

Since the concentrated sucrose solution has a lower water concentration than the cell contents, water passes out of the cell by osmosis. The living contents shrink and pull away from the fairly rigid cell wall. Cells in this state are said to be **plasmolysed**. However they are not dead. When immersed in water plasmolysed cells regain turgor by taking in water by osmosis.

KEY QUESTIONS

1 a) Give a named example of (i) an animal material, (ii) a plant material, in which the process of osmosis may be observed without the aid of a microscope.
b) Describe TWO changes undergone by the plant material when immersed in water for 24 hours.
c) Describe one change undergone by the animal material after immersion in salt solution for 2 days.

2 a) Using all the following words and phrases, describe the main features of osmosis: lower water concentration, membrane, water molecules, higher water concentration.
b) Rewrite the following sentence and complete the blanks. Since the movement of water molecules from a high to a low water concentration occurs through a __________, osmosis is said to be a special case of __________.

Extra Questions

3 a) When a membrane is described as being **selectively permeable**, what does this mean?
b) With reference to the terms selectively permeable membrane and concentraton gradient of water, explain how the process of osmosis occurs.

4 a) With reference to the water concentrations involved, explain why red blood cells burst when placed in water yet onion epidermal cells do not.
b) Why do red blood cells shrink when placed in concentrated salt solution?

5 a) Describe and explain the osmotic effect of a very concentrated sugar solution on onion epidermal cells.
b) What name is given to cells in this state?
c) How could such cells be restored to their normal turgid condition?

6 Investigating cell division

Cell division

Figure 6.1 shows the main events that occur during cell division in an animal cell. The nucleus divides first and the two daughter nuclei separate. Next the cytoplasm becomes pinched off between the two nuclei, forming two daughter cells.

Division in a plant cell is shown in figure 6.2. Since a cell wall is present the cytoplasm cannot pinch off in the middle. Instead, nuclear division is followed by the laying down of a new cell wall between the daughter cells.

In unicellular organisms such as *Amoeba* and *Pleurococcus*, cell division is a form of **reproduction** producing two new individuals. In multicellular organisms such as man and onion plant, cell division results in **growth** of the original organism. Cell division in both cases is the means by which the organism increases its number of cells.

Figure 6.3 Chromosomes

Chromosomes

Chromosomes are threadlike structures found inside the nucleus of every living cell (see figure 6.3).

Mitosis

This is the process by which the nucleus divides into two daughter nuclei each of which contains exactly

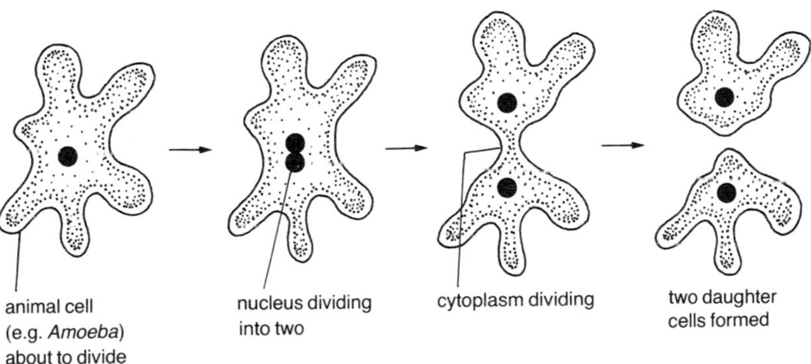

Figure 6.1 Cell division in an animal cell

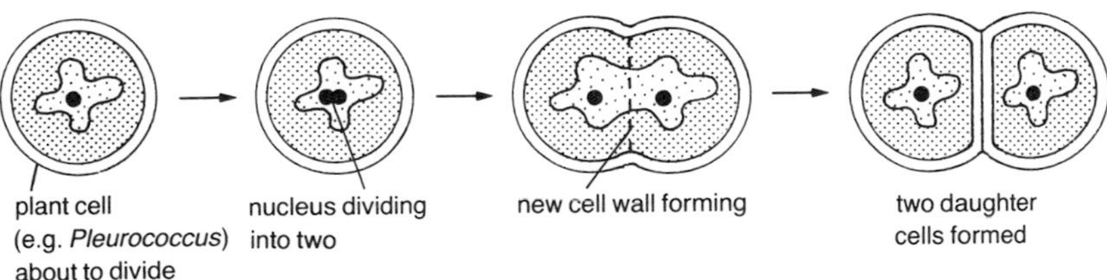

Figure 6.2 Cell division in a plant cell

the same number of chromosomes as the original nucleus. Figure 6.4 shows the sequence of events that occurs during mitosis.

As each chromosome becomes shorter and thicker it is seen to be a double thread. Each thread is called a **chromatid**. The two chromatids of a chromosome are joined together by a **centromere**. Soon a spindle forms and each chromosome becomes attached by its centromere to one of the spindle fibres at the 'equator'.

Next each centromere splits and one chromatid from each pair moves to the 'north' pole and one to the 'south' pole. Finally a nuclear membrane forms round each group of chromatids (now regarded as chromosomes) and nuclear division (mitosis) is complete.

Mitosis is followed by division of the cytoplasm to form two identical daughter cells. Each cell now undergoes a period of cell growth. During this time the

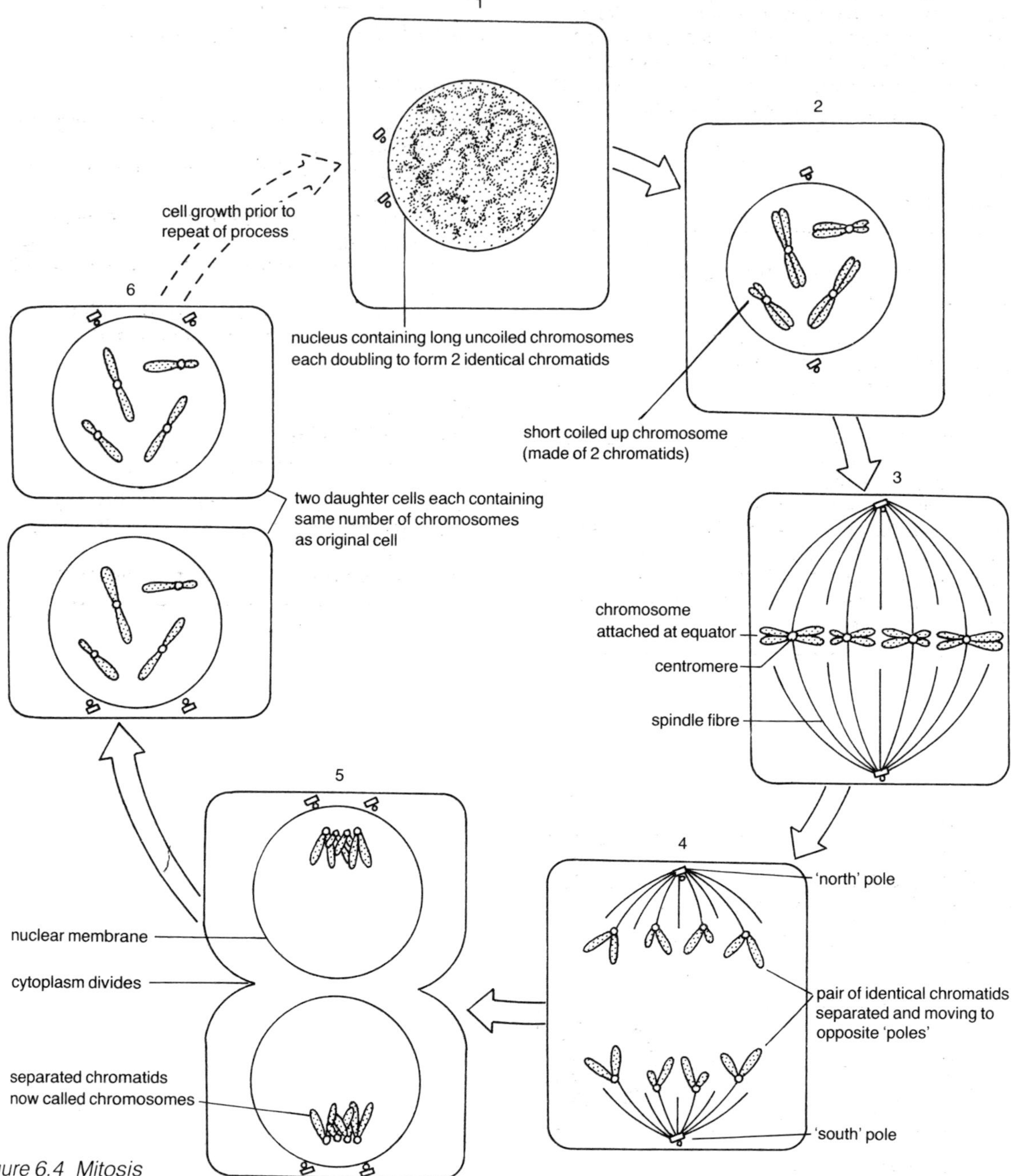

Figure 6.4 Mitosis

chromosomes in each nucleus cannot be seen and sometimes this stage (called interphase) is referred to as the resting period. However, the chromosomes are not really resting. Instead, each is completely uncoiled and busy making an exact double (replica) of itself in preparation for the next nuclear division.

Length of time

The time taken by a cell to go through the entire cycle of cell division from interphase to interphase varies from species to species. An onion root tip cell takes twenty-two hours to complete the full cycle at room temperature. Of this time, mitosis (nuclear division) only takes about 90 minutes. The remainder of the time is spent on cell growth and formation of chromatids.

Maintenance of chromosome complement

Every species of plant and animal has a definite and characteristic number of chromosomes (the chromosome **complement**) present in each cell (see table 6.1).

Chromsomes provide the main source of genetic information typical of a particular species of living thing. It is essential that each cell formed as a result of mitosis receives a full complement of chromosomes, so that during growth and development the cells of a multicellular organism will be able to endow the animal or plant with all the characteristics of its species.

Mitosis maintains this continuity of chromosome complement from cell to cell.

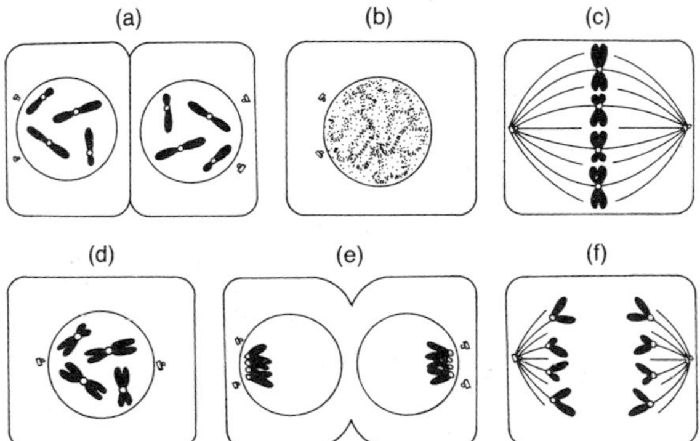

Figure 6.5 See question 5

species of living thing	chromosome complement
onion	16
cabbage	18
rice	24
fruitfly	8
frog	24
cat	38
man	46
horse	66

Table 6.1 Chromosome complements

KEY QUESTIONS

1 Describe the main events that occur during cell division in an animal cell.
2 How does cell division in a plant differ from that in an animal?
3 Which cell structure controls all cell activities including cell division?
4 Describe the process of mitosis (nuclear division) in a cell.
5 Arrange the stages of mitosis shown in figure 6.5 in the correct sequence beginning with (d).

6 a) Why is it important that the chromosome complement of daughter cells in a multicellular organism is maintained?
b) What process maintains this continuity?

7 Investigating enzymes

Effect of a catalyst

Look at the experiment shown in figure 7.1. The bubbles forming the froth in tube A are found to relight a glowing splint, showing that oxygen is being released during the breakdown of hydrogen peroxide (into water and oxygen).

In tube B, the control, the breakdown process is so slow that no oxgyen can be detected.

It is concluded, therefore, that manganese dioxide (which remains chemically unaltered at the end of the reaction) has increased the rate of a chemical reaction which otherwise would only proceed very slowly.

A chemical which alters the rate of a reaction and yet remains itself unaltered is called a **catalyst**.

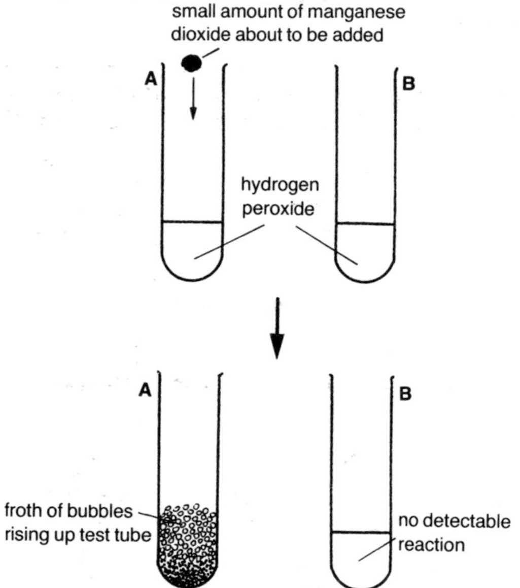

Figure 7.1 Effect of a catalyst

Effect of catalase on hydrogen peroxide

Look at the experiment shown in figure 7.2. The bubbles formed in tubes A and B are found to relight a glowing splint, showing that oxygen is being released during the breakdown of hydrogen peroxide. In the control tubes C and D there is no detectable reaction.

It is therefore concluded that some substance present in living cells has increased the rate of breakdown of hydrogen peroxide which would otherwise only proceed very slowly.

This substance is called **catalase**. It is made by living cells. Since it speeds up the rate of a chemical reaction in living cells yet remains itself unaltered, it is an example of a biological catalyst. Biological catalysts are called **enzymes**.

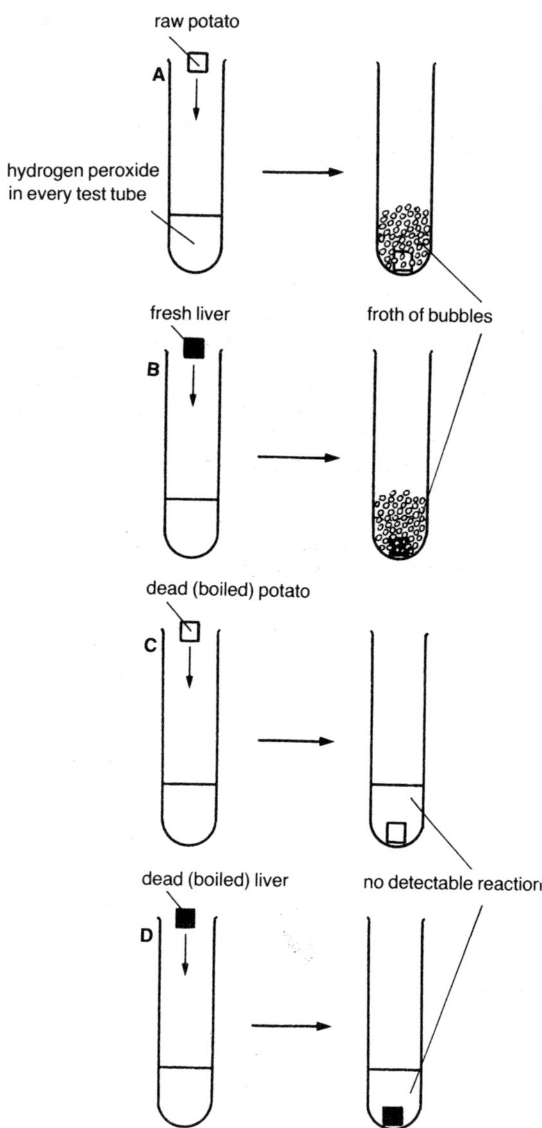

Figure 7.2 Effect of catalase on hydrogen peroxide

Enzymes

Catalase is only one example of an enzyme. Many different enzymes are made by and are present in all living cells. Each enzyme plays its role in speeding up a particular chemical reaction inside or outside the cells. All the chemical reactions that occur in living organisms are controlled by enzymes.

Importance of control

A **control** is a copy of the experiment in which all factors are kept exactly the same except the one being investigated in the original experiment. When the results are compared, any difference found between the two must be due to that one factor. For example, in the above experiments we can conclude that in the first one a catalyst and in the second one an enzyme increased the rate of breakdown of hydrogen peroxide. If controls had not been set up it would be valid to suggest that the breakdown of hydrogen peroxide would have proceeded rapidly whether a catalyst or an enzyme had been present or not.

More to do

Role of enzymes in cell chemistry

Normally the rate of a chemical reaction increases as temperature rises. For this reason chemical reactions involving non living substances are often carried out at very high temperatures to obtain the best results.

Most living cells function at fairly low temperatures (see table 7.1). Continued life on our planet depends on vital biological processes such as photosynthesis, respiration and growth, each of which consists of many biochemical reactions. Without the aid of biological catalysts, these reactions in living cells would proceed at too slow a rate to maintain life. Enzymes are needed to increase the rate of these biochemical reactions and are therefore essential to life.

living organism	typical temperature of body cells in organism's natural environment (°C)
trout	3
frog	15
spider plant	20
man	37
pigeon	42

Table 7.1 Typical body temperatures

KEY QUESTIONS

1 What is a catalyst?
2 a) By what general name are biological catalysts known?
 b) Give a named example of a biological catalyst.
 c) Where are enzymes found?
 d) What is the function of an enzyme?
3 Explain the importance of including a control in an experiment.

4 a) Within what approximate range of temperature are living cells found to function?
 b) Explain why enzymes are required in cell chemistry.

Effect of plant amylase (diastase) on starch

Look at the experiment shown in figure 7.3. From the results it is concluded that in tube A the enzyme plant **amylase** (diastase) has promoted the breakdown of

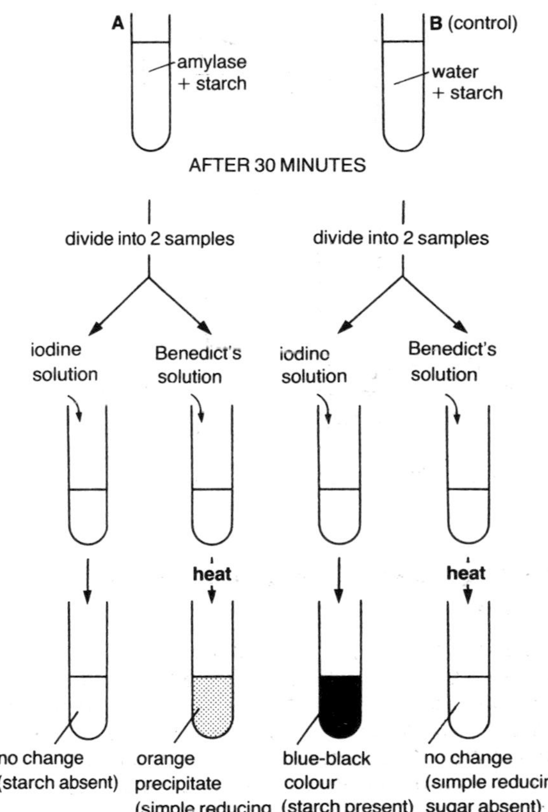

Figure 7.3 Action of plant amylase

starch to simple sugar (maltose). In tube B, the control, which lacks the enzyme no detectable reaction has occurred.

The substance upon which an enzyme acts is called the **substrate**. The substance produced as a result of the reaction is called the **end product**. The reaction being promoted in this experiment can therefore be summarised as in the following word equation:

starch (substrate	amylase ⟶ (enzyme)	simple sugar (maltose) (end product)

Salivary amylase and pancreatic amylase made in the human body similarly promote the breakdown of starch into simple sugar.

Action of potato phosphorylase on glucose-1-phosphate

A sample of potato extract is prepared by liquidising a mixture of fresh potato tuber and water and then centifuging the mixture until the supernatant (see figure 7.4) is starch-free.

This potato extract is added to an active form of glucose (called glucose-1-phosphate) in each of four dimples in row A of a tile as shown in figure 7.5. Rows B and C are controls. One dimple for each condition is tested at three minute intervals with iodine solution. Starch is found to be formed in row A only.

From this experiment it is concluded that a substance present in potato extract has promoted the building up (synthesis) of glucose into starch.

The substance present in the potato extract is an enzyme called **potato phosphorylase**. Its action can be summarised as follows:

glucose-1-phosphate (substrate)	potato phosphorylase ⟶ (enzyme)	starch (end product)

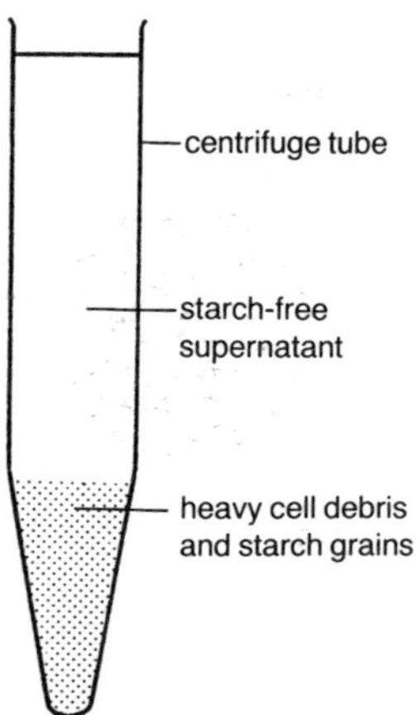

Figure 7.4 Preparation of potato extract

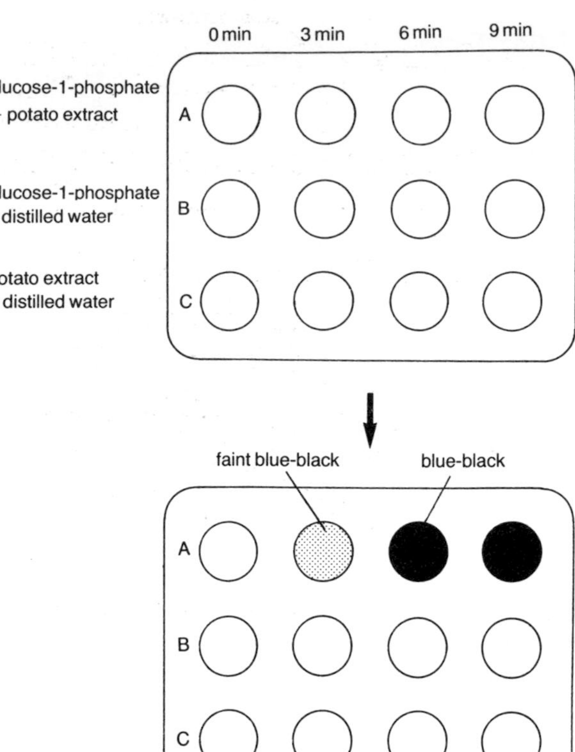

Figure 7.5 Action of potato phosphorylase

The above two experiments show that some enzymes (e.g. amylase) promote reactions which involve the breakdown (degradation) of complex molecules into simpler ones whereas other enzymes (e.g. potato phosphorylase) promote reactions which bring about the building up (synthesis) of complex molecules from simpler ones.

All enzymes are made of **protein**.

More to do

Specificity of an enzyme

Each enzyme acts on only one type of substance (the substrate). Amylase, for example, is only able to promote the breakdown of starch and no other substance. Each enzyme is said therefore to be **specific** to its one substrate. It is thought that the shape of a molecule of enzyme exactly matches the shape of a molecule of its substrate like a key which exactly fits a lock. This allows the two molecules to combine briefly bringing about the reaction. The **'lock-and-key'** theory of enzyme action is illustrated in figures 7.6 and 7.7.

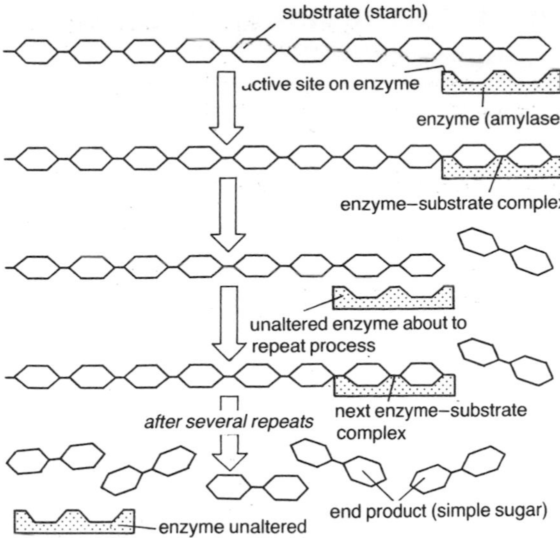

Figure 7.6 Lock-and-key model of enzyme action – amylase

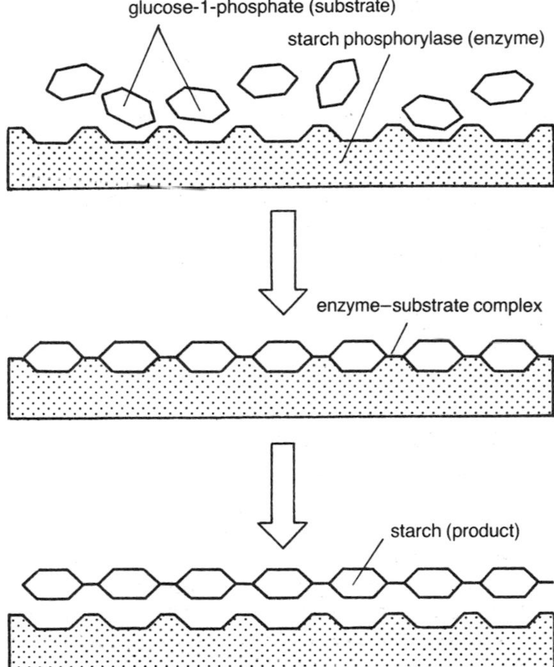

Figure 7.7 Lock-and-key model of enzyme action – phosphorylase

KEY QUESTIONS

1 a) Give a named example of an enzyme which promotes (i) the breakdown, (ii) the synthesis of a substance.
b) For each of these enzymes, summarise the reaction that it promotes in a word equation.

2 When preparing potato extract for the potato phosphorylase experiment, the mixture is centrifuged until the supernatant is starch-free. Explain the reason for this procedure.

3 Explain why rows B and C are included in the experiment shown in figure 7.5.

4 Of what substance are all enzymes composed?

Extra Questions

5 What is meant by the term **specific** as applied to enzymes?

6 In the models of enzyme activity shown in figures 7.6 and 7.7, which substance is equivalent to the 'lock' and which to the 'key'?

Starch agar

Starch agar is a jelly-like substance containing a uniform suspension of starch grains. Therefore, when a petri dish of starch agar is flooded with iodine solution, the contents of the dish turn blue-black as shown in figure 7.8.

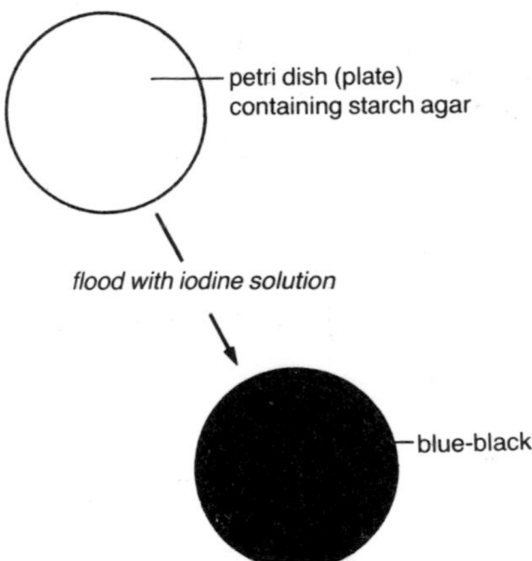

Figure 7.8 Starch agar

Effect of temperature on amylase activity

Plant amylase is an enzyme which promotes the breakdown of starch to simple sugar (see page 89). In the experiment shown in figure 7.9, a cork borer is used to make a hole in the centre of the five petri dishes of starch agar. The enzyme amylase is added to each hole and the plates kept at five different temperatures for twenty-four hours. When the plants are then flooded with iodine solution, those areas still containing starch turn blue-black but those lacking starch do not.

It is concluded from this experiment that an area lacking the blue-black colour is a region where the enzyme amylase has successfully promoted the breakdown of starch to sugar. The larger the diameter of the non blue-black zone, the more active the enzyme has been.

From the results it can be seen that amylase is most active at 30 °C. At high temperatures (e.g. 50 °C) no breakdown of starch has occurred. This is because enzymes are destroyed (denatured) at temperatures of 50 °C and above. At low temperatures (e.g. 4 °C) only a little breakdown of starch has occurred. This is because enzymes are fairly inactive (but not destroyed) at low temperatures.

One variable factor

An investigation is a fair test if at each stage only one difference (**variable factor**) is studied at a time. If several differences are involved at the same time then it is impossible to know which one is responsible for the results obtained.

The experiment shown in figure 7.9 is valid and fair because it tests only one variable factor (temperature). See also Appendix 8.

Effect of pH on catalase activity

In the experiment shown in figure 7.10, the action of catalase on hydrogen peroxide is investigated. The one variable factor is the pH of the hydrogen peroxide solution. Each of the different pH conditions is maintained by adding a suitable buffer solution (a special chemical which keeps an experiment at a required pH).

When an equal-sized piece of fresh liver is added to each cylinder, the results shown in the diagram are produced. This is because liver contains the enzyme catalase which promotes the reacton:

$$\text{hydrogen peroxide} \xrightarrow{\text{catalase}} \text{water} + \text{oxygen}$$

As oxygen is released, it produces a froth of bubbles. The height of the froth formed indicates the activity of the enzyme at each pH.

From the results it can be seen that catalase is most active at around pH 9 but works fairly well over a range of pH from about 7 to 11. This is called its working range. Catalase does not work well outside this range of pH.

Egg albumen

When glass tubes are filled with egg albumen (egg white) and heated, a solid rod of white albumen is produced inside each tube. This albumen is made of molecules of insoluble protein.

Figure 7.11 shows five rods of albumen (inside their glass tubes) placed against a scale.

Effect of pH on pepsin activity

Pepsin is an enzyme found in the human stomach which promotes the breakdown of insoluble protein molecules to soluble peptide molecules.

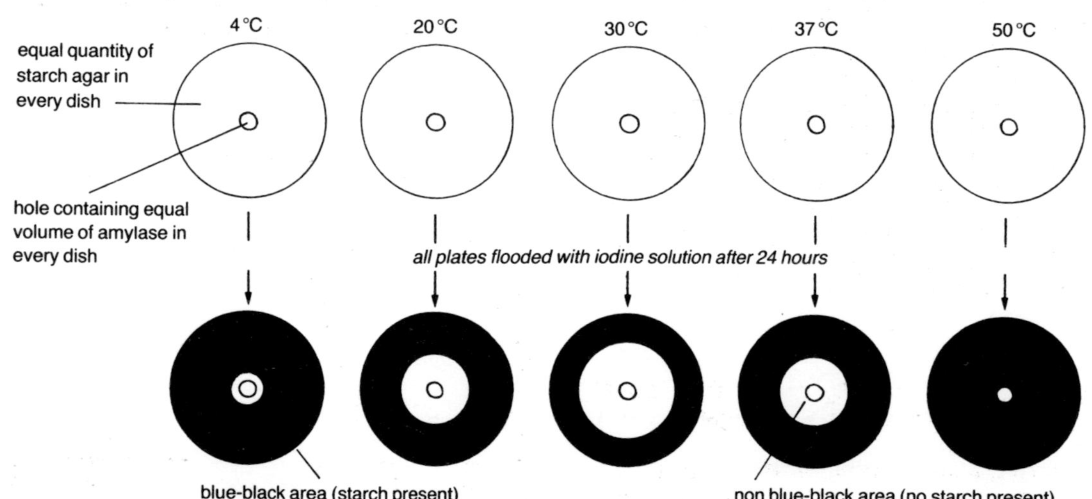

Figure 7.9 Effect of temperature on amylase activity

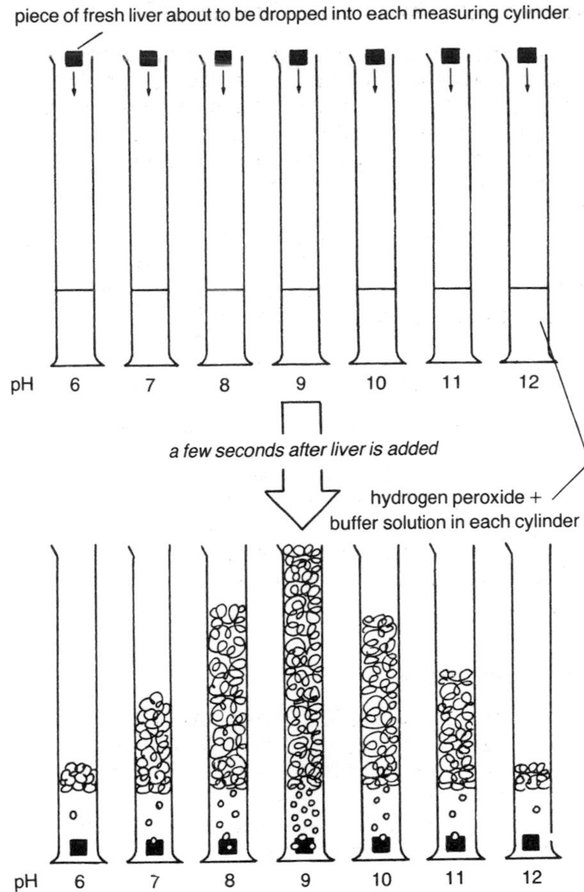

Figure 7.10 Effect of pH on catalase activity

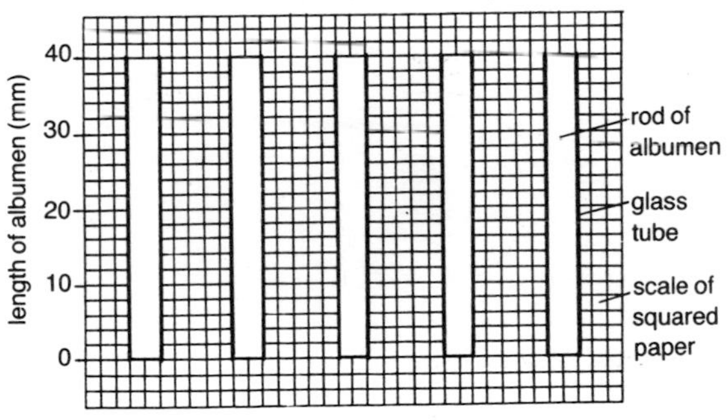

Figure 7.11 Albumen rods

In the experiment shown in figure 7.12, the action of pepsin on rods of egg albumen is investigated. The one variable factor is pH.

During the experiment, pepsin is able to enter the open ends of the glass tubes and under suitable pH conditions digest the insoluble protein to soluble products which pass into the solution. When the tubes

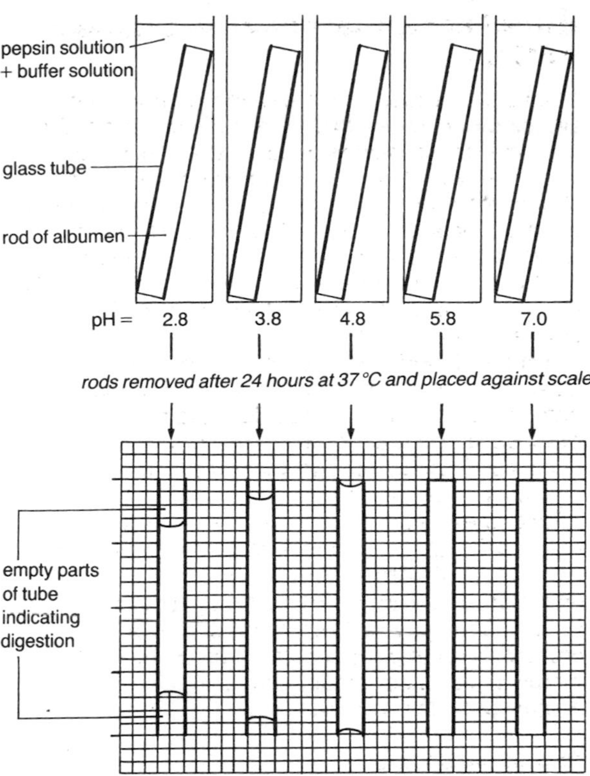

Figure 7.12 Effect of pH on pepsin activity

are placed against a scale at the end of the experiment, the amount of digestion is indicated by the decrease in length of each albumen rod.

From the results it is concluded that pepsin promotes the digestion of protein best in strongly acidic conditions (e.g. pH 2.8). Its activity decreases as pH increases. This is illustrated in figure 7.13 which shows the working pH range of pepsin.

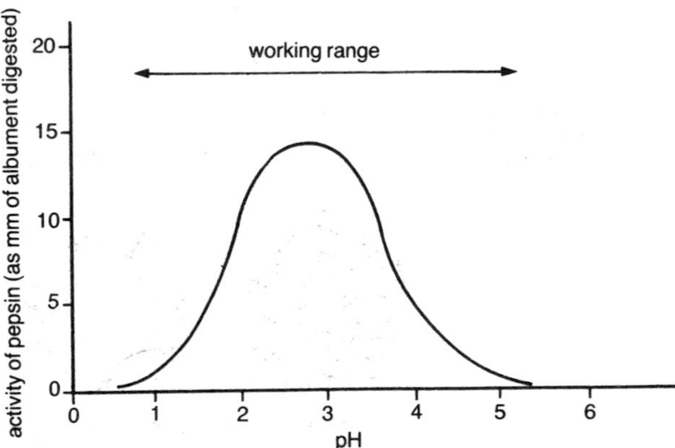

Figure 7.13 Working range of pepsin

Enzymes and pH

Each enzyme works best at a particular pH related to the conditions in which it normally operates. Pepsin is most active at pH 2–3 and is therefore ideally suited to the conditions of the stomach which are very acidic. However, most enzymes operate in fairly neutral conditions and are therefore found to work best at around pH 7.

More to do

Optimum condition

The above experiments show how enzyme activity is affected by various factors such as temperature and pH. The particular condition of a factor at which enzyme activity works best is called the **optimum** condition.

Since man is a warm-blooded animal, the optimum temperature for human enzyme activity, for example, is 37 °C as shown by the graph in figure 7.14. Many plant enzymes, on the other hand, have an optimum temperature several degrees Celsius below this.

The graph in figure 7.15 shows that for each enzyme there is a particular optimum condition of pH at which the enzyme is most active.

KEY QUESTIONS

1 a) Describe the effect of each of the following temperatures on the activity of plant amylase: (i) 4 °C, (ii) 30 °C, (iii) 60 °C.
b) At which of these temperatures is the enzyme permanently destroyed?

2 Explain why an equal volume of amylase must be added to each hole in the experiment shown in figure 7.9.

3 For each of the following enzymes, give the pH at which it would be most active: catalase, pepsin, plant amylase.

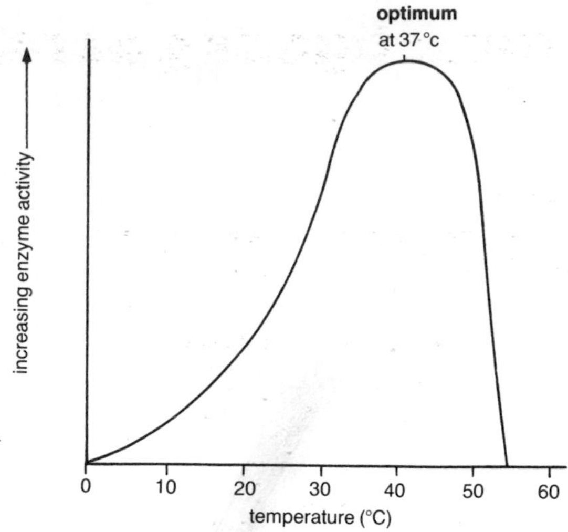

Figure 7.14 Effect of temperature on human enzyme activity

4 Explain the meaning of the term **optimum** condition as applied to the activity of an enzyme.

5 Look at table 7.2 and identify the optimum set of conditions for the activity of **(a)** catalase, **(b)** pepsin and **(c)** plant amylase.

temperature (°C)	pH
25	2.8
25	7.0
25	9.0
37	2.8
37	7.0
37	9.0

Table 7.2 See question 5

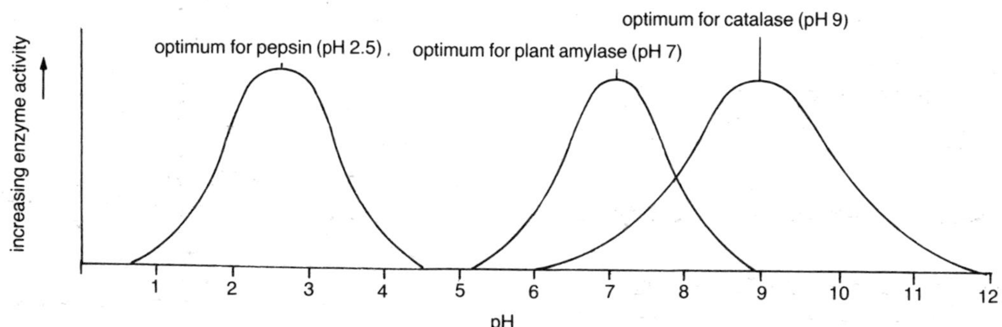

Figure 7.15 Effect of pH on activity of enzymes

8 Investigating aerobic respiration

Energy

Living cells need energy for a variety of reasons as shown in figure 8.1.

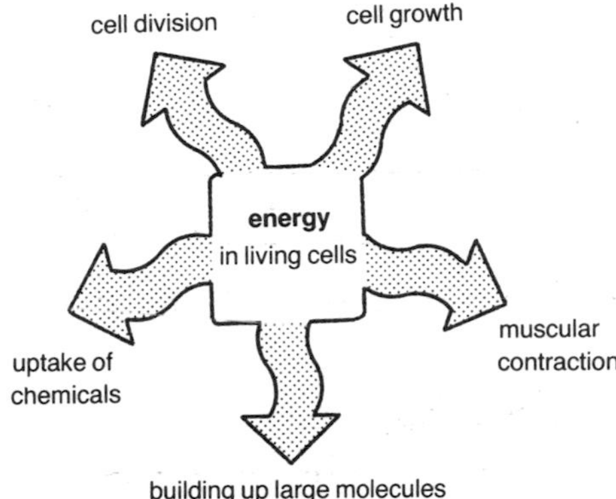

Figure 8.1 Uses of energy

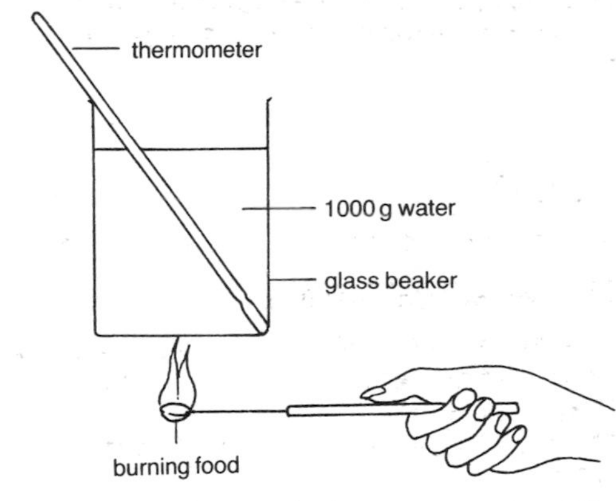

Figure 8.2 Measuring energy in food

Energy transformation

There are many forms of energy, e.g. chemical, heat, light and kinetic (movement) energy. Energy can be transformed (changed) from one form to another by living cells. The following energy transformation occurs in muscles during contraction:

chemical energy ⟶ kinetic energy + heat energy
(movement)

Photosynthesis in green plant cells involves light energy being changed to chemical energy (see page 91).

Food

Living organisms obtain the energy that they need from food. Green plants make their own food by photosynthesis; animals need a ready-made food supply (see page 15).

Measuring the amount of energy in food

The chemical energy ~~continued~~ CONTAINED in food can be transformed to heat energy and measured. In the experiment shown in figure 8.2, 1 g of peanut is ignited and held under a beaker containing 1000 g water. The temperature of the water rises slightly (by about 2 °C).

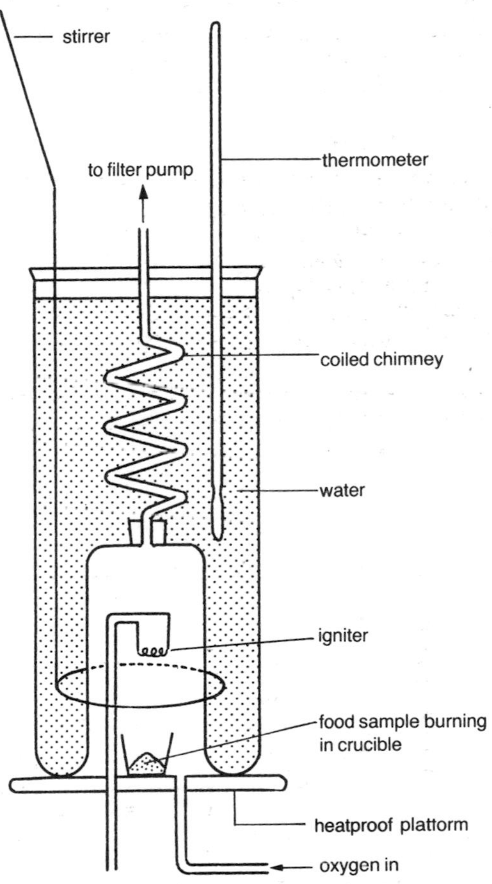

Figure 8.3 Food calorimeter

Energy is measured in **kilojoules** (kJ). 4.2 kJ is the amount of energy required to raise the temperature of 1000 g water by 1 °C. Thus 1 g of peanut has apparently released about 8.4 kJ.

However, much of the heat is lost to the surroundings, the heat that does reach the water is not evenly spread out and the peanut is not completely burned to ashes.

Food calorimeter

This apparatus (figure 8.3) is used to measure the energy content of a food and overcomes the difficulties described above. Since the food sample is enclosed, little heat loss occurs. The stirrer and coiled chimney bring about even distribution of heat and the oxygen supply ensures that the peanut burns completely.

This time the temperature of the water rises by about 6 °C, showing that 1 g of peanut releases about 25.2 kJ. The **calorific value** (amount of energy released when 1 g is burned) of various foods is given in table 8.1. From the table it can be seen that fats and oils contain more chemical energy per gram than either carbohydrates or protein.

food	main components (see also page 96)	calorific value (kJ)
peanut	protein and fat	25.5
butter	fat	33.5
olive oil	oil	37.7
bread	carbohydrate	10.1
meat	protein	13.6

Table 8.1 Energy content of foods

KEY QUESTIONS

1 State THREE reasons why living things need energy.

2 Give an example of an energy transformation that occurs in the cells of
(**a**) an animal, (**b**) a green plant.

3 Rewrite the following sentence to include only the correct words.

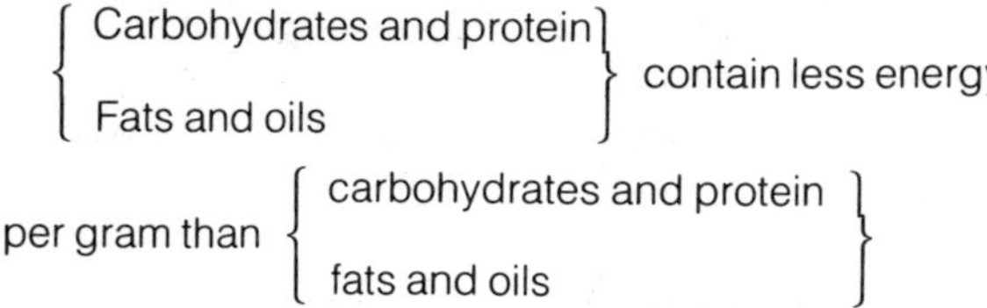
{ Carbohydrates and protein / Fats and oils } contain less energy per gram than { carbohydrates and protein / fats and oils }

Uptake of oxygen

Look at the experiment shown in figure 8.4. After three days a burning splint is plunged into each gas jar. It goes out immediately in gas jar A but continues to burn for a few seconds in B, showing that A contains less oxgyen than B. It is therefore concluded that germinating seeds have taken in oxygen.

Oxygen is needed by living cells to release energy from food during respiration.

Release of carbon dioxide

Green plant in darkness

Look at figure 8.5. Since lime water A remains clear, this indicates that all of the carbon dioxide in the original incoming air has been removed by the sodium

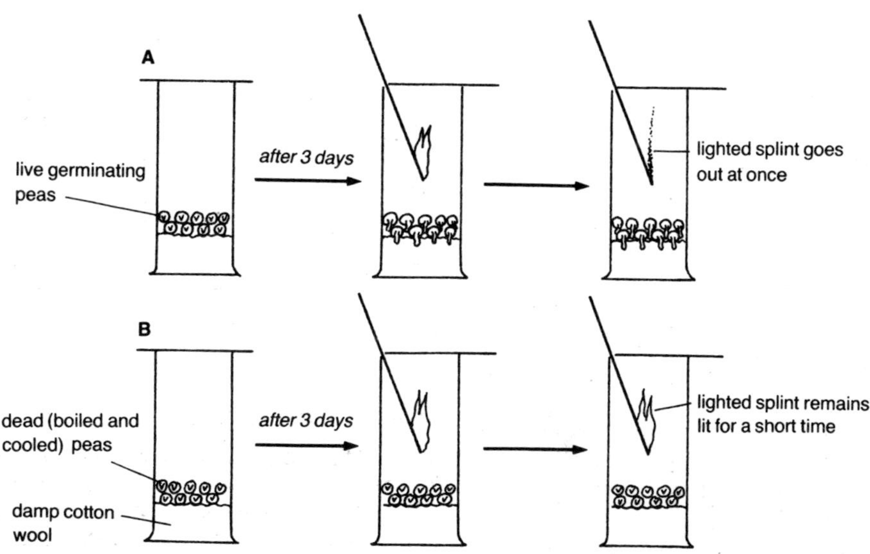

Figure 8.4 Oxygen uptake

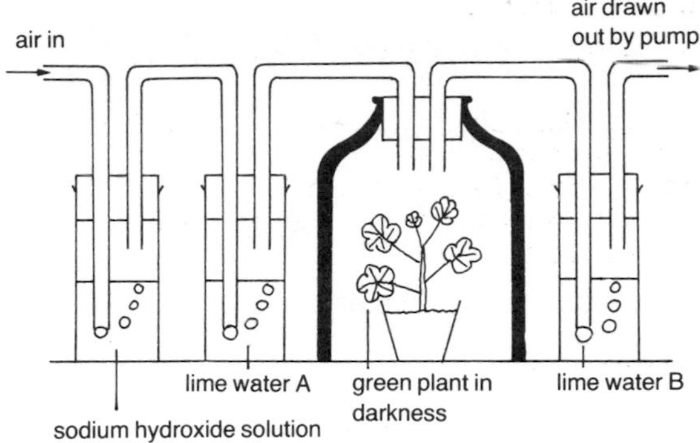

Figure 8.5 Release of CO_2 by green plant

hydroxide. Since lime water B turns milky, this shows that carbon dioxide is produced by a green plant during respiration. The plant is kept in darkness to prevent photosynthesis masking respiration.

Bicarbonate indicator and carbon dioxide

Bicarbonate indicator solution is a chemical which varies in colour depending on the concentration of carbon dioxide (CO_2) that it contains. Table 8.2 shows the relative CO_2 concentrations indicated by the various colours.

Look at the experiment shown in figure 8.6. After a few hours, the bicarbonate indicator in tubes A, B and C turns from red to yellow showing that each organism has given out CO_2 during respiration. Tube D, the control, which remains unchanged shows that the results are valid (and not due to some other factor such as the air already present in each test tube).

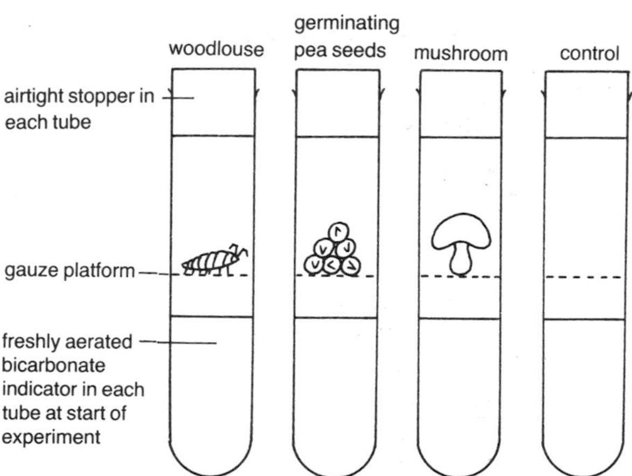

Figure 8.6 Release of CO_2 by living organisms

relative CO_2 concentration	colour of bicarbonate indicator solution
high (above atmospheric)	yellow
medium (atmospheric)	red
low (below atmospheric)	purple

Table 8.2 Bicarbonate indicator range

Universal indicator and pH

Look at figure 8.7. Although the initial colour of the water in the test tube is yellow and therefore neutral (see table 8.3), it changes to orange after a few hours showing that it has become slightly **acidic**. When CO_2 dissolves in water it forms a weak acid. It is therefore concluded that the acidic conditions in the test tube have been produced by the snails giving out CO_2 during respiration.

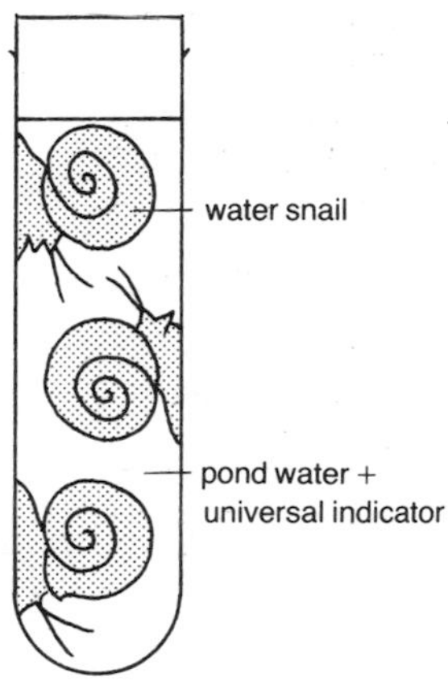

Figure 8.7 CO_2 release and pH

pH condition	colour of universal indicator
strongly acidic	red
weakly acidic	orange
neutral	yellow
weakly alkaline	green
strongly alkaline	blue

Table 8.3 Universal indicator range

Source of carbon dioxide

Look at the experiment shown in figure 8.8. The food is burned and then the contents of the gas jar tested with lime water. Since the lime water turns milky, it is concluded that the food has released CO_2 during burning.

Carbon dioxide is released from food when it is broken down by living cells during respiration.

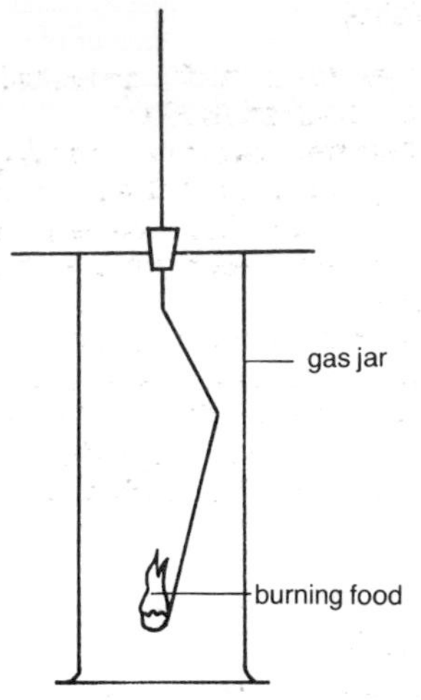

Figure 8.8 Source of CO_2

Release of heat energy

Heat production by germinating peas

Look at figure 8.9. After a few days the temperature is found to have risen, showing that germinating pea seeds release heat energy formed during respiration.

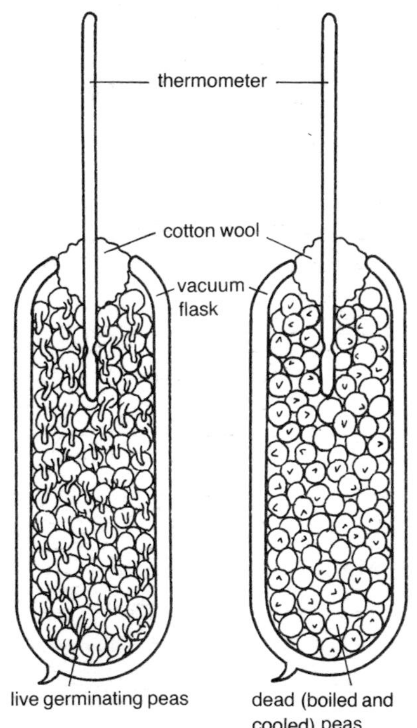

Figure 8.9 Release of heat energy by respiring plant

A control flask containing boiled and cooled seeds shows no temperature increase in the first few days since the peas are dead. However, if the pea seeds have not been washed in disinfectant prior to the experiment, then a little heat energy is released in the control flask after a few more days. This is produced by fungi which begin to attack the dead seeds and respire using the food present in the seeds.

Heat production by animals during respiration

Look at the apparatus set up in figure 8.10. Although the manometer levels are equal at the start of the experiment, after a few minutes X drops and Y rises. It is therefore concluded that the animal is giving out heat energy formed during respiration. This heat, unable to escape, gradually heats up the air trapped in the nearby test tube causing the air to expand and depress level X.

Since the only difference between sides A and B is the presence or absence of the respiring animal, side B acts as a control.

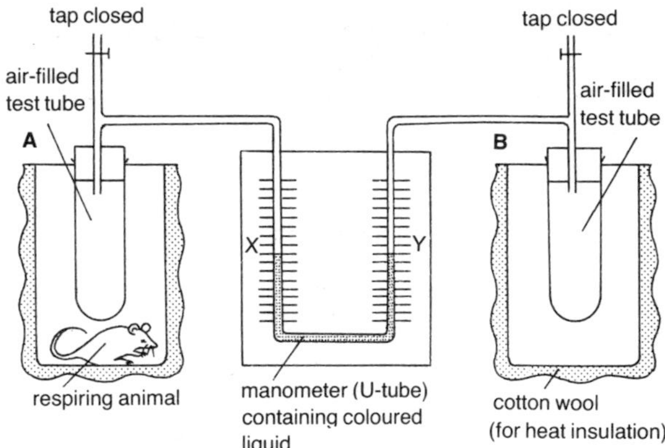

Figure 8.10 Release of heat energy by respiring animal

Aerobic respiration

Respiration involving the use of oxygen is called **aerobic** respiration. The experiments in this chapter have shown that during this process:

cells need oxygen to release energy from food,
carbon dioxide is given off by living cells,
carbon dioxide is derived from food and
heat energy may be released from cells.

In addition, water is formed during aerobic respiration. These facts are often summarised in a word equation as follows:

food + oxygen ⟶ carbon dioxide + water + energy
(e.g. glucose) (e.g. heat)

More to do

Metabolism

Metabolism is the sum of all the chemical processes that occur in a living organism and keep it alive. All metabolic reactions within a living cell take place in a watery solution and are controlled by enzymes.

There are two types of metabolic reaction. The first involves the breaking down of a substance and the release of energy (e.g. the breakdown of glucose into carbon dioxide and water during aerobic respiration). The second type involves the building up of a chemical substance and requires energy to do so (e.g. the synthesis of protein from amino acids).

Thus, aerobic respiration in living cells is a metabolic reaction of the first type which is important because it releases the energy needed by the second type of reaction to build up the materials essential for life. This relationship is summarised in figure 8.11.

In reality, metabolic processes proceed as a large number of small enzyme-controlled steps. Energy is therefore released in small usable amounts rather than in vast quantities which might kill the cell.

KEY QUESTIONS

1 Which gas is **(a)** taken in, **(b)** given out by living cells during aerobic respiration?
2 Rewrite the following sentence choosing only the correct word(s) from the brackets in each case. Cells need (carbon dioxide/oxygen) to release (oxygen/energy) from (food/carbon dioxide) during aerobic respiration.
3 Suggest a suitable control for each of the experiments shown in figures 8.7 and 8.8.
4 Name the form of energy that is released by germinating seeds during respiration.
5 Give the word equation of aerobic respiration.

6 What is meant by the term **metabolism**?
7 Why is the energy released from food during aerobic respiration important to the metabolism of cells?

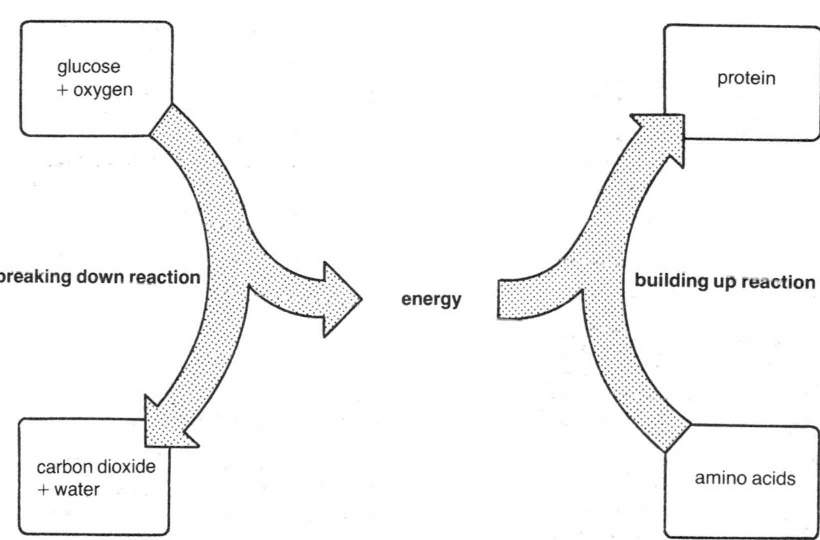

Figure 8.11 Metabolic reactions

PROBLEM SOLVING

1 Read the passage and answer the questions which are based on it.

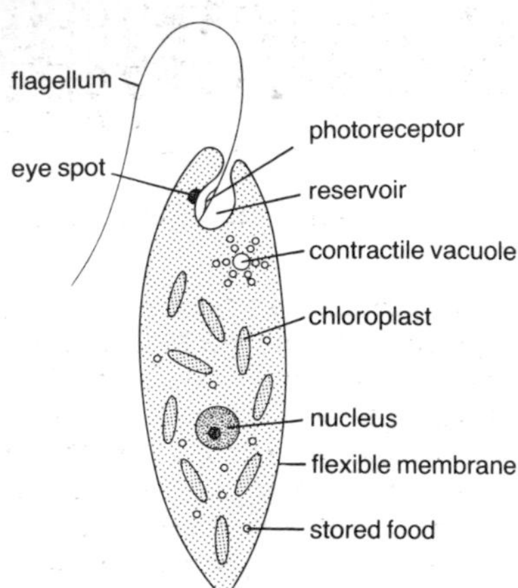

***Euglena viridis*, shown in the diagram, is a unicellular organism found living in stagnant pondwater. It swims by means of its long whip-like flagellum, movements of which draw the organism forwards. It is able to feed both by photosynthesis and by taking in organic substances present in the water. It uses its contractile vacuole to discharge unwanted water into the reservoir and then to the outside of the cell. Its eye spot and photoreceptor are used to guide it towards light.**

a) What is the function of *Euglena*'s flagellum?
b) *Euglena* is often found living in water containing rotting organic matter. Suggest why.
c) Why is it of benefit to *Euglena* to be able to swim towards light?
d) Identify ONE structural feature possessed by *Euglena* that is often found in plant cells but never in animal cells.
e) What structural feature typical of all plant cells does *Euglena* lack?
f) In what way does *Euglena* contradict the idea that every living thing can definitely be classified as either a plant or an animal?

Extra Question

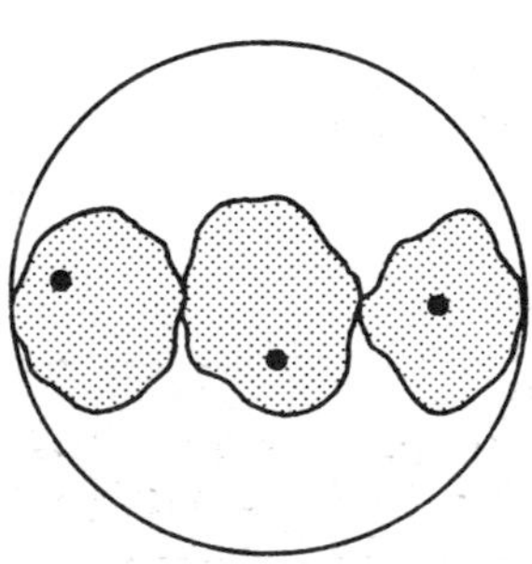

2 The accompanying diagram shows three human cheek cells under the high power of a microscope. The diameter of the field of vision is 0.03 mm. What is the average diameter of a cheek cell in micrometres?
(1 millimetre = 1000 micrometres)

3 Three identical cylinders of fresh beetroot were immersed in the liquids shown in the diagram for 24 hours.

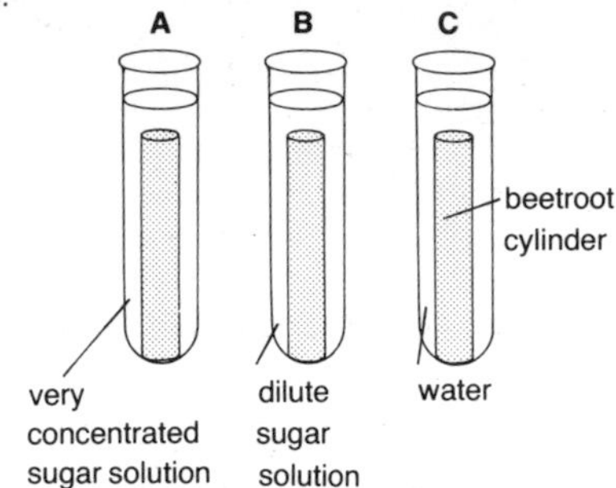

Each was then removed and pinned to a cork as shown below.

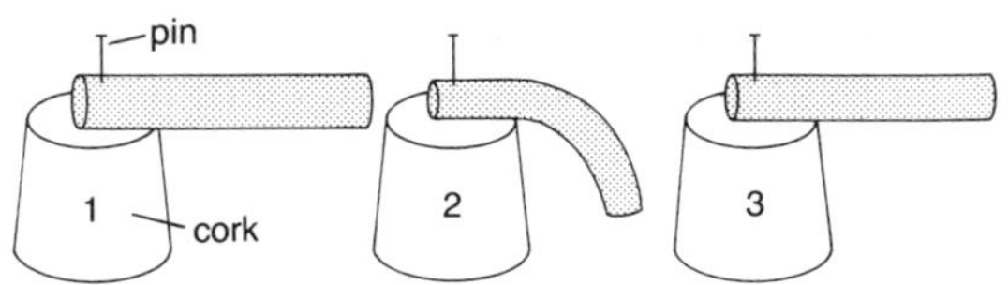

a) Match each of the diagrams 1–3 with the cylinders A, B and C.
b) Explain your answer to (a).

Extra Questions

4 Several potato cylinders of similar length were weighed and then immersed in water for 30 minutes. They were then reweighed and their increase in mass recorded in the following table.

potato cylinder	increase in mass (g)
1	0.113
2	0.109
3	0.121
4	0.118
5	0.124
6	0.105

Calculate the average increase in mass.

5 *Paramecium*, shown in the diagram, is a unicellular animal that lives in pond water. In its natural environment it gains water from its surroundings by

osmosis.
Unwanted excess water is removed by two contractile vacuoles. Each rhythmic expansion and contraction of a contractile vacuole is called a pulsation.

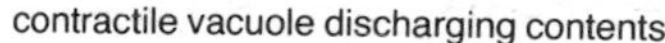

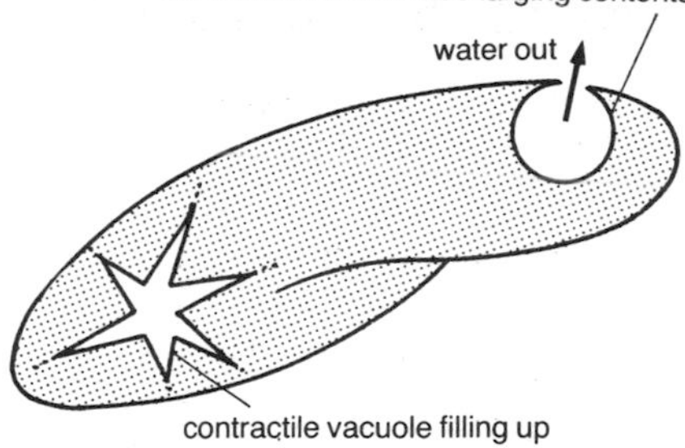

When the animal is placed in different liquids the following results are obtained.

bathing liquid	number of pulsations
1% salt solution 0.5% salt solution distilled water	fewest ↕ most

Give an explanation of the results in the table.

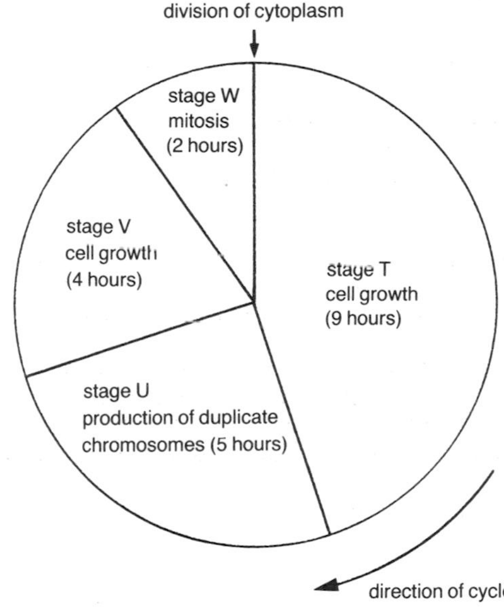

6 The above diagram shows a complete cycle of the stages that occur before, during and after mitosis in a certain type of animal cell.
a) Which of the following bar graphs best represents the above information?
b) How much time is spent on cell growth in one complete cycle?
c) Express your answer to b) as a percentage of the total time required for one cycle.

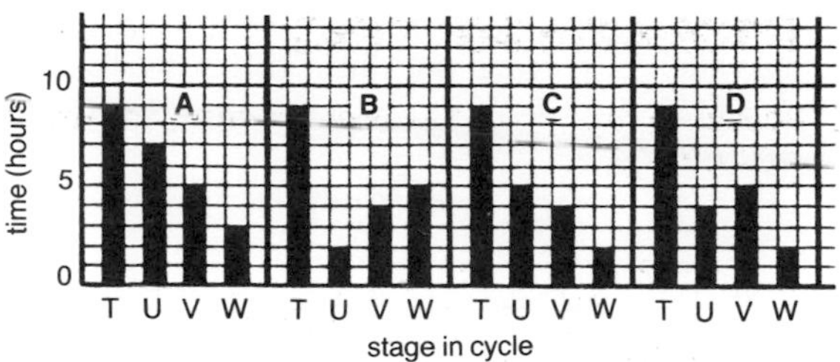

7 Read the passage and answer the questions that are based on it.

When a few cells are removed from a living organism and grown in a nutrient solution, the group of cells formed is called a tissue culture. Dividing cells from tissue cultures examined under the microscope show that the time taken by a cell to divide varies from species to species. An onion root tip cell, for example, takes 22 hours to undergo the full cycle, whereas a human fibrocyte (a type of connective cell) takes 18 hours from interphase to interphase. Out of this time only 45 minutes is devoted by the fibrocyte to mitosis.

The rate of cell division is found to be affected by factors such as nutrition and temperature. Cells from a grasshopper embryo require 3½ hours at 38 °C but 8 hours at 26 °C. Some bacteria divide every 20 minutes at 37 °C but take many hours at low temperatures. Human cells cease to divide at temperatures below 24 °C and above 46 °C.

a) By what means can a supply of cells be readily obtained for studying cell division?
b) A human fibrocyte takes 18 hours to undergo the full cycle of cell division. How much of this time is spent preparing for the actual process of mitosis?
c) Describe the effect of raising the temperature from 26 °C to 38 °C on the rate of division of grasshopper embryo cells.
d) Imagine a single bacterial cell placed in nutrient solution at 37 °C. How many bacteria would be present after two hours?
e) The table refers to four flasks containing human fibrocyte cells about to be cultured.

flask	A	B	C	D
temperature (°C)	17	27	37	47

In which flask(s) would
(i) cell division occur?
(ii) cell division occur at the fastest rate?
(iii) cell division not occur?

f) Apart from temperature, what other factor affects the rate of cell division? Suggest why.

8 A pupil set up the experiment shown in the diagram in an attempt to investigate the effect of snail's gut on flour.

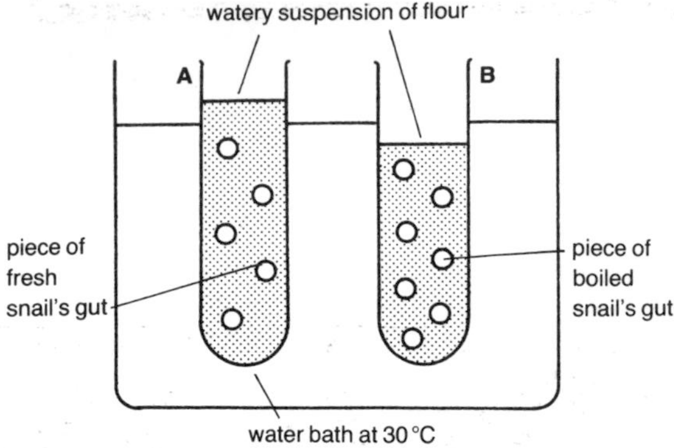

a) State TWO ways in which the experimental set-up needs to be altered in order to make it a fair test. These two changes were made in a second attempt at the experiment. The pupil then tested the flour suspension from each test tube for starch and reducing sugar at the start and after thirty minutes. She tabulated her results as follows.

	tube A		tube B	
	starch test	sugar test	starch test	sugar test
at start	+	–	+	–
after 30 min	–	+	+	–

(+ = positive result, – = negative result)

b) In which tube did the starch disappear?
c) In which tube did sugar appear?

Extra Question

d) Find out the meaning of the term hypothesis by reading Appendix 10, and then construct a hypothesis to account for the results in the above experiment.

9 The graph below shows the rate of uptake of oxygen by a plant in darkness at different temperatures.
a) At what temperature was a volume of 60 cm^3 of oxygen per day being taken up?
b) What volume of oxygen was taken up at 15 °C?
c) At what temperature was the highest volume of oxygen taken up?

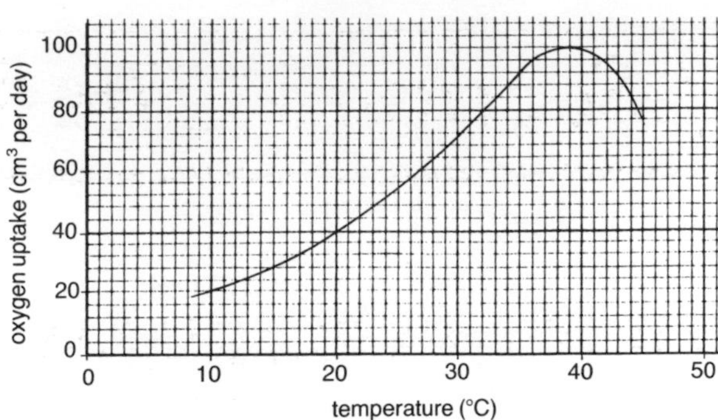

Extra Questions

d) By how many times was the volume of oxygen taken up at 30 °C greater than that taken up at 10 °C?
e) What increase in temperature was required to increase the volume of oxygen taken up at 20 °C to double that amount?

10 A boy set up the following experiment to investigate gas exchange in a locust.

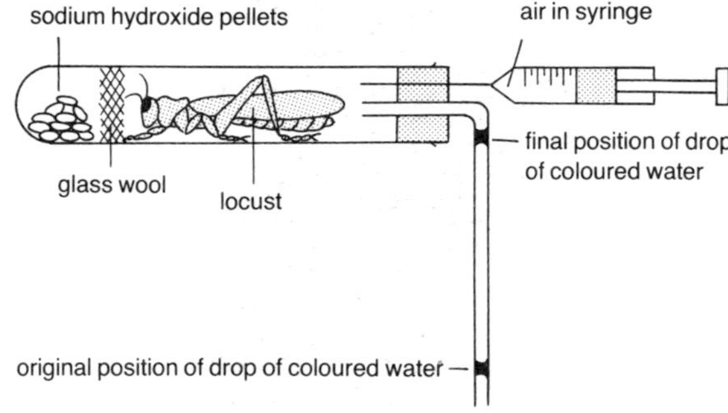

a) What gas, breathed out by the locust, is absorbed by the sodium hydroxide pellets?
b) The boy claimed that the experiment shows that a locust breathes in oxygen. Explain why he was justified in making this claim.

Extra Questions

c) Using only the apparatus shown, suggest how the boy could measure the volume of oxygen taken in by the locust during a given period of time.
d) Why is it essential to keep the entire apparatus at constant temperature throughout the experiment?

Section 3 The World of Plants

9 Introducing plants

Variety of plants

The Earth possesses an enormous **variety** of plants. These range from the Giant Redwood tree (figure 9.1) which may be 100 metres in height to tiny microscopic algae (simple water plants) smaller even than the full stop at the end of this sentence. Plants live nearly everywhere on Earth and in many different forms, from the slimy seaweeds of coastal waters to the leafy oak trees of the forest and the prickly cacti of the desert (see figure 9.2 and Appendix 1).

Figure 9.1 Giant Redwood tree

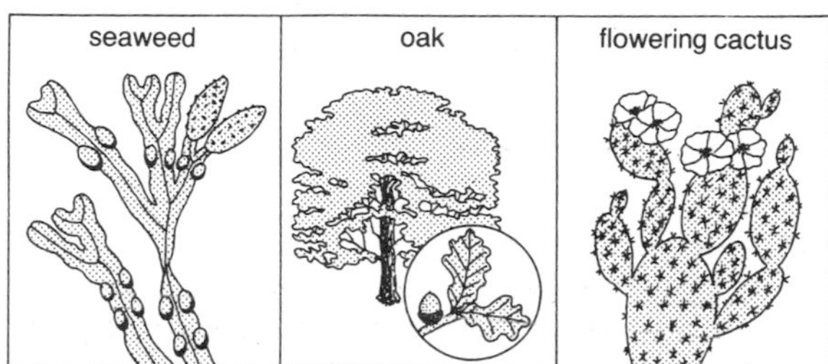

Figure 9.2 Different plant forms

Advantages to man

Food

The wide variety of plants provides man with a vast range of **edible** products. All food on Earth comes directly or indirectly from plants. Although no one plant can alone supply man with a balanced diet (the correct amount of each class of food), many plants make different contributions. A few are given in table 9.1.

Medicine

The wide variety of plants present in the world provides man in turn with an enormous range of **medicinal** substances. About 25 per cent of all medicines contain one or more active ingredients derived originally from flowering plants. A few examples of medicinal plants are shown in figure 9.3 and described in table 9.2.

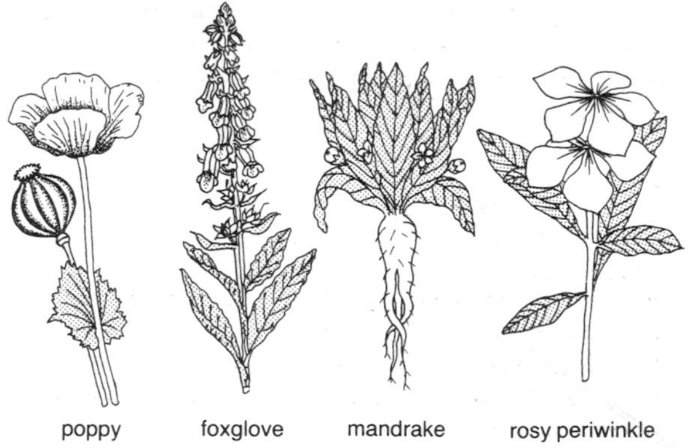

Figure 9.3 Medicinal plants

plant	plant organ eaten	class of food (all values per 100 g of edible food) protein (g)	fat (g)	carbo-hydrate (g)	calcium (mg)	iron (mg)	vitamin A (mg)	vitamin C (mg)
cabbage	leaf	1.5	0	5.8	65.0	1.0	0	60.2
carrot	root	0.7	0	5.4	50.0	0.7	2	6.1
orange	fruit	0.8	0	8.5	42.8	0.3	0	50.1
pea	seed	6.0	0	10.7	14.2	1.7	0	25.0
potato	tuber	2.1	0	20.1	7.1	0.7	0	17.8
soya	bean seed	41.0	23.9	13.5	210.7	7.1	0	0
wheat	seed grain	11.7	1.7	69.6	35.7	3.9	0	0

Table 9.1 Composition of foods from plants

common name	poppy	foxglove	mandrake	rosy periwinkle
scientific name	*Papaver somniferum*	*Digitalis lanata*	*Mandragora officinarum*	*Catharanthus roseus*
drug extracted	morphine	digitoxin	hyoscyamine	vincristine
use of drug	relief of pain	treatment of heart disease	sedation of patient before operation	treatment of leukaemia

Table 9.2 Medicinal plants

Figure 9.4 Clearing the rain forest

More to do

Ecological loss

Tropical rain forests cover only about 6 per cent of the Earth's land surface, yet they are estimated to contain half or more of all the existing species of plants, animals and micro-organisms. When this type of forest is cleared (see figure 9.4), not only is a **natural resource** destroyed but the ecosystem and its wide variety of interdependent species are lost forever.

For every plant species that becomes extinct through man's activities, several animal species that were dependent on the plant for food and shelter are threatened and may also become extinct. Animals nearing extinction include Brazil's muriqui monkey, the flightless Kagu bird of New Caledonia and the birdwing butterfly of Papua New Guinea.

One possible solution to the problem is 'debt-payment-for-nature' exchange. This involves donations from conservationists being used to settle developing countries' international debts in exchange for binding promises that these countries will create permanent reserves of tropical rain forest.

Loss of future applications of plants

Food plants

Only about twenty of the world's plant species are used to produce approximately 90 per cent of people's food; yet the Earth is thought to possess at least 75000 species of edible plant. A few examples of underexploited food plants are shown in figure 9.5 and described in table 9.3.

plant	natural habitat	useful part of plant
yeheb nut bush	dry land of Somalia (Africa)	pod containing seeds rich in protein and fat
eelgrass	coast of Mexico	grain seeds containing flour suitable for bread-making
buffalo gourd	dry land of Mexico	seeds rich in protein and oil
naranjilla	forest of South America	tomato-like fruit with taste like mixture of pineapple and strawberry
amaranth	dry mountainous regions of South America	grain seeds containing flour rich in protein
Chinese gooseberry	rain forest of South East Asia	fruit containing juice 15 times richer in vitamin C than oranges

Table 9.3 Underexploited food plants

Figure 9.5 Underexploited food plants

Medicinal plants

More than 40 species of flowering plants have already provided mankind with medicines. Tropical rain forests are the Earth's main source of plants containing chemicals of possible medicinal use in the future. It is thought that at least 2000 different plant species may contain **anti-cancer** properties. However, most of these plants have not yet been studied in detail. Clearing rain forests and the subsequent loss in the variety of species means losing forever these plants and their potential medicinal value in combating disease.

Genetic storehouse

Over 70 per cent of present day American maize plants are genetically identical. If a new disease arrives they could all be affected. A wild form of maize has been discovered in Mexico that is perennial (grows for many years without being replanted) and is resistant to many diseases. Geneticists are trying to introduce these useful characteristics into modern maize by **cross-breeding**.

In the future it will not be possible to introduce new beneficial characteristics into crop plants from wild varieties if they are lost by man destroying their habitats for quick commercial gain. It is essential therefore that we conserve a wide variety of strains of each plant species to provide a **genetic storehouse** of useful characteristics for the future.

KEY QUESTIONS

1 Name TWO very different types of plant and state two ways in which they differ from one another.
2 No one plant species supplies mankind with all the essential foods needed for a balanced diet. Why is this not a problem?
3 State ONE other way in which man benefits from the wide variety of plants present in the world.

Extra Questions

4 In what TWO ways does the clearing of plants from a rain forest ecosystem affect animals?
5 Explain why it is important that the world's underexploited edible plants are saved from extinction.
6 Why will mankind lose out in the long term if wild varieties of crop plant species are allowed to become extinct?

Uses of plants

Plants and plant products are put to many everyday uses. Three of these are shown in figure 9.6.

Specialised uses of plants

Rape

The rape plant is a member of the cabbage family. Its leaves provide forage for farm animals such as sheep. Its tiny black seeds contain oil used in lubricants, metal tempering and foodstuffs.

Cotton

A cotton seed bears a mass of 'hairy' fibres which act like a parachute during wind dispersal of the seed. A group of seeds with their fibres entangled in a fibrous mass is called a cotton boll (figure 9.7).

The seeds are separated from their fibres. Oil and protein for animal feed are extracted from the seeds. The fibres are spun (twisted to give strength) into yarn (thread). The yarn is then woven into cloth (cotton).

Cotton is a cheap hardwearing absorbent material. It is particularly suitable for mixing with synthetic fibres to make a wide variety of useful fabrics.

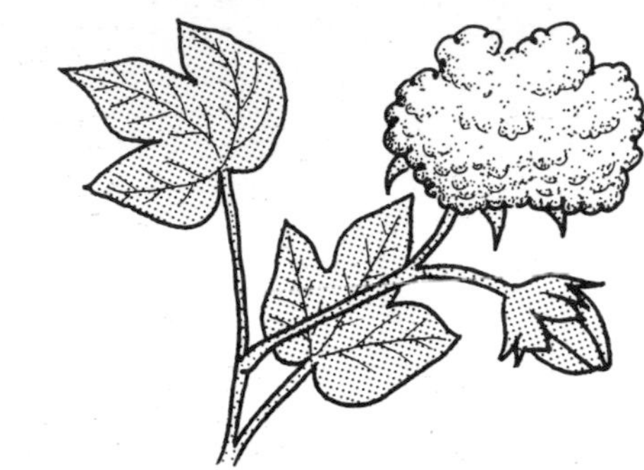

Figure 9.7 Cotton boll

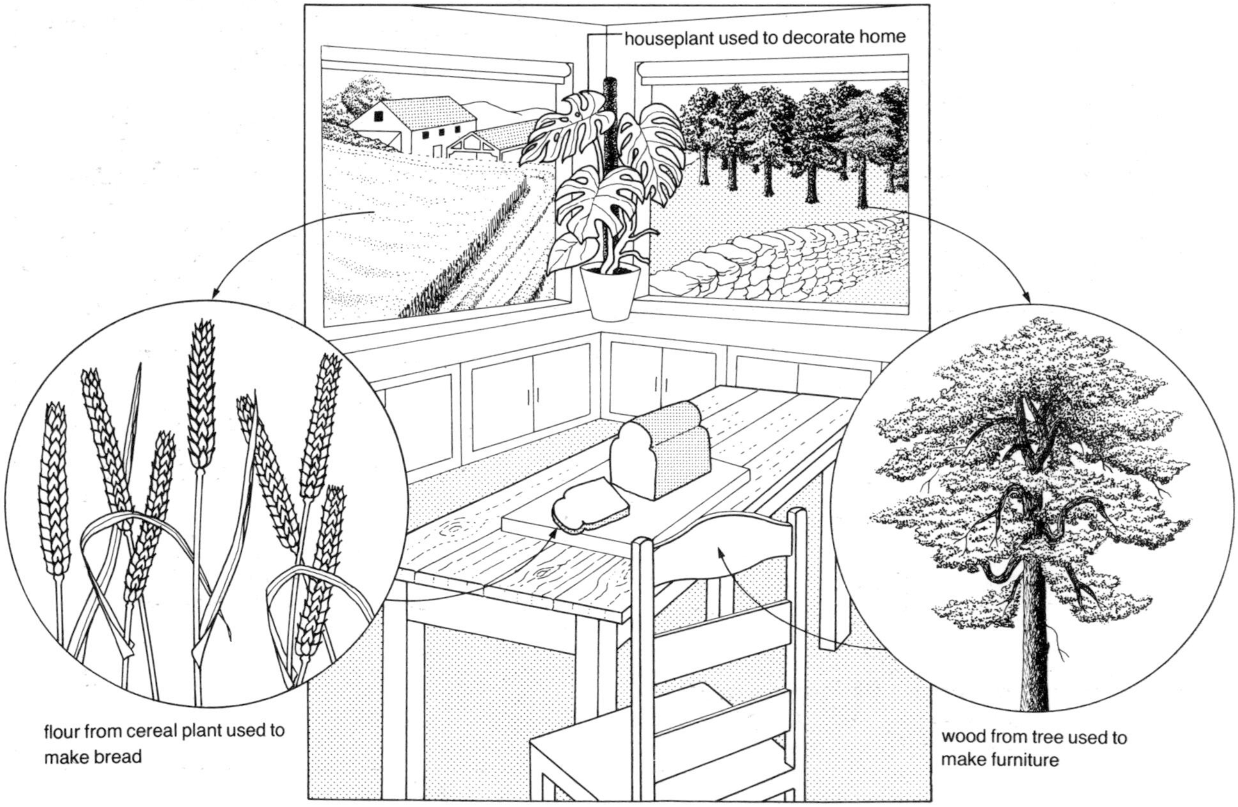

houseplant used to decorate home

flour from cereal plant used to make bread

wood from tree used to make furniture

Figure 9.6 Uses of plants

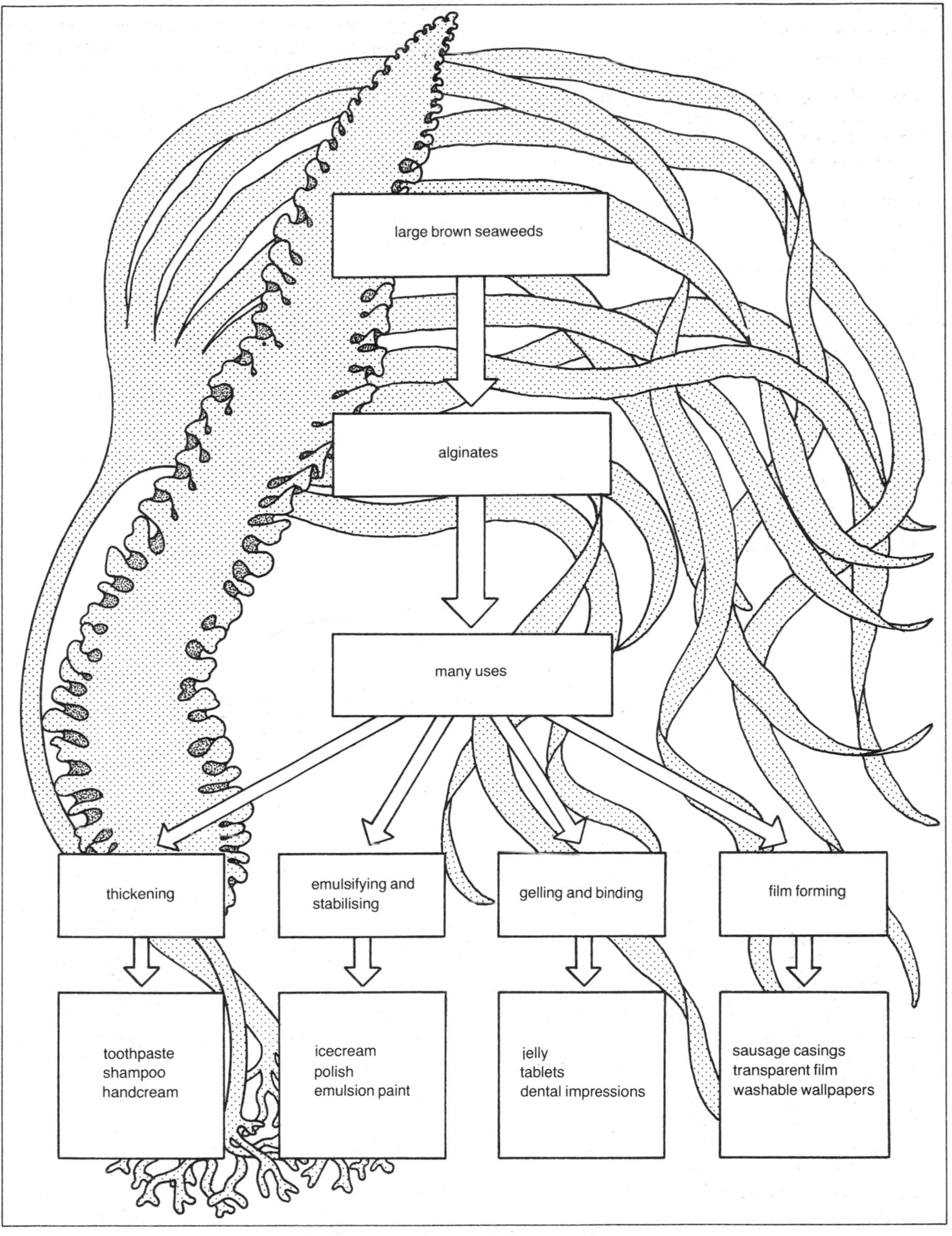

Figure 9.8 Uses of alginates

Alginates
Alginic acid is a chemical extracted from large brown seaweeds. The salts of alginic acid, **alginates**, are used in about 150 different products. Some of these are shown in figure 9.8.

Timber production
In Scotland vast forests of conifer trees are grown for the commercial value of their timber. Timber production involves the following stages.

Preparation of the land
High sloping land that is too poor to support food crops is often suitable for the low demands of conifer trees. However, this land must normally be opened up by digging drainage trenches (see figure 9.9) to remove excess water and allow air into the soil.

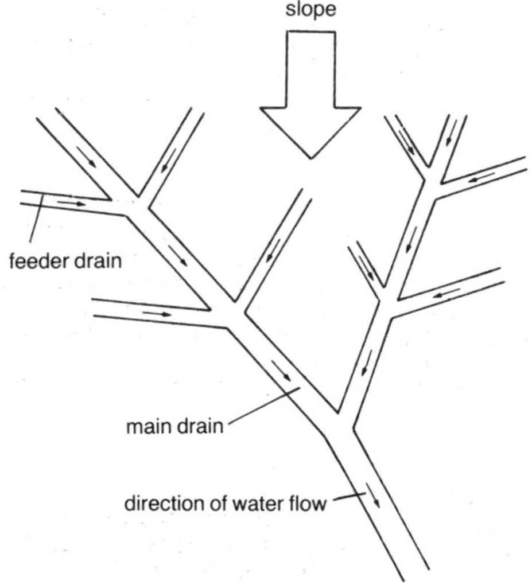

Figure 9.9 Drainage ditches

Planting and managing the forest
Young trees are brought from a nursery and planted by hand in rows in the turf beside the trenches. Weeds are cleared and animals such as hares which would kill the young trees are kept out.

As the years go by, poorer trees are removed during thinning to give the best trees room to grow. Once the trees have gained a height of about ten metres, their lower branches die because they are in permanent shade. These are pruned off by the forester using a debranching machine so that the tree trunks will be suitable later for sawing into long straight lengths.

The forest is inspected regularly for disease and treated with pesticide when necessary.

Harvesting and marketing
After a long time scale of about fifty years, the trees are mature and ready to be felled (harvested) using a chain saw. Strip or block cutting (figure 9.10) is often

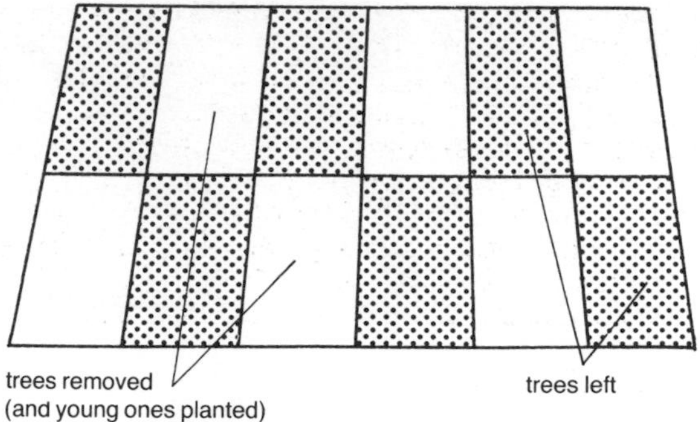

Figure 9.10 Block cutting

done to prevent soil erosion and sudden exposure of neighbouring farmlands to extremes of climate.

The trees are sent to the sawmill to be cut up and prepared for marketing. The wood from conifers is used for many purposes. Scots pine provides timber for telegraph poles, railway sleepers, mine props, chip-board and paper pulp. Norway spruce timber is light and strong and ideal for interior joinery and furniture-making.

More to do

Potential uses of plants or plant products

Alternative source of fuel
More than half of Brazil's 200 million ton annual sugar crop is converted into sugar cane **alcohol** for use as motor fuel instead of petrol. One disadvantage of running a car on alcohol is that it needs a tropical climate to run efficiently. At below 25 °C a car engine fuelled on alcohol will not start without a squirt of petrol to enable the engine to fire.

Research is also being done on the possible use of castor oil and soya oil as motor fuels. However, motor manufacturers have to be persuaded that a vegetable-based fuel oil is a realistic alternative to traditional fossil fuels before they are willing to invest the amount of money needed to make the necessary adjustments to current motor engine designs.

New medicines
People have appreciated the medicinal value of certain plants for thousands of years. The ancient Chinese used **opium** from poppies as a painkiller. The discovery this century of the **antibiotic** penicillin (extracted from the fungus *Penicillium*) opened an exciting new chapter in medicine.

Today medical researchers are ever on the lookout for new plant products. In particular new antibiotics from fungi are being discovered and developed. However, it may take ten or more years to develop a new medicine and require vast sums of money to be invested. Figure 9.11 shows some of the stages involved in the development of a new antibiotic drug and the difficulties that can arise. Even if the chain of events is successfully completed, the new antibiotic may become ineffective within a few years if a strain of micro-organism that is resistant to it appears.

KEY QUESTIONS

1 Name THREE plants or plant products that are put to everyday use by humans.
2 State THREE specialised uses of plants or plant products and for each describe how the plant or product is useful.
3 Write a short account of timber production.

4 Describe how Brazil is overcoming its lack of fossil fuels.
5 Write a short paragraph discussing the difficulties involved in developing a new antibiotic drug.

VARIETY OF SUBSTANCES EXTRACTED FROM FUNGI

↓

SUBSTANCES ABLE TO PREVENT GROWTH OF BACTERIA IDENTIFIED

↓

POTENTIALLY USEFUL SUBSTANCE (ANTIBIOTIC DRUG) SELECTED AND PRODUCED ON A LARGER SCALE

↓

DRUG TESTED ON HUMAN CELLS IN TISSUE CULTURE (difficulty: effects on single cells may not be the same on tissues in multicellular organism)

↓

DRUG TESTED ON ANIMALS (difficulties: (a) it may not be possible to do all required tests on animals without breaching government rules on use of animals in research or incurring adverse publicity which could affect sales of company's other products (b) effects on animals may not be the same on humans)

↓

FOLLOWING GOVERNMENT APPROVAL DRUG SUBJECTED TO CLINICAL TRIALS ON THOUSANDS OF VOLUNTEERS TO ENSURE THAT IT PREVENTS BACTERIAL GROWTH WITHOUT HARMING HUMANS (difficulty: it may not be possible to test a complete range of people, e.g. those who are pregnant)

↓

DRUG MASS-PRODUCED AFTER RECEIVING FURTHER GOVERNMENT APPROVAL (difficulty: it may not be possible to mass-produce the drug at a price cheap enough to compete with other antibiotics already on the market)

↓

DRUG LAUNCHED (difficulty: making doctors aware of the drug's existence and then persuading them to prescribe it as a valid alternative to existing drugs)

Figure 9.11 Development of a new drug

10 Growing plants

Examining the structure of a broad bean seed

Figure 10.1 shows the three main parts of a seed and states their functions. Germination is the development of a plant embryo into an independent plant with green leaves (see figure 10.2). During germination the food store present in the seed leaves is digested to give the young plant food and energy for growth.

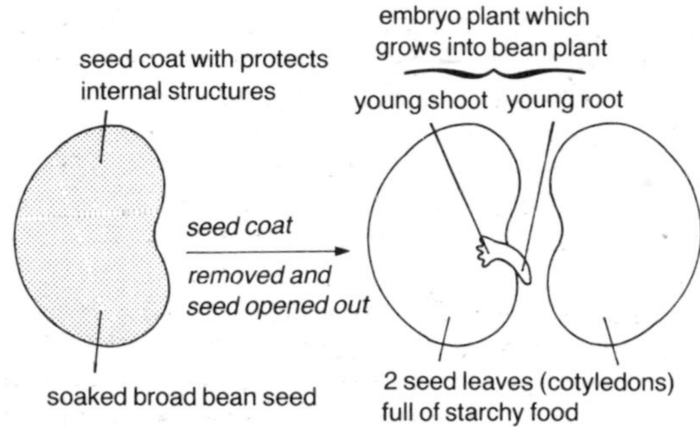

Figure 10.1 Structure of broad bean seed

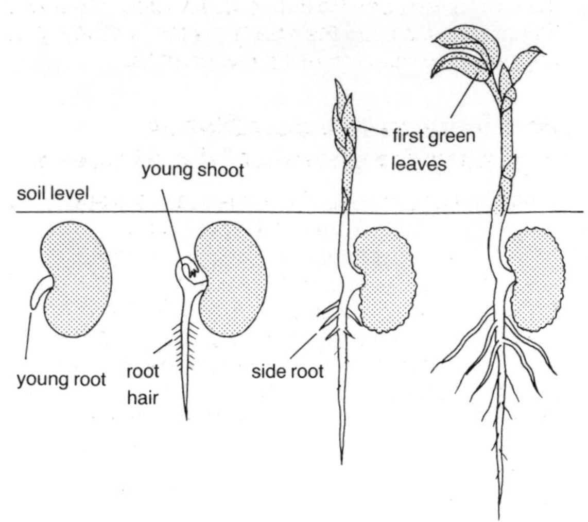

Figure 10.2 Germination of broad bean seed

Rules when performing a biological investigation

In biology the results of an investigation are valid if:

1. at each stage only **one variable factor** is studied at a time because if several are involved then it is impossible to know which is responsible for the results (see also Appendix 8);
2. **many organisms** are used because if only a few are used then perhaps these were unusual and not typical of the species in general;

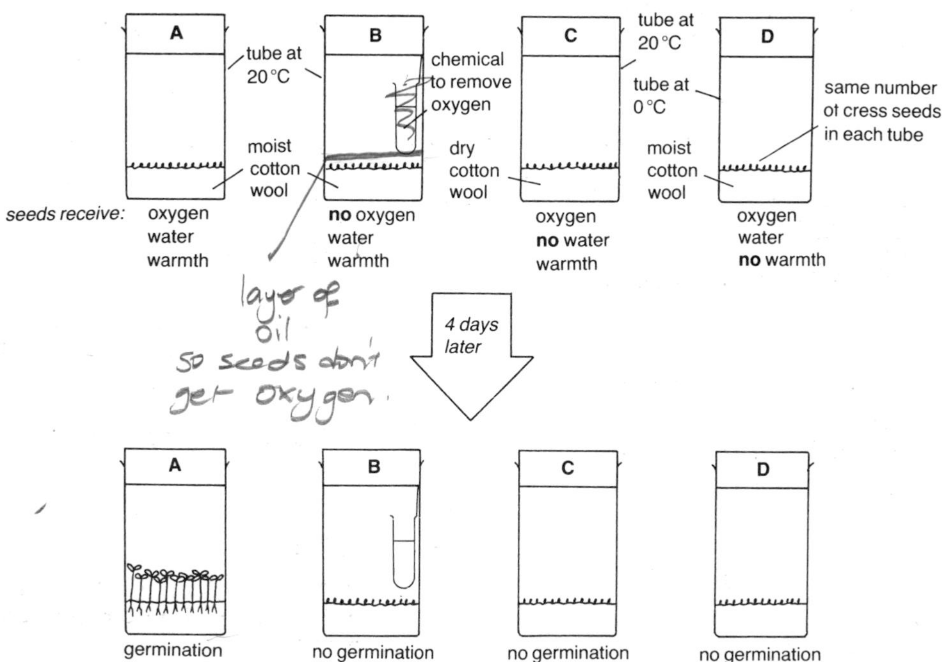

Figure 10.3 Conditions necessary for germination

3. the experiment can be successfully repeated **many times** because if not then perhaps the outcome just happened to result from a lucky chance.

Investigating the conditions necessary for germination of seeds

To satisfy the above rules the investigation is set up as shown in figure 10.3. Tubes B, C and D each differ from tube A by only one factor. The same large number of cress seeds is used in each tube to allow for a few seeds being unusual or dead. The whole experiment is repeated by several groups of pupils.

The results show that seeds need **oxygen**, **water** and **warmth** for germination. Germinating seeds need: oxygen for respiration to give energy for growth; water to allow chemicals called enzymes to digest stored food for the growing embryo; and warmth to give a suitable temperature for enzymes to act.

Effect of temperature on germination

Look at the experiment shown in figure 10.4. The oat grains are first rinsed in disinfectant (to remove surface micro-organisms) and then jammed between the wall of the beaker and the cardboard 'collar'. Six similar containers are kept at the different temperatures indicated in table 10.1. The table also gives a typical set of results (where emergence of a young root is regarded as germination).

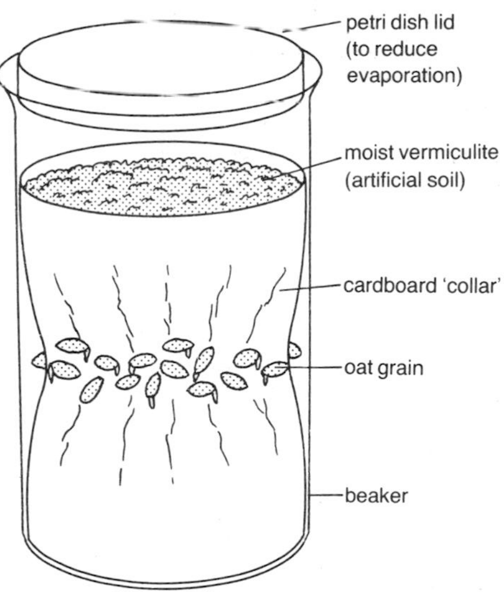

Figure 10.4 Investigating effect of temperature

temperature (°C)	location	% germination
below 0	freezing compartment of fridge	0
5	chiller compartment of fridge	4
18	room	60
27	seed propagator	80
34	incubator	10
40	oven	0

Table 10.1 Germination at different temperatures

When these results are graphed (see figure 10.5), it can be seen that very few or no grains germinate at very low temperatures. As the temperature increases so does the percentage number of grains germinating until a maximum is reached at about 27 °C. This is called the **optimum** (most favourable) condition of the one variable factor (temperature) being investigated. As the temperature increases beyond the optimum, the percentage number of grains germinating decreases eventually reaching zero at around 40 °C.

A possible explanation of the results is that germination depends on the activities of certain enzymes which in oat grains work best within a temperature range of 25–30 °C. To obtain a more accurate measurement of optimum temperature, it would be necessary to repeat the experiment using a narrower but more detailed range of temperatures e.g. 24, 25, 26, 27, 28, 29, 30 and 31 °C.

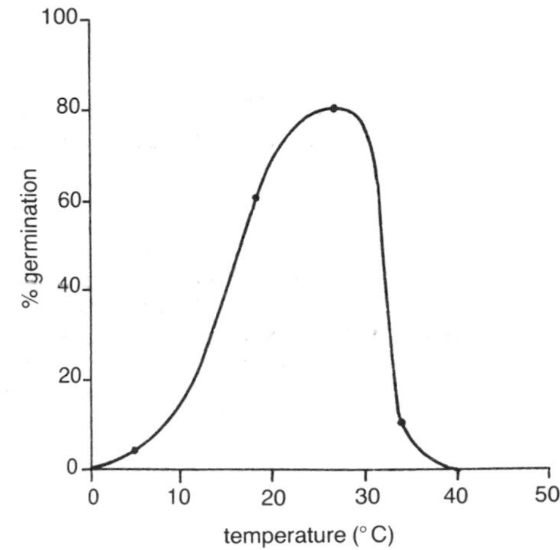

Figure 10.5 Graph of germination results

KEY QUESTIONS

1 What is the function of a broad bean's seed coat?
2 What is the function of the food store present in the seed leaves?
3 State TWO conditions other than water required by a seed for germination.

Extra Questions

4 Describe the effect of varying temperature on the percentage germination of oat grains.
5 What is meant by an **optimum** condition of a factor?
6 What was the optimum temperature found to be for the germination of oat grains in the experiment graphed in figure 10.5?

Examining the parts of a flower

Reproduction is the production of new members of a species. In flowering plants the **flowers** are responsible for reproduction. It is here that the seeds are made. Although flowers often appear to be very different from one another, they are all built to the same basic plan. Usually the male and female parts are both present in the same flower. Figure 10.6 shows the parts of a typical flower and states the function of each part.

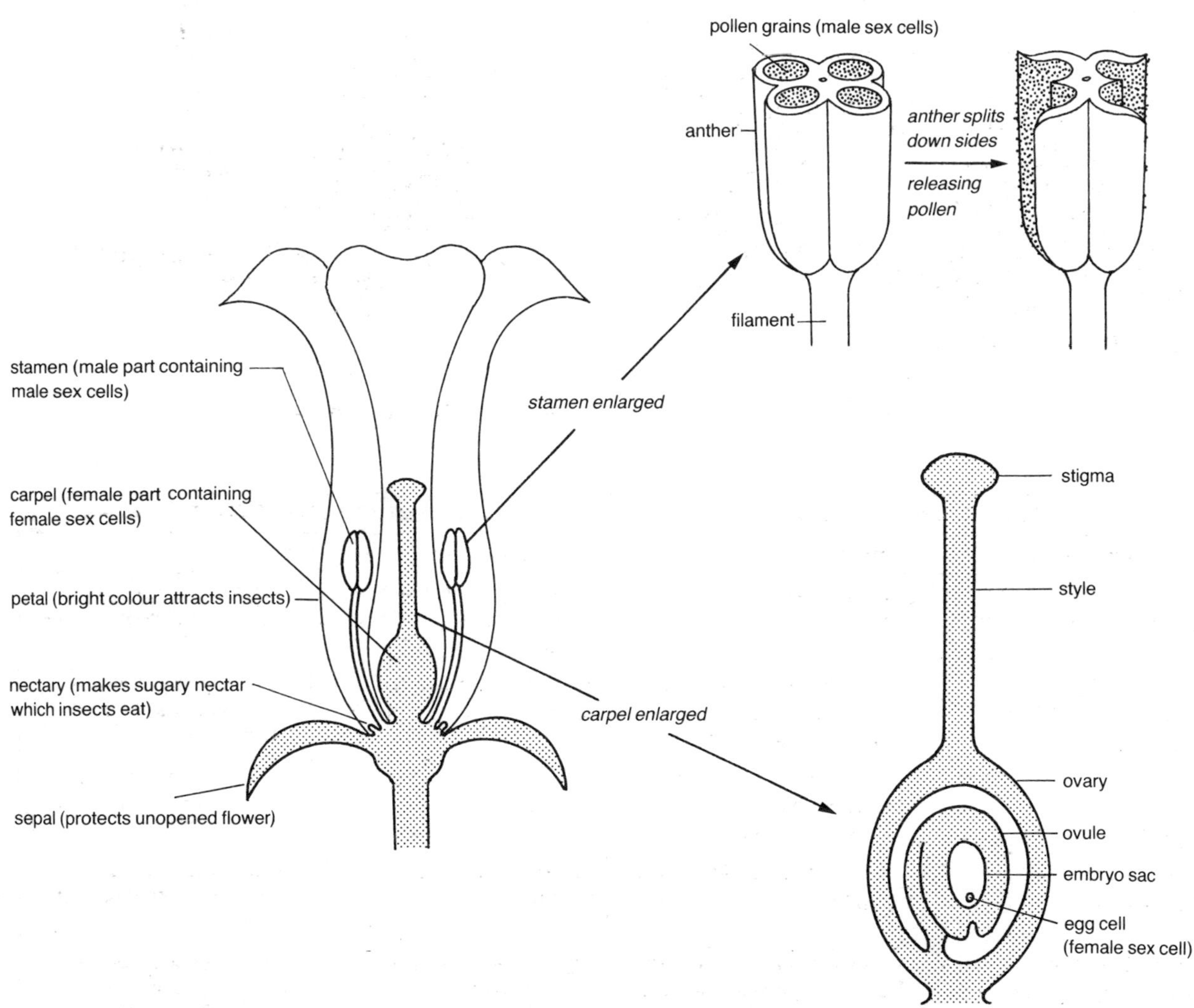

Figure 10.6 Parts of a flower

Pollination

This is the transfer of pollen grains from an anther to a stigma. **Self-pollination** is the transfer of pollen from an anther to a stigma in the same flower or in another flower on the same plant. **Cross-pollination** is the transfer of pollen from an anther of one flower to a stigma in a flower on another plant of the same species.

Insect pollination

Pollen is carried from one flower to another by insects as shown in figure 10.7. Roses and primroses are examples of insect pollinated plants.

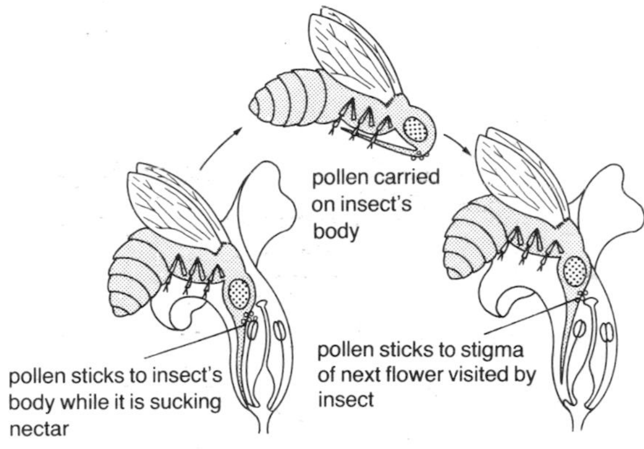

Figure 10.7 Insect pollination

Wind pollination

Pollen is carried from one flower to another by wind as shown in figure 10.8. Grasses and cereals are examples of wind pollinated plants.

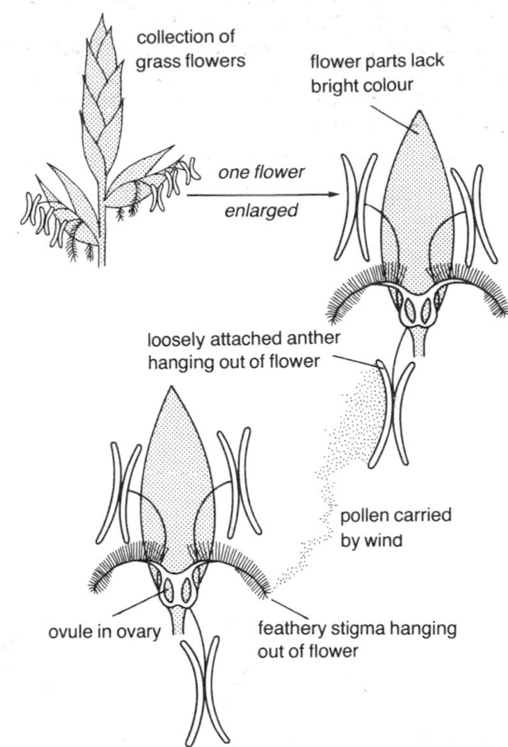

Figure 10.8 Wind pollination

WIND POLLINATED FLOWER		INSECT POLLINATED FLOWER	
structural feature	reason	structural feature	reason
flower small, lacking bright colour, scent and nectar	no visit from an insect required	flower large, with brightly coloured petals, scent and nectar	to attract insects which will eat nectar and collect pollen
anthers loosely attached and hanging out of flower	to enable them to be shaken and to allow pollen to be carried away by wind	anthers firmly attached inside flower	to be in a position where insects are likely to brush against them
pollen grains light and smooth	to enable them to be carried in air currents without sticking together	pollen grains sticky or with rough surface	to enable them to stick on to insect's body easily
large quantity of pollen produced	low chance of pollen reaching stigma	small quantity of pollen produced	insect greatly increases chance of pollen reaching stigma
stigmas hanging out of flower and feathery in structure	to be in a good position and present a large surface area for trapping pollen	stigmas inside the flower with sticky surface	to be in a position where insect is likely to brush against them and leave pollen stuck to them

Table 10.2 Comparison of wind and insect pollinated flowers

More to do

Comparison of wind and insect pollinated flowers

Table 10.2 relates the structure of wind and insect pollinated flowers to their agents of pollination.

Fertilisation

Once a pollen grain has landed on a stigma, it absorbs sugar from the stigma and forms a **pollen tube** which grows towards the ovule (see figure 10.9). When the nucleus of the male sex cell (gamete) reaches the nucleus of the female sex cell (gamete), they fuse forming a zygote. This process is called **fertilisation**. The zygote divides by cell division many times and grows into an embryo plant. This new individual found inside a seed inherits some features from one parent and some from the other and is therefore different from them both.

The method of reproduction described above involving gametes and fertilisation is called **sexual** reproduction. Since two different parents contribute towards the offspring, **new varieties** are formed by this method. Some plants can also reproduce asexually (see page 78).

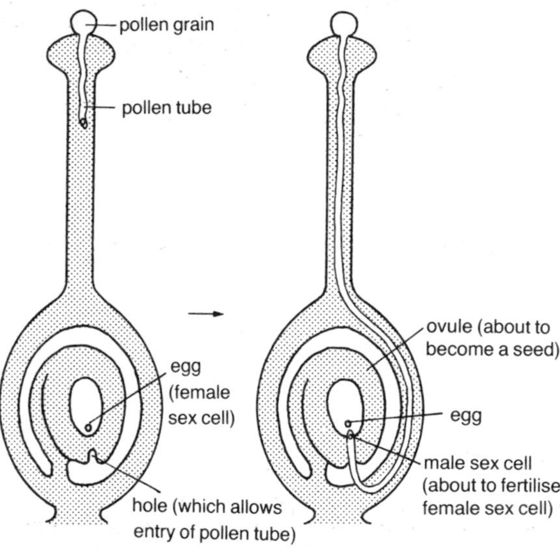

Figure 10.9 Fertilisation

More to do

Seed and fruit formation

Following fertilisation, an ovule becomes a **seed** which contains an embryo plant and a food store. Its outer wall becomes harder and thicker, forming the seed coat.

Fertilisation also affects the rest of the carpel. The stigma and style normally wither away but the ovary containing the seed(s) continues to develop into the **fruit**. The fruit coat (previously the ovary wall) often becomes soft and fleshy as in grape, cherry and tomato (figure 10.10). However the fruit wall does not always become edible. It may become a hard and dry pod (e.g. bean) or nut (e.g. hazel).

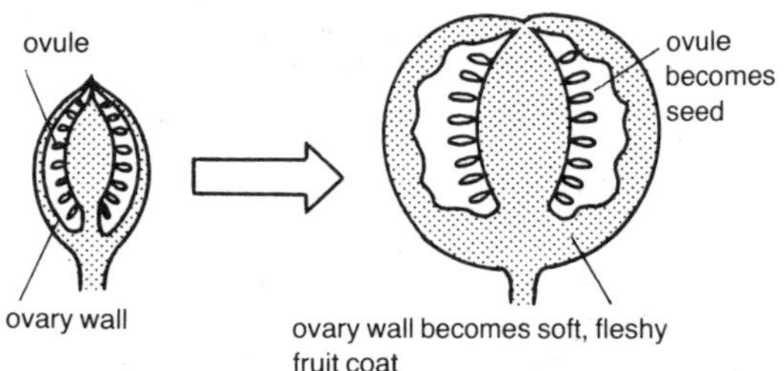

Figure 10.10 Tomato seed and fruit formation

KEY QUESTIONS

1 State the function of each of the following parts of a rose flower:
   a) sepal;
   b) petal;
   c) nectary.
2 Name the type of sex cells present in an anther.
3 What does the term **pollination** mean?
4 Describe how an insect brings about pollination.
5 Name the other agent of pollination.
6 Describe how a male gamete travels from the stigma to the female sex cell.
7 What name is given to the fusion of the nuclei of the two gametes to form a zygote?

8 Why do wind pollinated flowers have loosely attached anthers and make many smooth light pollen grains?
9 Explain why insect pollinated flowers have sticky stigmas positioned inside the flower whereas wind pollinated flowers have feathery stigmas hanging out of the flower.

10 Briefly describe how a tomato fruit is formed.
11 Name a fruit which is normally referred to as a vegetable.

Seed and fruit dispersal

A **seed** is an ovule after fertilisation which contains an embryo plant and a food store. A **fruit** is an ovary after fertilisation which contains one or more seeds. Plants need to scatter their seeds over a wide area. They do this in a variety of ways. Figure 10.11 shows three different seed dispersal mechanisms.

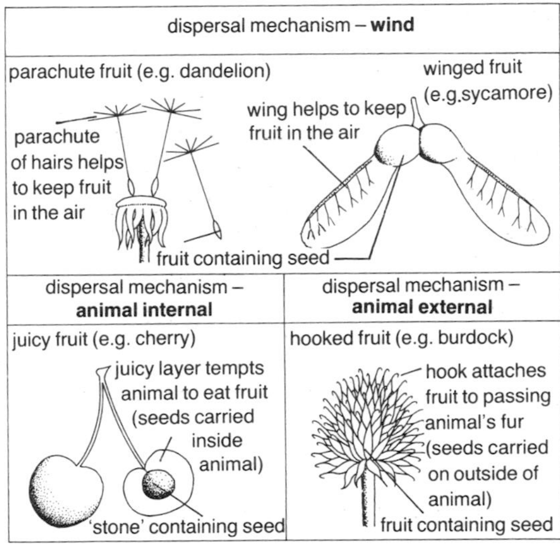

Figure 10.11 Seed dispersal mechanisms

More to do

Dispersal

The structure of each type of fruit shown in figure 10.11 is related to its mechanism of dispersal.

A **wing** or **parachute** offers resistance to airflow, making the fruit remain airborne for a longer time. This increases its chance of being carried away some distance from the parent plant before landing.

A **brightly coloured, juicy, edible layer** entices a hungry animal to eat the fruit. There is a good chance that the seeds will be spat out. However, if they are swallowed their tough seed coats protect them from the animal's digestive juices and they pass out unharmed in faeces well away from the parent plant.

Similarly, a **hooked** seed may drop off the animal's coat far away from the parent plant.

Further examples of seed dispersal mechanisms are shown in figure 10.12.

A **pod** that bursts open ejects the seeds to a location further away from the parent plant than they would have reached by simply falling off the parent.

A **fibrous coat** which traps air under a waterproof outer layer allows the fruit to be transported by ocean currents to a new site many miles away from the parent plant.

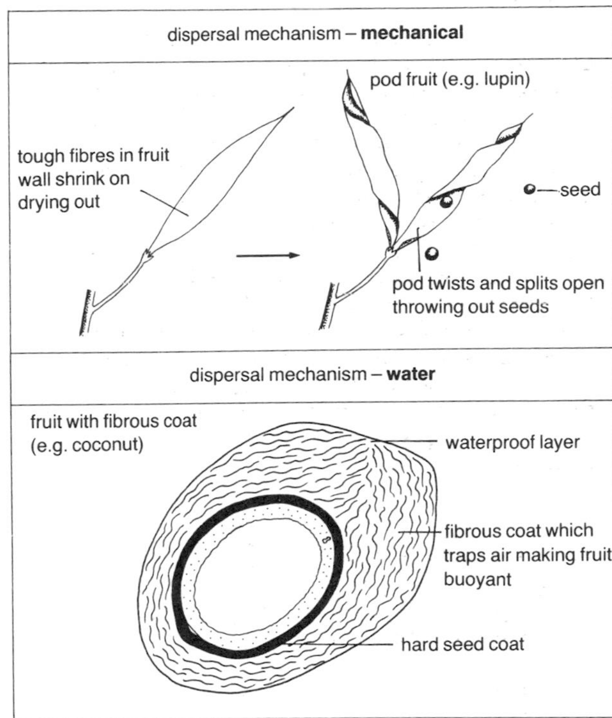

Figure 10.12 Further seed dispersal mechanisms

Survival

If a plant's seeds are well dispersed the chance of survival of the species is increased because (a) there is a good chance that at least some seeds will land in a favourable habitat, (b) the species may be able to colonise new habitats and increase its geographical range (thus reducing the chance of its extinction by a local disaster such as a volcanic erruption, flood, pollution, etc.) and (c) the offspring will not be competing with the parent (and one another) for light, water, space, minerals,etc.

KEY QUESTIONS

1 Name a plant that makes (a) a winged fruit, (b) a hooked fruit.
2 Describe how (a) a wing (b) a hook, help to disperse the fruit.
3 Describe a further example of a seed dispersal mechanism.
4 Beginning with germination of seed, rewrite the following stages in the life cycle of a plant in the correct order:

fertilisation; flowering; germination of seed; growth of young plant; pollination; seed dispersal; seed formation.

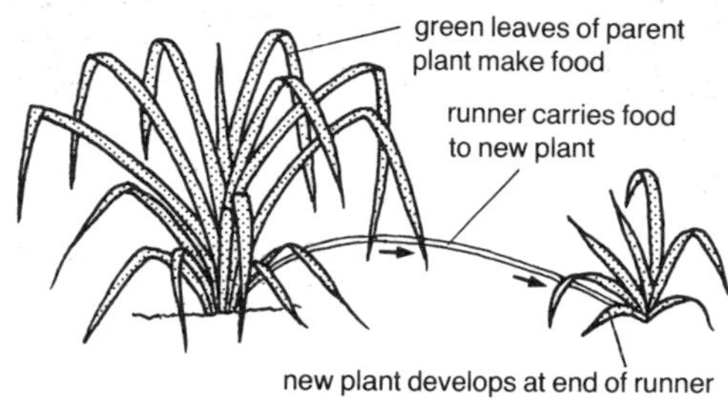

Figure 10.13 Runner formation in spider plant

Extra Questions

5 Explain why it is of advantage to a cherry to be eaten by an animal even if the cherry is swallowed.
6 Explain how the structure of (a) a lupin pod and (b) a coconut is related to its seed dispersal mechanism.
7 Give TWO reasons why good seed dispersal increases a plant species' chance of survival.

Asexual reproduction

During **asexual** reproduction, new plants are produced by a single parent plant without involving sex cells or fertilisation. Since no variation or change is introduced, the new plants formed by this method are **genetically identical** to one another (and to the parent plant) and are said to make up a **clone**.

Asexual reproduction in plants is also called vegetative propagation.

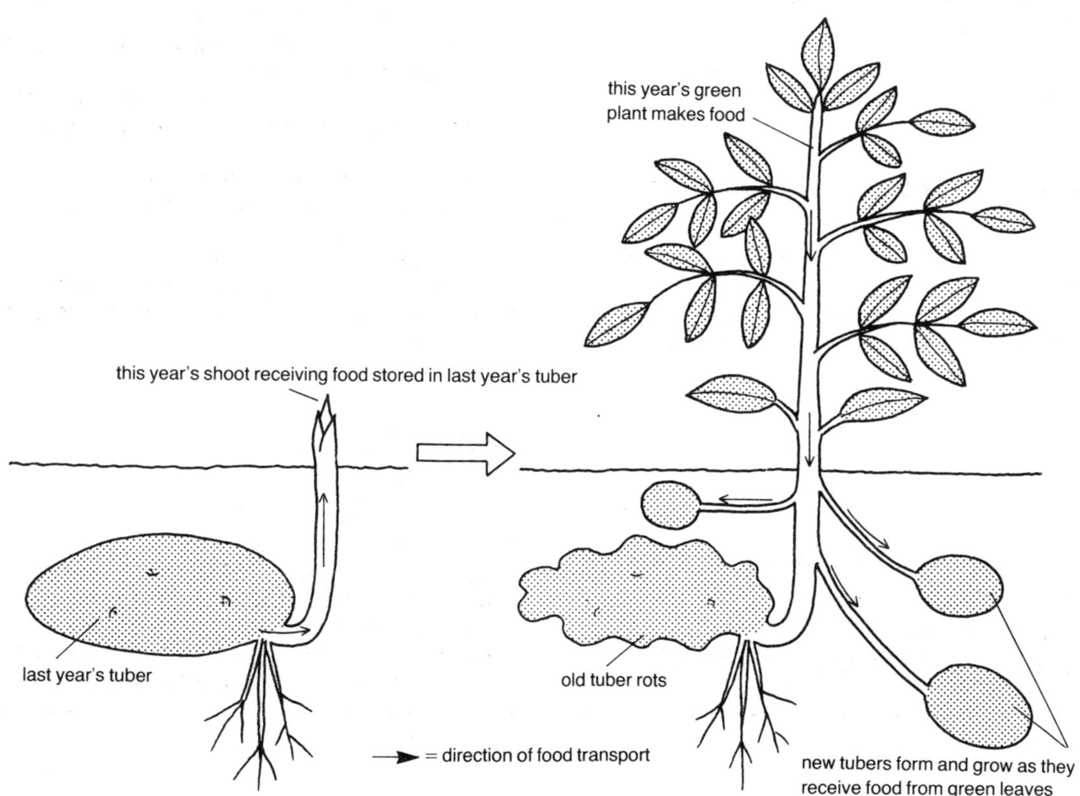

Figure 10.14 Tuber formation in potato

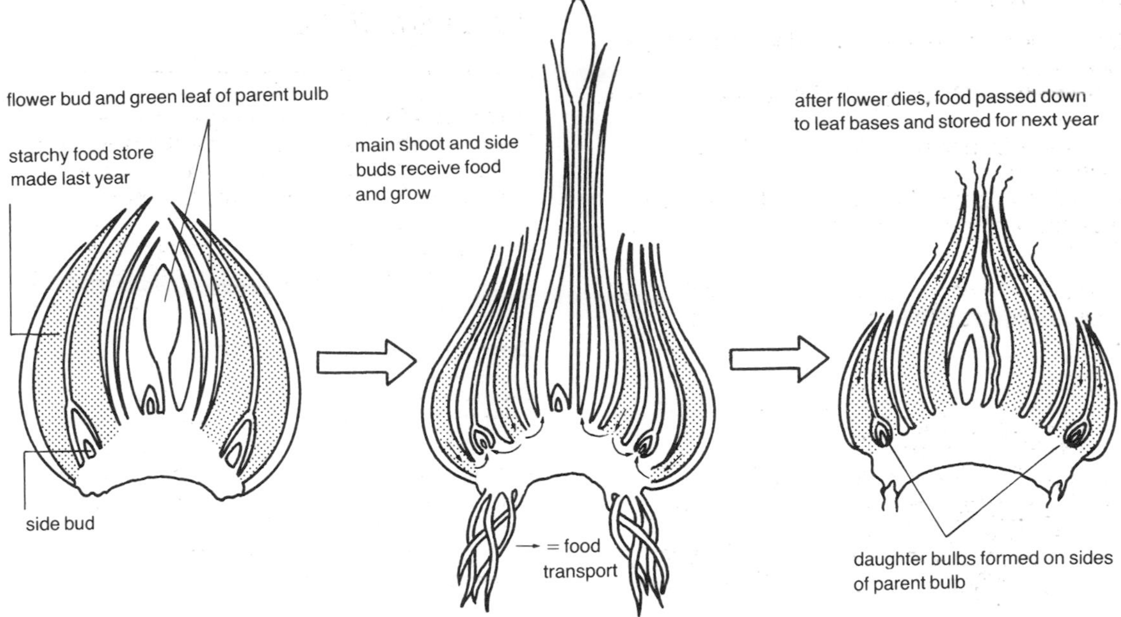

Figure 10.15 Bulb formation in daffodil

Methods of natural vegetative propagation

Many plants are able to reproduce asexually by forming offspring from their stems or buds. Some examples are shown in figures 10.13–15.

Advantages of asexual reproduction to plants

The offspring are in contact with the parent plant from which they receive food and water. The young plants are produced in an environment which suits growth of the parent plant and will also suit them when they become independent.

KEY QUESTIONS

1 Name the method of natural vegetative propagation employed by a daffodil plant.
2 Describe the method of asexual reproduction that occurs in a potato plant.
3 Arrange the following pieces of information about a spider plant's method of asexual reproduction into the correct order:
new plant becomes independent;
parent plant grows a side branch (runner);
runner rots away;
bud grows into a new plant;
bud at end of runner gets food from parent.
4 Describe two advantages of asexual reproduction to plants.

More to do

Comparison of asexual and sexual reproduction

Table 10.3 explains the advantages and disadvantages of both sexual and asexual reproduction to plants.

Extra Questions

5 Explain why each of your answers to question 4 above is of advantage to the plant.
6 a) State TWO disadvantages of asexual reproduction to a plant.
b) For each of these explain how the problem is overcome by sexual reproduction.

ASEXUAL		SEXUAL	
advantage to plant	**reason**	**advantage to plant**	**reason**
young plant has its own food store or receives food and water from parent	young plant able to grow quickly ahead of competitors	some characteristics inherited from one parent, some from the other, thus great variation amongst offspring	whatever disaster occurs in the environment (e.g. disease) there is a good chance that some plants will survive
young plant produced in an environment which suits growth of parent plant	environment will also provide conditions needed by young plants for growth	plants well distributed	(a) wide geographical range increases chance of survival (b) less competition for light, space, etc., amongst offspring
pollination plays no part	plant not dependent on wind or insects for success		
disadvantage to plant	**reason**	**disadvantage to plant**	**reason**
all plants are identical	all plants susceptible to same disaster (e.g. disease)	requires two sex cells (gametes) to meet and fuse	transfer of male sex cell (pollen) depends on wind or insects which may fail to bring about pollination
plants poorly distributed	(a) whole population or even species could be wiped out by disaster (e.g. pollution) (b) members of population forced to compete for light, water, etc.		

Table 10.3 Comparison of asexual and sexual reproduction

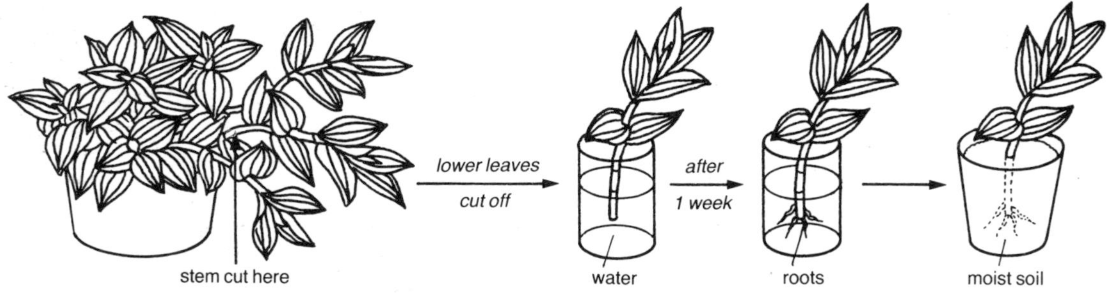

Figure 10.16 Taking a wandering sailor cutting

Methods of artificial propagation

When a plant lacks its own natural method of vegetative propagation, man often employs an artificial method of reproducing it asexually.

Taking cuttings

Sometimes the cutting is allowed to form roots in water before being planted into moist soil, as shown in figure 10.16. In other cases (figure 10.17) the cutting is put directly into soil and enclosed in a polythene bag to prevent the soil from drying out during early root formation. Heel 'cuttings' are torn rather than cut off the parent hardwood plant (see figure 10.18). All cuttings grow into plants with features identical to the parent plant.

Grafting

A cutting (**scion**) from a cultivated (and often delicate) variety of fruit tree or rose bush is grafted onto the **stock** of a hardy (and often disease-resistant) variety by cleft or bud grafting as shown in figure 10.19. The two cut surfaces bond together, healing the wound. Grafting is necessary because on its own the cultivated variety would not grow well. If the stock

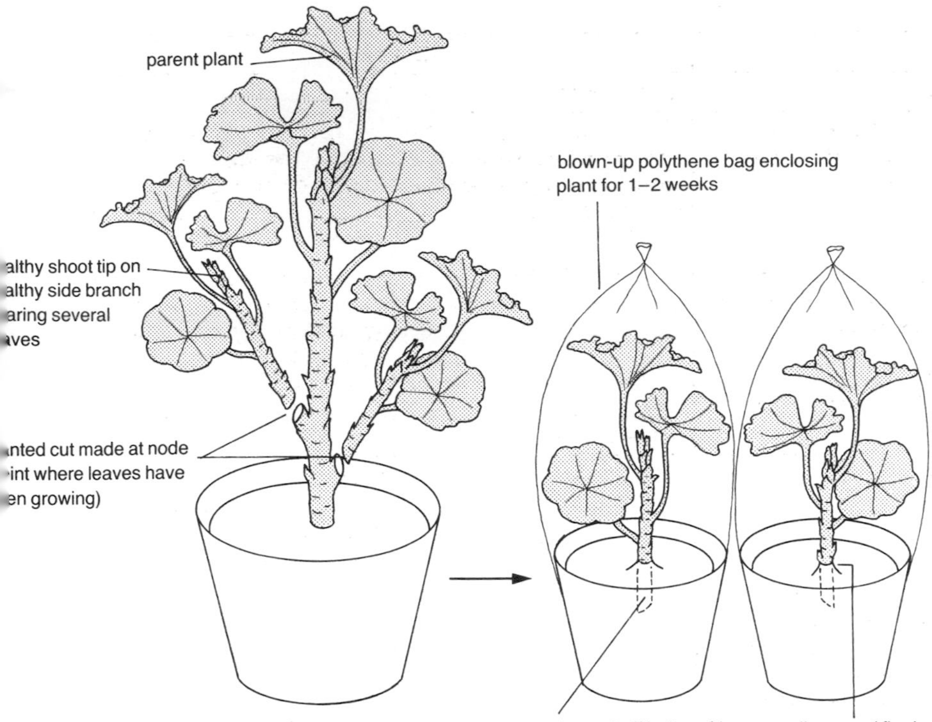

parent plant
althy shoot tip on
althy side branch
aring several
aves
inted cut made at node
int where leaves have
en growing)
blown-up polythene bag enclosing plant for 1–2 weeks
leaves trimmed off bottom ⅓ of cutting before it goes into soil
soil pressed firmly round cutting

ure 10.17 *Taking geranium cuttings*

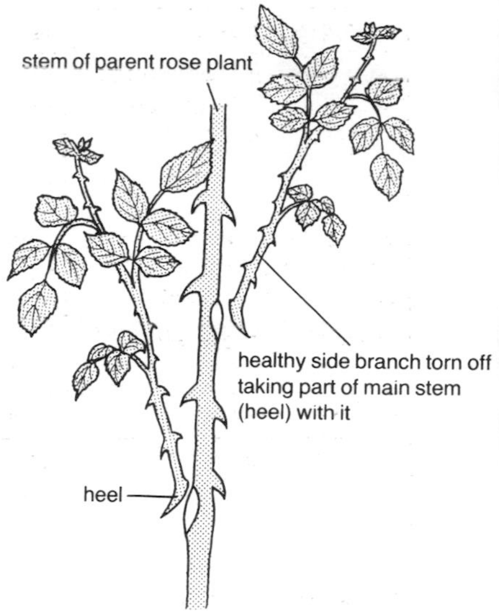

stem of parent rose plant
healthy side branch torn off taking part of main stem (heel) with it
heel

Figure 10.18 Taking rose cutttings

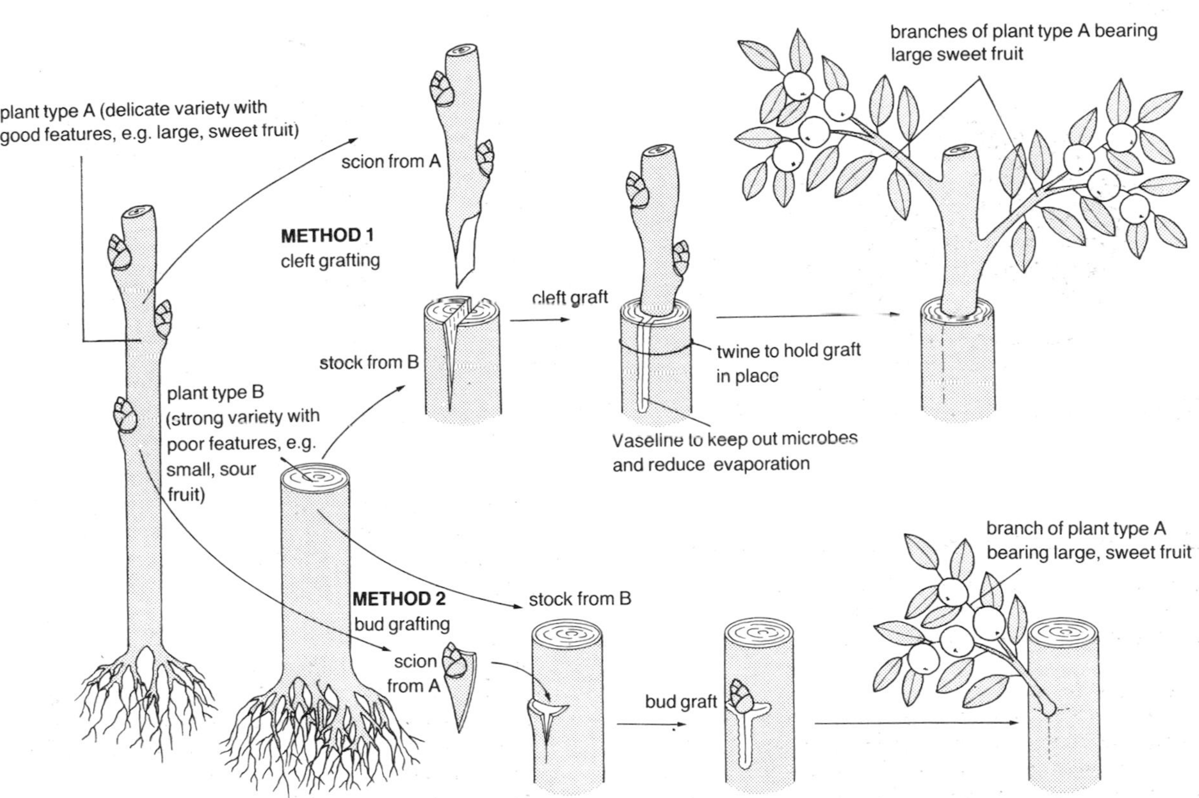

Figure 10.19 Two methods of grafting

develops wild side shoots, these unwanted 'suckers' are cut off to prevent them depriving the part of the plant above the graft of water and essential minerals.

Layering
This method of artificial propagation is often used to increase a supply of carnation plants (see figure 10.20).

Tissue cultures
This modern technique (figure 10.21) is also used by commercial growers to produce thousands of identical offspring from plants such as pineapple, orchids and rubber plants which are difficult to propagate by other means.

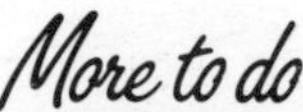

Advantages to man of artificial propagation

Guarantee of uniformity
Once a commercially desirable variety of plant (e.g. rose, fruit tree) has been produced by sexual reproduction, this plant can be artificially propagated by asexual reproduction (e.g. grafting) to produce a huge supply of offspring all possessing the desirable characteristic (e.g. beautiful flower, delicious fruit, etc.).

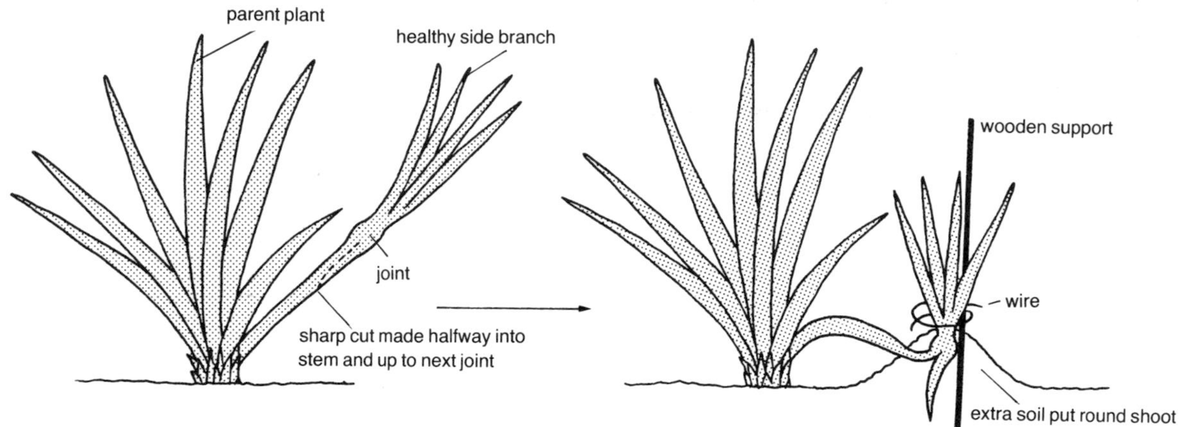

Figure 10.20 Layering

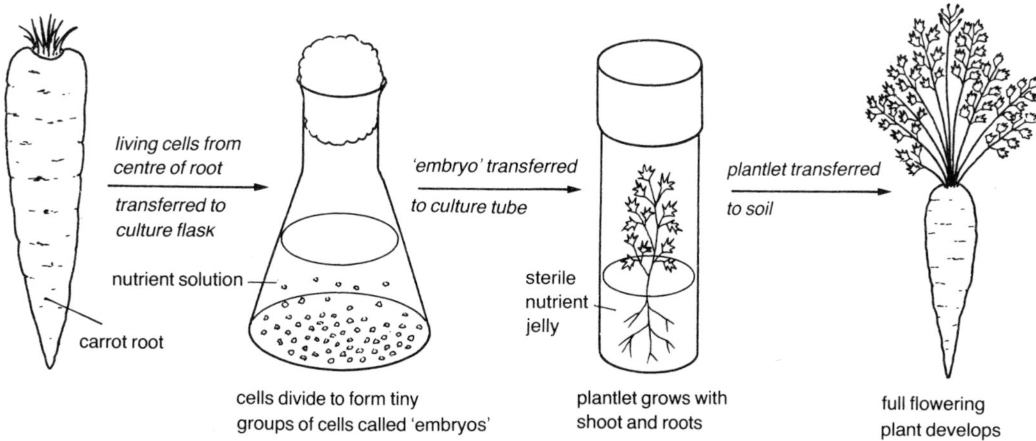

Figure 10.21 Tissue culturing

Reproduction of sterile varieties
Many cultivated plants, although producing flowers, do not produce seeds and are therefore sterile. By employing artificial propagation, man is able to produce vast crops of plants such as seedless grapes and citrus fruits.

Conservation
Artificial propagation allows man to save rare plants (of possible future economic value) that are threatened with extinction in their natural habitat.

KEY QUESTIONS

1 Rewrite and complete the following sentence by making one choice from the answers given.
Plants produced from cuttings are:

A identical to one another but different from the parent plant
B different from one another but identical to the parent plant
C identical to one another and identical to the parent plant
D different from one another and different to the parent plant

2 Describe a suitable method of taking a geranium cutting.
3 State TWO methods by which a scion from an apple tree could be grafted to the stock of a closely related variety.
4 Name a plant that can be artificially propagated by layering.
5 What name is given to a group of identical offspring that have been produced from one parent plant?

Extra Questions

6 Give TWO reasons why artificial propagation of flowering plants is advantageous to man.
7 Explain why the members of a clone are genetically identical.

11 Making food

Need for transport systems

All living plant cells need **water** and **sugar** to stay alive. A land plant's roots are buried in the ground where they absorb water. The green leaves are above ground where they make sugar by photosynthesis. The plant needs transport systems (see figure 11.1) so that water can travel up to the leaf cells and sugar can pass down to the root cells.

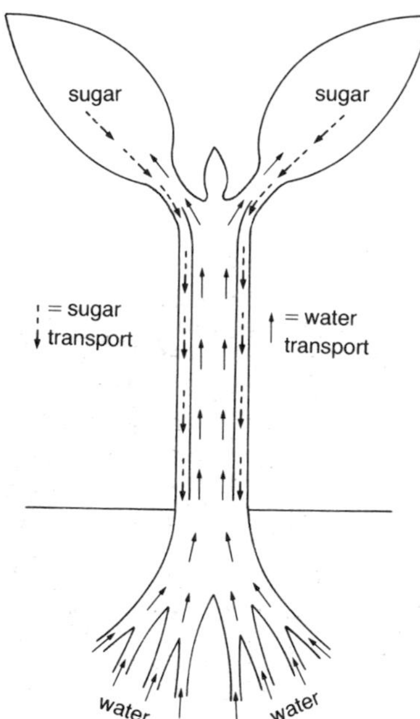

Figure 11.1 Plant transport systems

Structure of a one-year-old stem

Figure 11.2 shows where xylem and phloem tissue are found in a one-year-old stem of a plant such as privet.

Demonstrating the site of water movement in a plant

The cut end of a leafy shoot is placed in red dye for about an hour. A portion of stem is then cut out and examined as shown in figure 11.3. Red dye is found to be present only in the **xylem** vessels showing that xylem is the site of water transport in a plant.

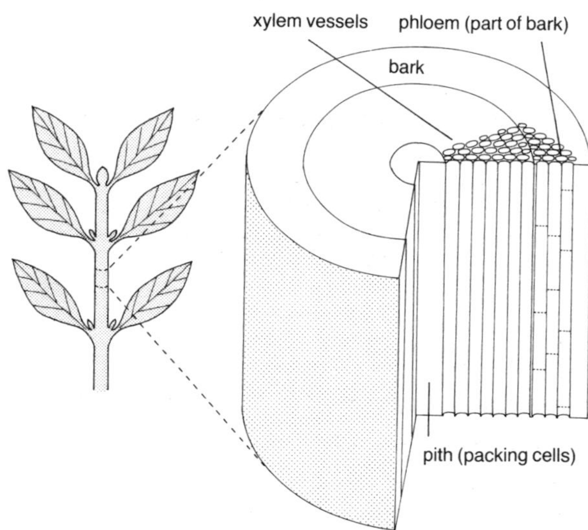

Figure 11.2 Site of xylem and phloem in a one-year-old stem

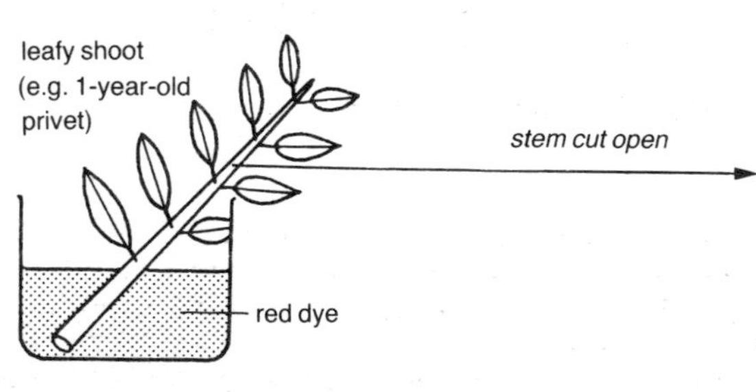

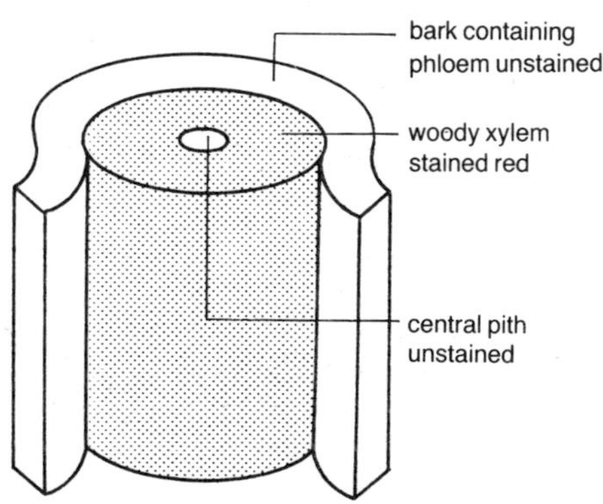

Figure 11.3 Red dye experiment

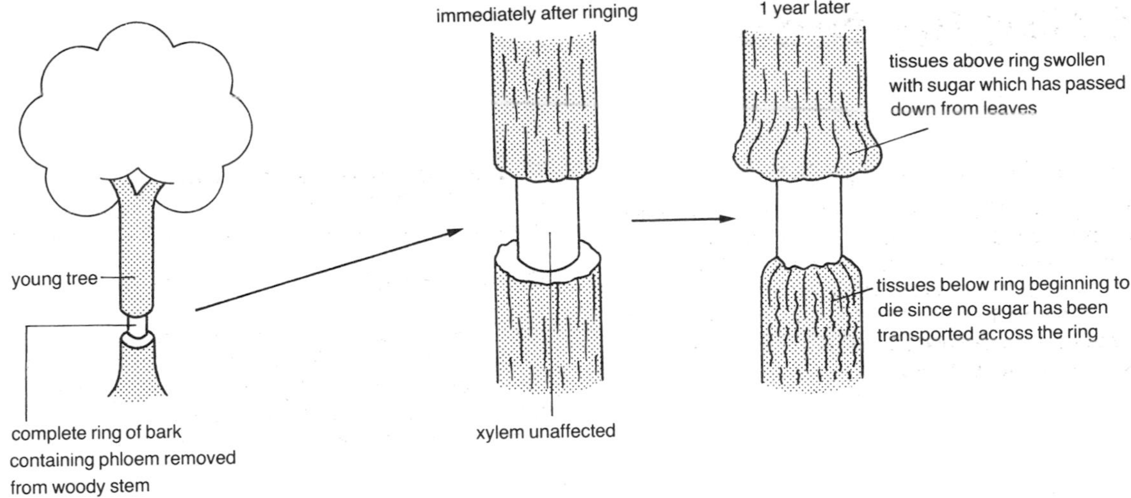

Figure 11.4 Ringing experiment

Identifying the site of sugar movement in a plant
The experiment shown in figure 11.4 demonstrates that the removal of a complete ring of bark (containing the phloem tissue) prevents sugar transport in the plant. It is therefore concluded that **phloem** tissue is the site of sugar transport (translocation).

Transport of materials in phloem and xylem

Phloem
The microscopic structure of phloem tissue is shown in figure 11.5. In addition to movement down a plant, sugar is also transported up to new growing points. The mechanism of translocation is still not fully understood but it is known from experiments that sugar transport is dependent upon the phloem tissue being alive. Translocation comes to a halt when the phloem is treated with respiratory poisons which stop cell respiration (energy release) and so kill the cells.

Xylem
The microscopic structure of xylem tissue is shown in figure 11.6 The movement of water in xylem is in an upward direction only. Xylem vessels are hollow dead tubes. Water transport is unaffected by the xylem being treated with respiratory poisons as the cells are already dead.

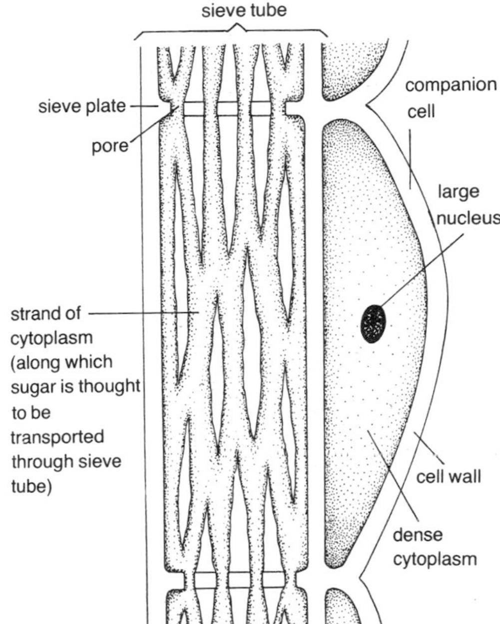

Figure 11.5 Structure of phloem

Other functions of transport system
Since xylem vessels make up a series of strong **lignified tubes** running from root to leaves, they help to support the plant.
In addition to transporting water, xylem is the route by which essential **mineral elements** absorbed from the soil solution are transported to all parts of the plant.

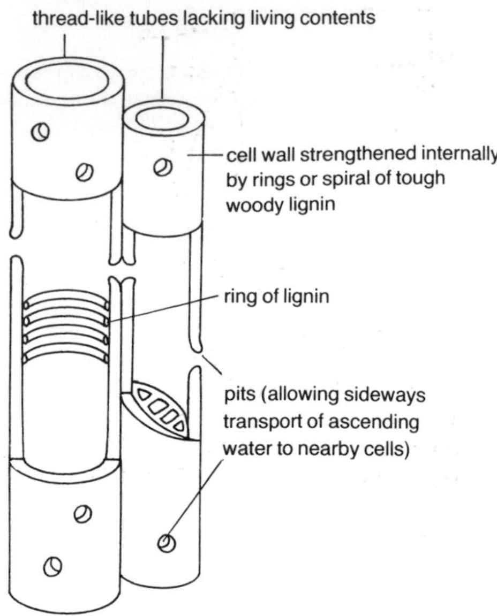

Figure 11.6 Structure of xylem

KEY QUESTIONS

1 Explain why a land plant needs two different transport systems.
2 Name the tissue which transports water.
3 Name the tissue which transports sugar.
4 With reference to direction, state one way in which the movement of water differs from the movement of sugar in an oak tree in midsummer.

Extra Questions

5 Draw up a table to compare transport of materials in phloem and xylem with respect to (**a**) type of material transported, (**b**) direction taken by material and (**c**) state of tissue (dead or alive).
6 Name TWO other functions of xylem in addition to water transport.

Stomata

Stomata are tiny pores found on the surfaces of a leaf (see figure 11.7). Plants take in carbon dioxide from the air through these stomata.

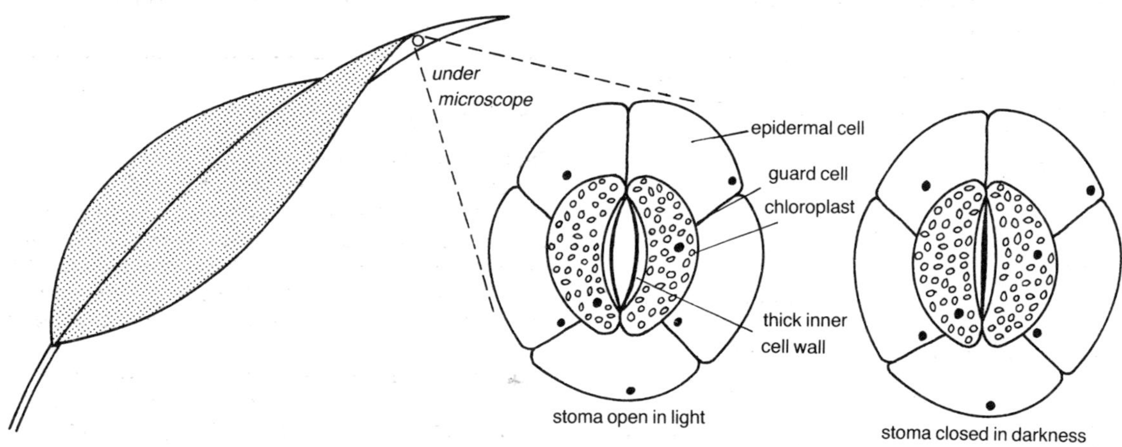

Figure 11.7 Stomata

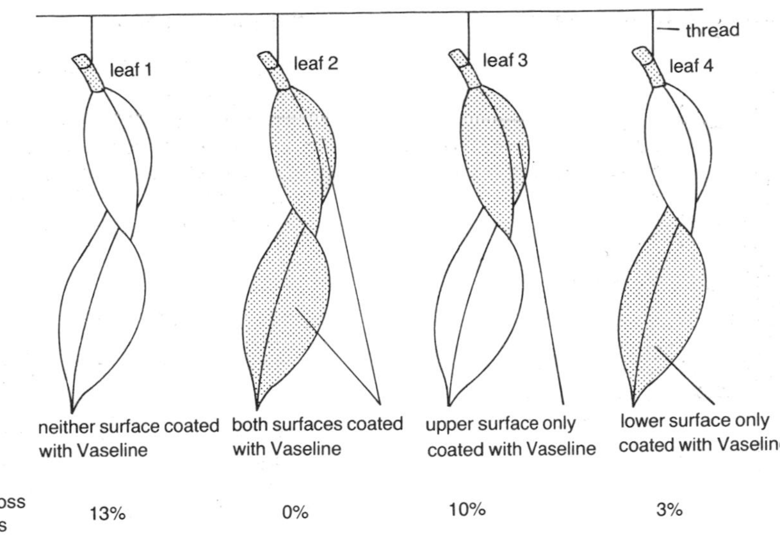

Figure 11.8 Location of stomata experiment

Location of stomata
Water vapour passes out of a leaf by the stomata. This is prevented if the stomata are blocked with Vaseline. Look at the leaves in the experiment shown in figure 11.8. Leaf 1 loses most weight because none of its stomata are clogged with Vaseline. Leaf 2 loses no weight because all of its stomata are blocked. Since leaf 3 loses more weight than leaf 4, it is concluded that leaf 3 has fewer blocked stomata and that most stomata must therefore occur on the under surface of this type of leaf.

More to do

Structure of a leaf (e.g. privet)

A leaf is structurally suited to the function of gas exchange, as shown in figure 11.9.

KEY QUESTIONS

1 Name the tiny pores found on a leaf surface.
2 State which gas enters a leaf by these tiny holes.
3 Why is this gas normally unable to enter the leaf during darkness?
4 Which surface of a privet leaf has more stomata?

Extra Question

5 Describe (a) the external features and (b) the internal structures possessed by a privet leaf that make it ideally suited to its function of gas exchange.

Photosynthesis

Testing a leaf for starch
The series of steps shown in figure 11.10 is carried out to test a leaf for starch (revise the test for starch using iodine solution by referring to Appendix 6). From the experiment it is concluded that this green plant has made its own food in the form of starch. Such food production by a green plant is called **photosynthesis**.

Dependence of food webs on photosynthesis
Without green plants there would be no life on Earth because only plants can produce high-energy compounds such as sugar and starch by photosynthesis. These energy-rich foodstuffs are the starting point in food chains and webs. Animals depend directly or indirectly on green plants for their energy as it passes along food chains and webs (see page 16). Humans are no exception to this rule; we also depend on plants for much of our food (see page 15).

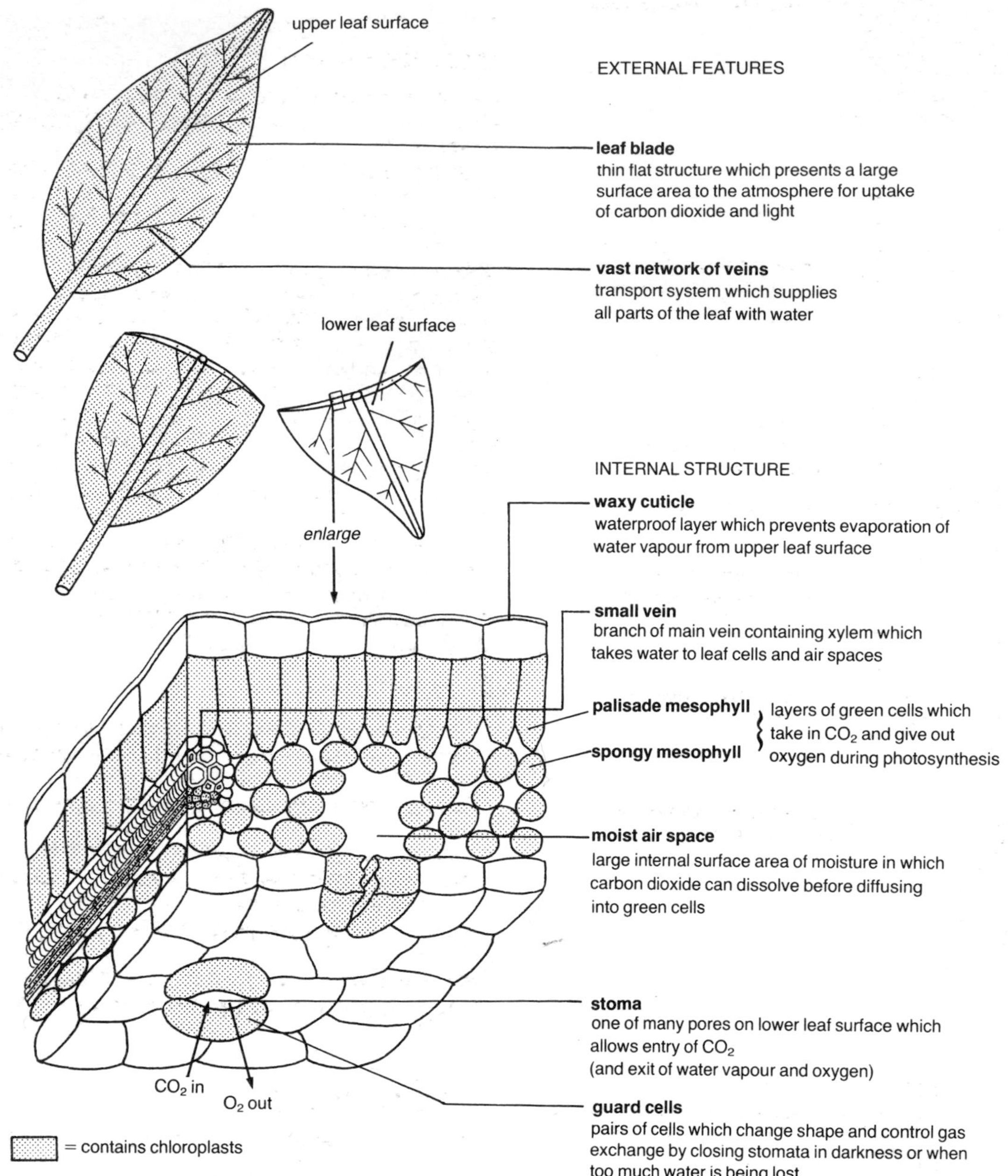

Figure 11.9 Structure of a leaf

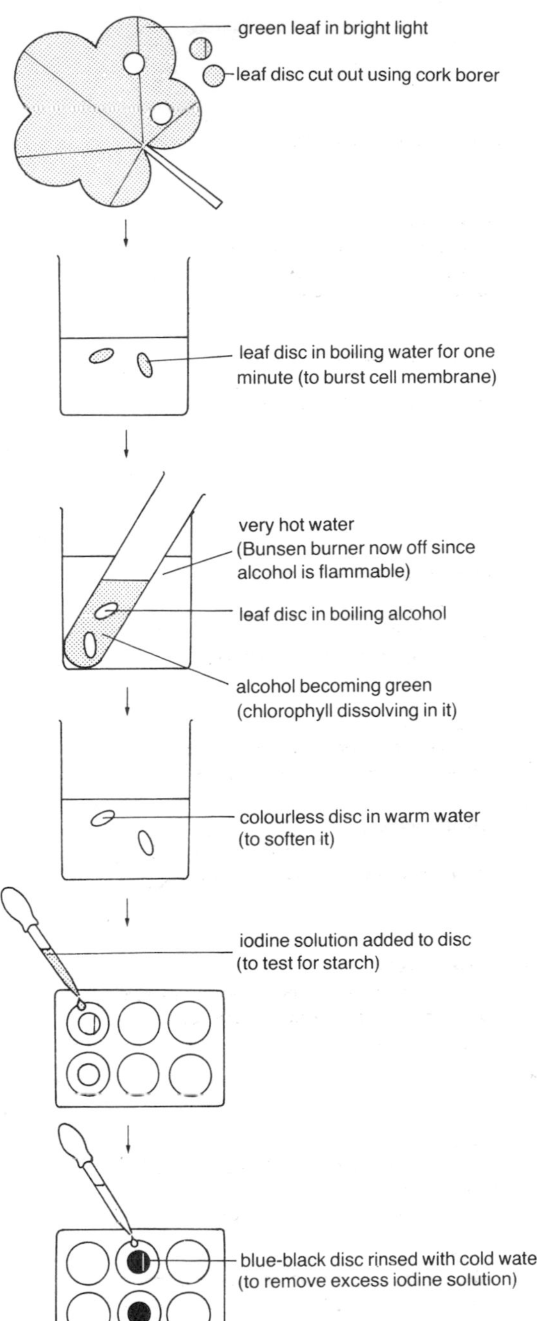

Figure 11.10 Testing a leaf for starch

More to do

Carbohydrates

A carbohydrate is a compound containing the elements **carbon** (C), **hydrogen** (H) and **oxygen** (O) combined together.

During photosynthesis molecules of carbon dioxide (CO_2) combine with molecules of water (H_2O) to form simple carbohydrates such as glucose ($C_6H_{12}O_6$).

Building up complex carbohydrates

As a plant grows it continues to make sugar by photosynthesis. Some of this sugar is broken down again when required, to provide the plant with energy for growth and reproduction. The remaining sugar molecules are linked into long chains and packed together into spherical starch grains found in a cell's cytoplasm (see figure 11.11). This starch is the plant's store of food and can be converted back to sugar for energy when needed. Starch is therefore called a **storage** carbohydrate.

Other sugar molecules are built into long chains of cellulose. These are gathered together to form ribbon-like fibres used to build the cell wall. Cellulose is therefore called a **structural** carbohydrate.

KEY QUESTIONS

1 What chemical reagent is used to test for starch?
2 What colour results when this chemical reacts with starch?
3 Why are the leaf discs boiled in water prior to this test?
4 Why are the discs next boiled in alcohol?
5 Name the foodstuff that this experiment shows to be present in the leaf discs.
6 At which position in a food chain is a green plant found?
7 What would happen to the animals on our planet if all the green plants became extinct? Explain why.

Extra Questions

8 What is a carbohydrate?
9 Give TWO differences between starch and cellulose.

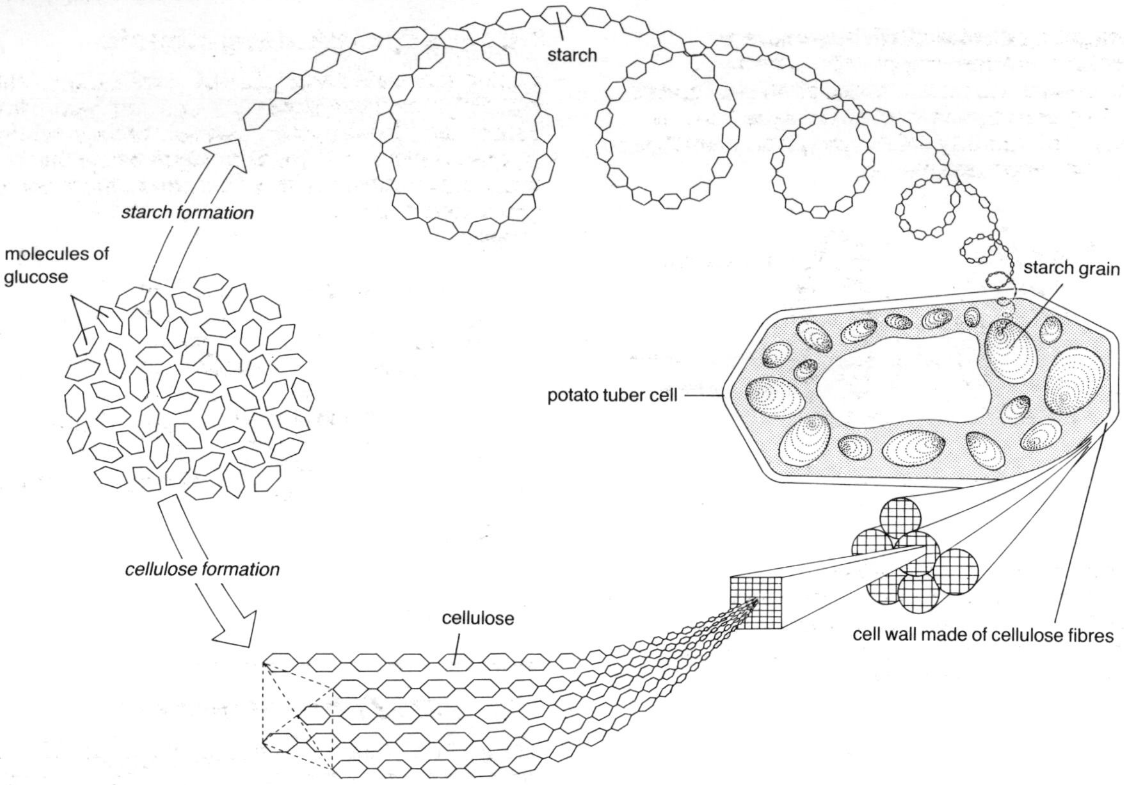

Figure 11.11 Formation of storage and structural carbohydrates

Requirements for photosynthesis

Investigating if light is necessary

The experiment shown in figure 11.10 is repeated using leaf discs from a plant that has been in darkness for two days. The discs do not turn blue-black when tested with iodine solution, showing that **light** is necessary for photosynthesis.

Investigating if carbon dioxide is necessary

Before being used in the experiment shown in figure 11.12, the two plants are kept in darkness for two days to ensure that at the start of the experiment they do not contain starch. In the experiment they are left for two days in bright light and then leaf discs from each plant are tested for starch. Those from A (deprived of carbon dioxide) fail to give a positive result with iodine solution but those from B turn blue-black. It is therefore concluded that **carbon dioxide** is essential for photosynthesis.

Importance of control

A **control** is a copy of the experiment in which all factors are kept exactly the same except the one being investigated in the original experiment. When the results are compared, any difference found between the two must be due to that one factor.

In the above experiment, if bell jar B (the control) had not been set up, it would be valid to suggest that the plant in bell jar A was unable to photosynthesise for some reason other than the lack of CO_2 (e.g. photosynthesis does not occur when the plant is enclosed in a confined space).

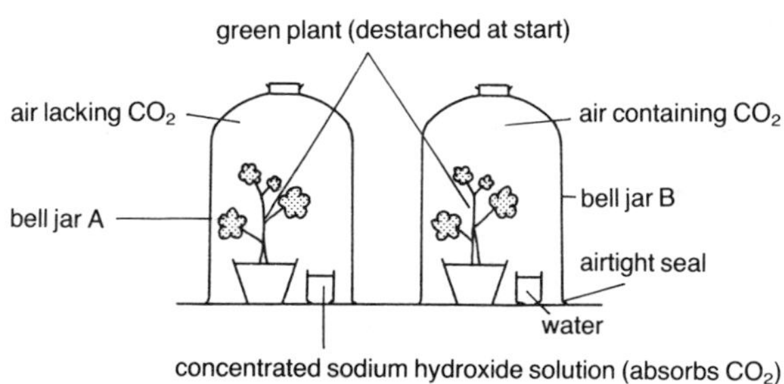

Figure 11.12 Need for carbon dioxide

Investigating if chlorophyll is necessary

When the starch test is applied to a leaf that is variegated (has two colours, one of which is green), only the green regions react positively with iodine solution (see figure 11.13) showing that **chlorophyll** is necessary for photosynthesis.

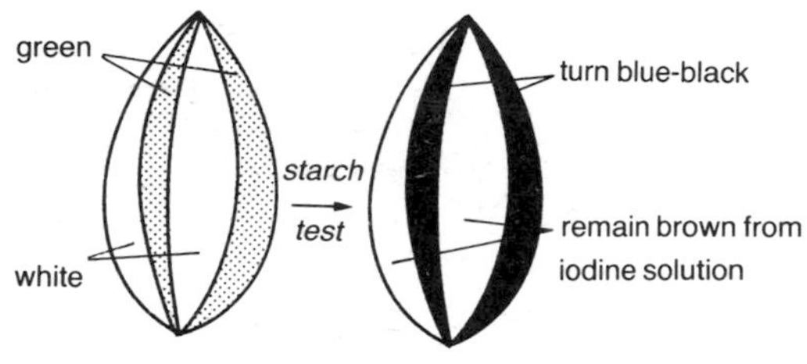

Figure 11.13 Need for chlorophyll

By-product of photosynthesis

The gas given off by the waterweed in the experiment shown in figure 11.14 is found to relight a glowing splint. This shows that in addition to food, **oxygen** is produced during photosynthesis.

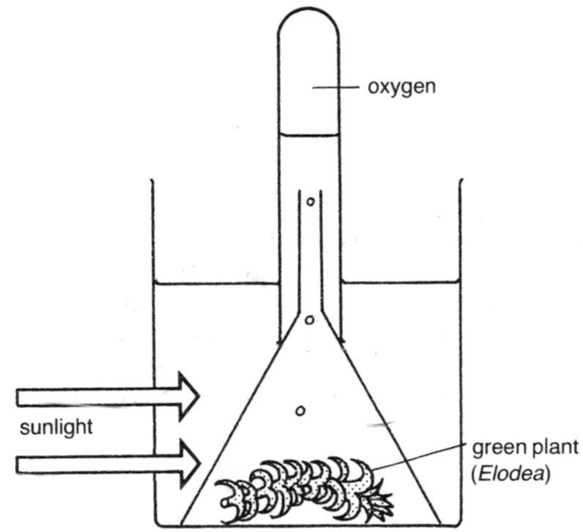

Figure 11.14 Oxgyen release

Summary of photosynthesis

Photosynthesis is the process by which green plants make high-energy foods (e.g. sugar and starch) from carbon dioxide and water using light energy trapped by green chlorophyll. Thus, during photosynthesis green leaves convert light energy to **chemical** energy (contained in food).

Equation:

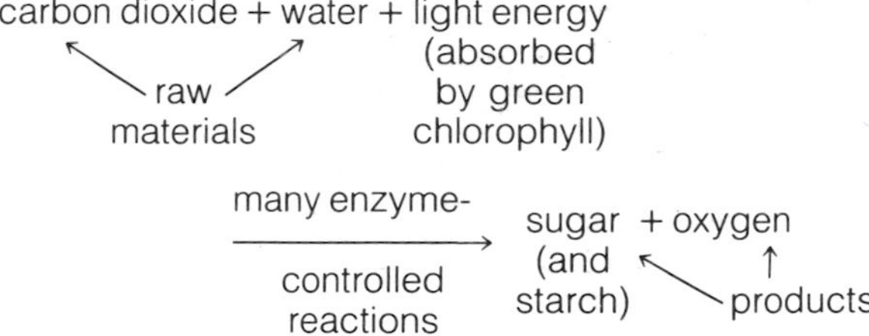

More to do

Elodea bubbler experiment

Figure 11.15 shows an investigation into the effect of light intensity on photosynthetic rate. The number of oxygen bubbles released per minute by the cut end of an *Elodea* stem indicates the rate at which photosynthesis is proceeding. At first the lamp is placed exactly 100 cm from the plant and the number of oxygen bubbles released per minute counted. The lamp is then moved to a new position (say 60 cm from the plant) and the rate of bubbling noted (once the plant has had a short time to become acclimatised to this new higher light intensity). The process is repeated for lamp positions even nearer the plant as shown in table 11.1.

distance from plant (cm)	units of light (calculated using mathematical formula)	number of oxygen bubbles/min
100	4	4
60	11	10
40	25	19
30	45	24
25	64	25
20	100	25

Table 11.1 Elodea *bubbler results*

When this typical set of results is displayed as a graph (figure 11.16), it can be seen that as light intensity increases, photosynthetic rate also increases until it reaches a maximum of 25 bubbles per minute at around 64 units of light.

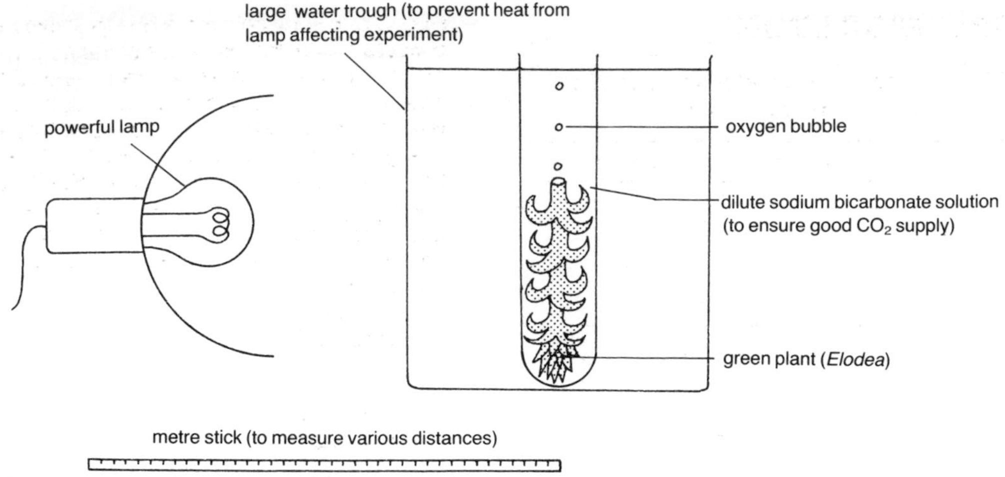

Figure 11.15 Elodea *bubbler experiment*

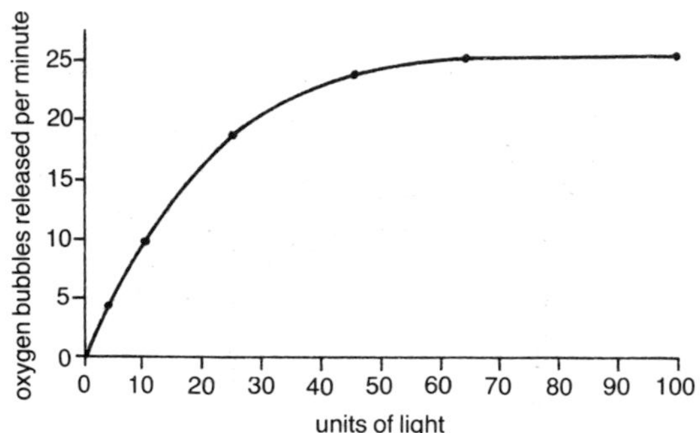

Figure 11.16 Graph of Elodea *bubbler results*

ingredients needed to make one loaf	Baker A's stock	Baker B's stock
500 g flour 30 g fat 10 g yeast 5 g sugar	5 kg flour 60 g fat 50 g yeast 40 g sugar	5 kg flour 300 g fat 50 g yeast 15 g sugar

Table 11.2 Factors limiting bread making

Limiting factors

Consider the information in table 11.2. Although Baker A has plenty of flour, yeast and sugar he can only make two loaves because he has a limited supply of fat. Baker B has plenty of fat but he can only make three loaves because his stock of sugar limits production. A **limiting factor** is a factor which holds up a process because it is in short supply.

Similarly, limiting factors can hold up photosynthesis. In the *Elodea* bubbler experiment described above, further increase in light intensity above sixty-four units does not increase photosynthetic rate. This is because shortage of CO_2 is now holding up the process. CO_2 concentration is acting as a limiting factor but when more CO_2 is supplied (as sodium bicarbonate solution) the rate of bubbling increases again.

KEY QUESTIONS

1 State THREE requirements for photosynthesis.
2 What energy conversion occurs during photosynthesis?
3 Describe the process of photosynthesis using all of the following words and phrases in your answer: green chlorophyll, oxygen, water, light energy, carbon dioxide, food.

Extra Questions

4 What is meant by the term **limiting factor**?
5 Name TWO factors which could limit the process of photosynthesis.

PROBLEM SOLVING

1 The table below gives the world production of major cereal plants grown in the year 1969.

cereal	millions of metric tonnes
wheat	305
corn	233
rice	181
barley	110
oats	51

Calculate the average amount of cereal produced in 1969.

2 The bar graph shows the fat and protein content of several foods from plants.

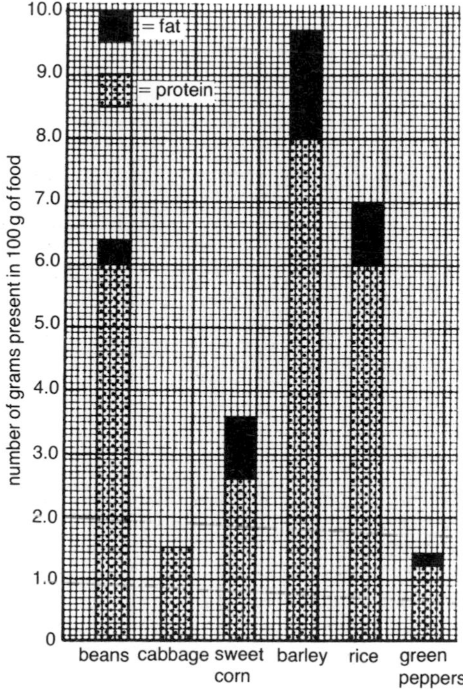

a) Name TWO foods that contain the same amount of protein.
b) Which TWO foods contain an equal amount of fat?
c) Which food contains no fat?
d) Which food contains least protein?
e) How many foods contain more fat than beans?
f) How many foods contain less protein than barley?
g) State the percentage fat content of barley.
h) By how many percent is the protein content of sweet corn greater than that of cabbage?
i) How many more grams of protein would be present in 1 kilogram of barley compared with 1 kilogram of rice?

3 During a biology field trip a group of five pupils collected different examples of various types of plants. Pupil 1 found 6 flowering plants, 2 seaweeds, 3 mosses and a fern. Pupil 2 collected 3 flowering plants and a piece of cone-bearing evergreen tree. Pupil 3 brought back 1 moss, 10 flowering plants and 2 seaweeds. Pupil 4's collection consisted of 3 ferns, 1 seaweed and 2 fungi. Pupil 5 collected 4 mosses, 1 fungus and 1 flowering plant. There were no duplicate specimens.
a) How many different non flowering plants were collected by the group?

Extra Question

b) Present all the above data in a table that shows at a glance what each pupil found and what was contained in the group's overall collection.

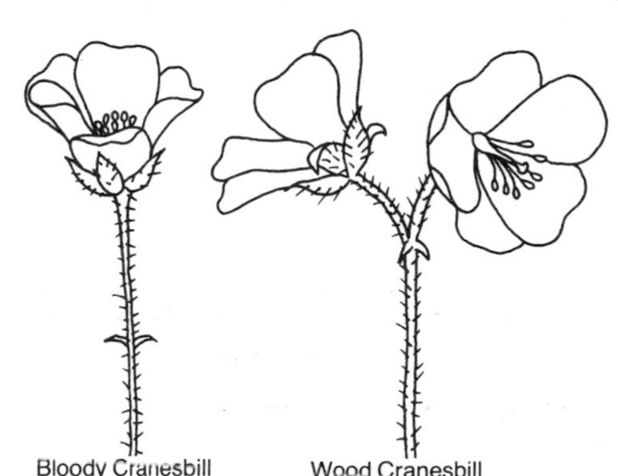

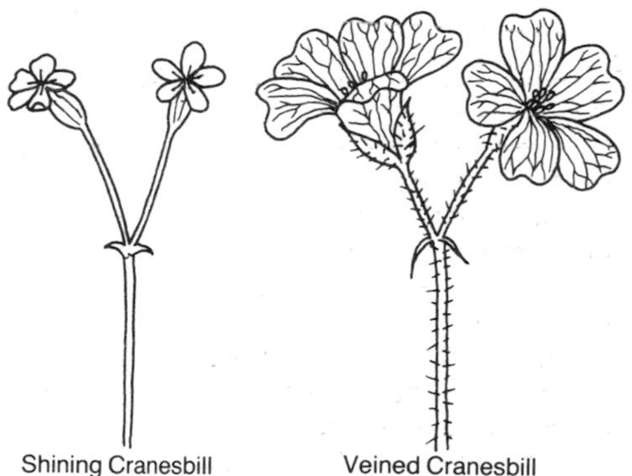

4 Carefully compare the four species of *Geranium* drawn to scale in the above diagram and then construct a simple branched key that could be used to identify each species.

5 The following table shows the results of an experiment with cress seeds.

number of seeds planted	300
number of seeds germinatng	240

Calculate the percentage germination failure.

6 The experiment shown in the diagram was set up to demonstrate that germinating seeds require oxygen.

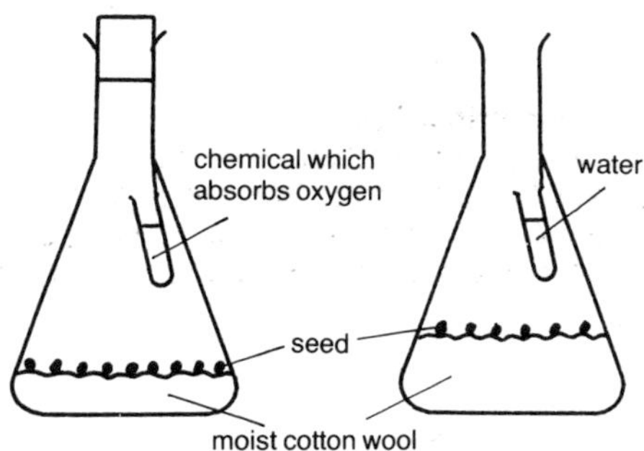

State THREE ways in which the experiment needs to be altered in order to make it a fair test.

7 The following experiment was set up in order to demonstrate that a green plant requires an atmosphere containing carbon dioxide for photosynthesis to occur.

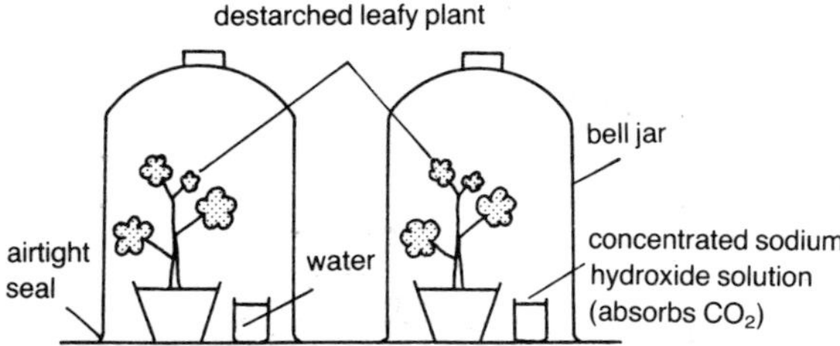

Name TWO environmental factors that must be kept the same for both bell jars to ensure that the experiment is fair.

8 Assume that to produce 1 unit of sugar by photosynthesis, each of the plants referred to in the following table must receive 3 units of CO_2, 3 units of water and 6 units of light energy.

plant	units of CO_2 available to plant	units of water available to plant	units of light available to plant
A	12	12	12
B	6	12	24
C	12	24	12
D	12	12	24
E	24	12	12

a) How many units of sugar will plant A be able to make?

Extra Questions

b) Which factor is in short supply and is holding up plant A's photosynthesis?
c) Which plant will be able to make the greatest number of sugar units?
d) If plant E is given an unlimited supply of light energy, how many units of sugar will it be able to make under the conditions given?

9 Cobalt chloride paper is blue when dry and pink when dampened by water vapour. In the experiment shown below the strip of paper was blue when it was taken out of the desiccator (drying chamber). The paper was taken to a plant on a window sill at the other side of the laboratory and placed in the enclosed area of leaf surface as shown.

After 10 minutes in contact with the leaf, the paper had turned pink. It was therefore concluded that 10 minutes is the time needed for the leaf surface to give out enough water vapour to affect the paper.

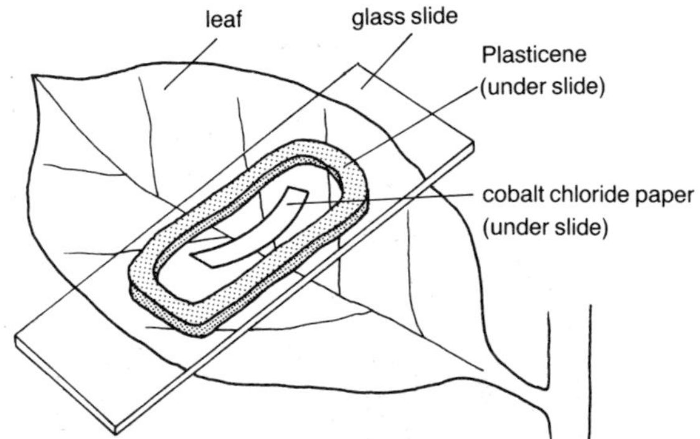

a) What colour change would be observed when moist cobalt chloride is placed in dessicator for several hours?

Extra Questions

b) Identify a possible source of experimental error in the above experiment.
c) Suggest a method of minimising the source of error you identified in (b).

Section 4 Animal Survival

12 The need for food

Food

Food is required for a variety of reasons as shown in figure 12.1. It provides the body with **fuel** for energy and supplies it with **building materials** for growth and tissue repair. Food also **protects** the body from illness and enables it to fight disease.

Chemical elements in food

For good health, man's diet must include **carbohydrate**, **fat** and **protein**. Each of these classes of food contains the chemical elements **carbon** (C), **hydrogen** (H) and **oxygen** (O). Protein also contains **nitrogen** (N).

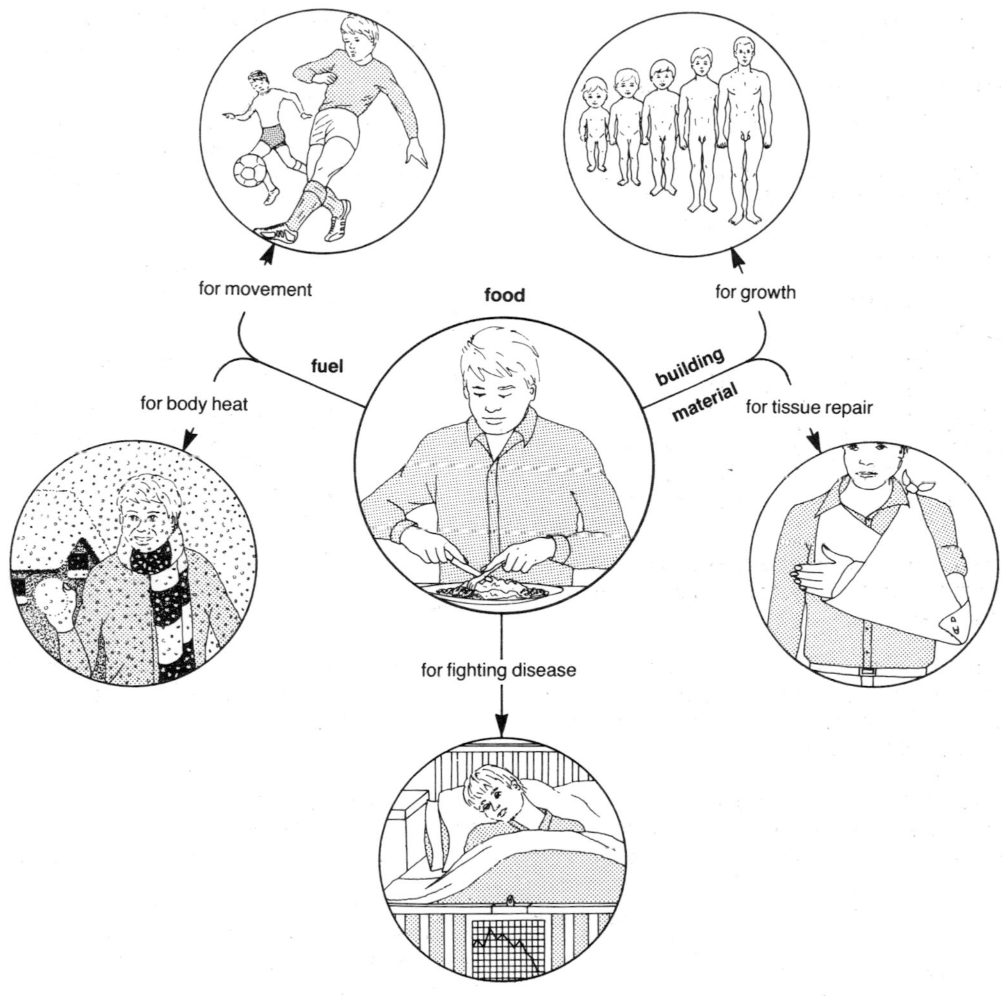

Figure 12.1 Need for food

Carbohydrates
These are energy-rich compounds often referred to as 'fuel' foods. Complex carbohydrates are built up from repeating units of simpler ones as shown in figure 12.2.

In plants excess **glucose** (which is soluble) is stored as insoluble **starch**. In animals (e.g. man) excess glucose is stored as insoluble **glycogen** in the liver.

Cellulose is a complex carbohydrate made of thousands of glucose molecules arranged in long chains which group together into cellulose fibres (the basic framework of plant cell walls). Although people are unable to digest cellulose, it is an essential part of their diet because it gives bulk to faeces. This stimulates the muscular action of the large intestine and prevents constipation. Plant foods rich in cellulose are called roughage.

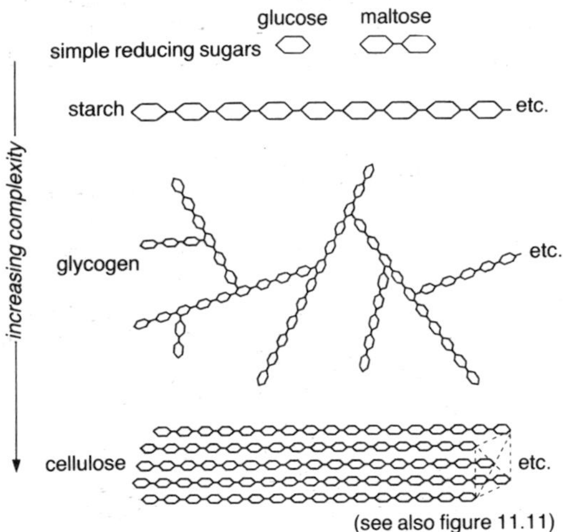

Figure 12.2 Carbohydrates

Fat
Since fat releases about twice as much energy per gram as carbohydrate, it is also a 'fuel' food. A molecule of fat is composed of **fatty acids** and **glycerol** as shown in figure 12.3.

Excess fat is stored round man's kidneys and under the skin, where it is especially effective in acting as a layer of insulation.

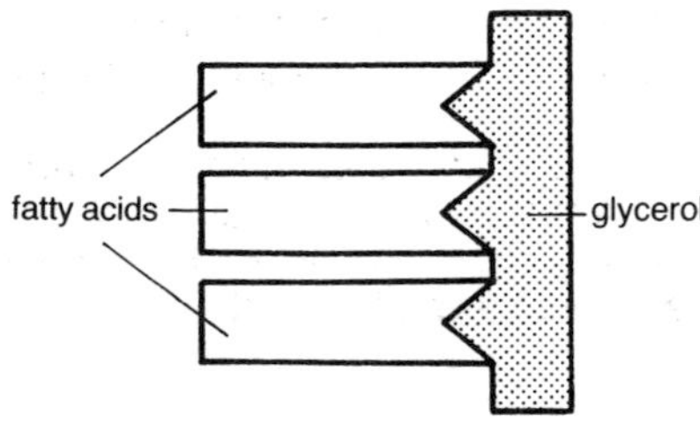

Figure 12.3 Fat molecule

Protein
Each molecule of protein is made of many, many subunits called **amino acids**, of which there are about twenty different types (see figure 12.4).

Since excess protein cannot be stored by man, an adequate daily intake (about 80 g) is required for body growth, formation of new cells and tissue repair. Though not normally a 'fuel' food, protein can release energy in a crisis (e.g. starvation).

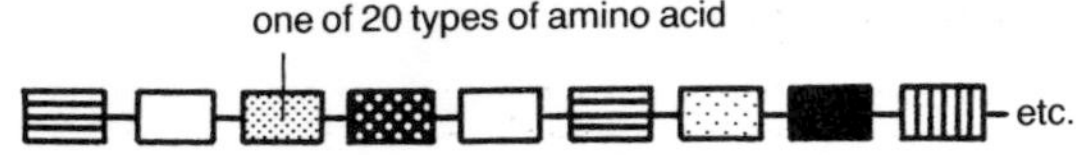

Figure 12.4 Small part of a protein molecule

Need for digestion

Every living cell in an animal's body needs a constant supply of food for the uses shown in figure 12.1. This food is carried to the cells by the bloodstream. In order to be absorbed into the bloodstream and then transported in the blood, food must first be digested.

Digestion is the breakdown of large particles of food into smaller particles that can be absorbed into the bloodstream through the wall of the small intestine (see figure 12.5).

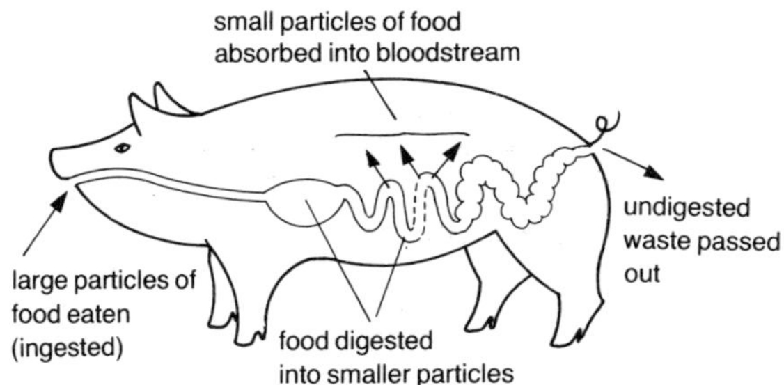

Figure 12.5 Feeding and digestion

More to do

The process of digestion

Figures 12.6 and 12.7 show how digestion involves the gradual breakdown of large insoluble molecules of food into smaller soluble molecules.

KEY QUESTIONS

1 Give THREE reasons why animals need food for survival.
2 a) Name THREE chemical elements always found in carbohydrates, fats and proteins.

b) Name a fourth chemical element found only in proteins.

3 By what means is food transported to living cells in the human body?

4 a) What must happen to large particles of food before they can be absorbed into the bloodstream?

b) What name is given to this process?

5 a) Describe the structure of glycogen in terms of simple sugar molecules.

b) With the aid of a simple diagram, describe the structure of a molecule of protein.

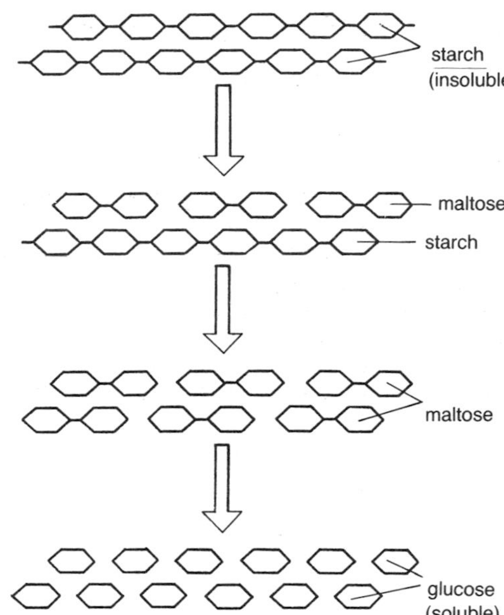

Figure 12.6 Digestion of carbohydrate (starch)

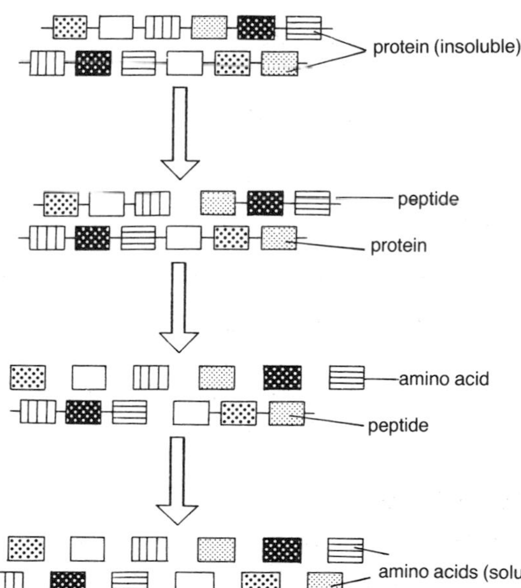

Figure 12.7 Digestion of protein

6 Using the words soluble and insoluble in your answer, explain what happens to food during digestion.

Action of teeth on food

Mammals use **teeth** to mechanically break food down into small fragments suitable for being swallowed and digested. The number and types of teeth present in an animal's dentition are directly related to its diet.

Omnivore

An **omnivore** (e.g. man) eats both plants and animal material. To suit this mixed diet, the four types of teeth (see figure 12.8) are all approximately the same size.

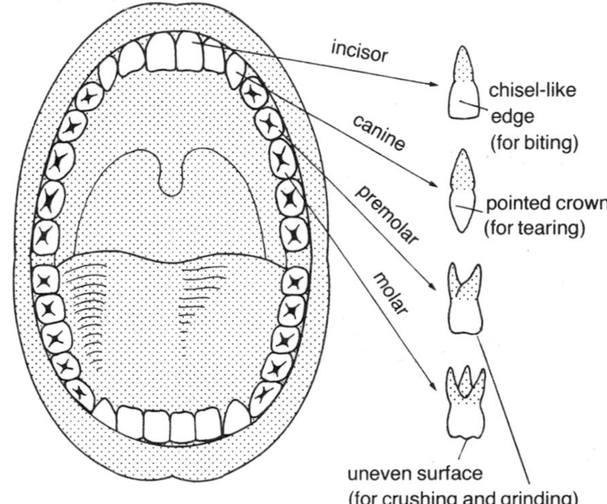

Figure 12.8 Dentition of adult human (omnivore)

Carnivore

A **carnivore** (e.g. dog) normally eats only animal material such as flesh. Its dentition is specialised to suit such a diet, as shown in figure 12.9.

The long backward-curved **canine** teeth are especially suited to stabbing and holding prey. The **premolars** and **molars** have sharp cutting edges. Since the lower jaw bites inside the upper jaw, a shearing action results on contraction of the powerful jaw muscles.

The massive **carnassial** teeth gain leverage by being at the back of the mouth. They are such effective shears that they can even slice tendons and crack bones as well as cutting flesh.

Herbivore

A **herbivore** (e.g. sheep) eats only plant material and its dentition (figure 12.10) is suited to this diet.

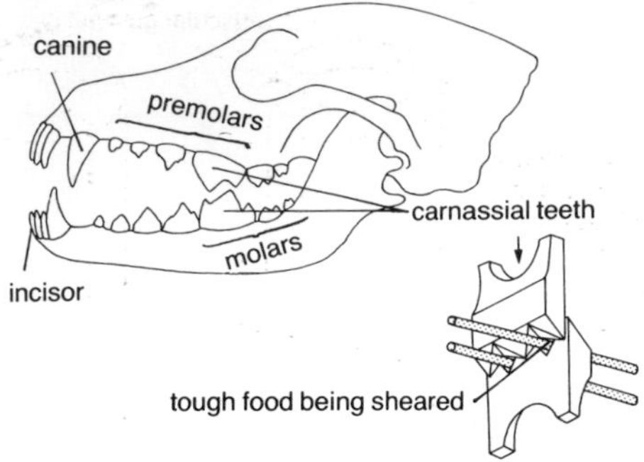

Figure 12.9 Dentition of dog (carnivore)

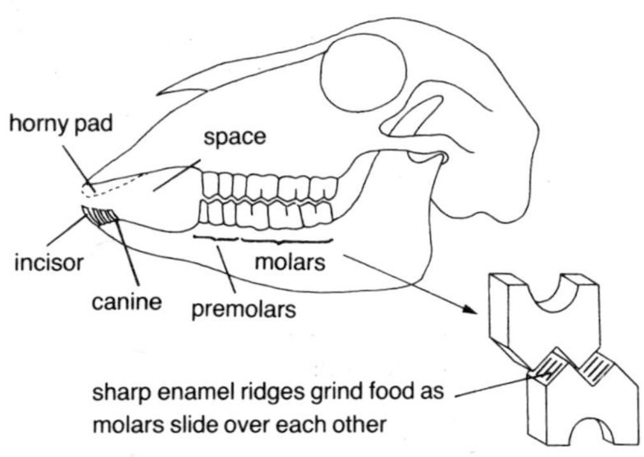

Figure 12.10 Dentition of sheep (herbivore)

The **incisor** teeth in the lower jaw of a sheep, crop grass by biting against a **horny pad** on the upper jaw. In place of large canines, there is a space where food collects before being pushed back by the tongue.

The herbivore also moves its lower jaw from side to side. This ensures that tough plant material is efficiently ground down by **enamel ridges** on the flat surfaces of its back teeth as they slide over one another.

Action of digestive system on food

The digestive system is made up of the **alimentary canal** (figure 12.11) which is a long muscular tube running from mouth to anus. The **salivary glands**, **liver** and **pancreas** (known as the associated organs) are connected to the alimentary canal by tubes called **ducts**.

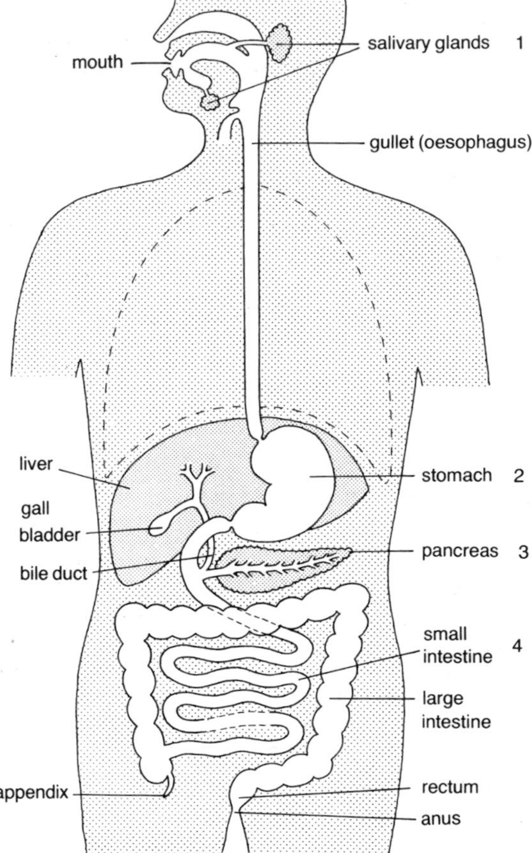

Figure 12.11 Alimentary canal (gut)

More to do

Digestive juices

These are made at various sites along a mammal's gut. The main ones are given in table 12.1

site of production	numbered part of figure 12.11	digestive juice produced
salivary glands	1	saliva
gastric glands in stomach wall	2	gastric juice
pancreas	3	pancreatic juice
glands in wall of small intestine	4	intestinal juice

Table 12.1 Sites of digestive juice production

Peristalsis

The alimentary canal is basically a muscular tube. When food is swallowed, muscular contractions of the gullet wall force the food down into the stomach. This muscular activity is called **peristalsis**. It is the means by which food is squeezed down the gullet, through the stomach and all the way along the intestines.

Mechanism of peristalsis

Figure 12.12 shows peristaltic activity in a region of the gut (e.g. gullet). Part of the gut wall is made of **circular muscle**. When this contracts behind a portion of food, the central hole of the tube narrows and the food is pushed along. At the same time, the circular muscle in front of the food becomes relaxed, making the central hole enlarge and allowing the food to slip along easily.

Peristalsis is a wave-like motion since it results from the alternate contraction and relaxation of the gut wall muscle along the entire length of the alimentary canal.

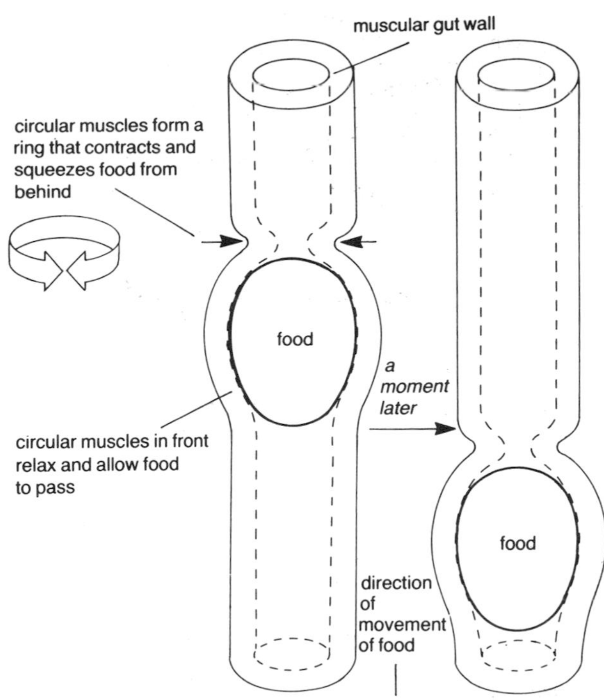

Figure 12.12 Peristalsis

Stomach

The wall of the stomach contains layers of muscle which contract and become relaxed in turn, causing **churning** of the stomach contents as shown in figure 12.13. During churning, food is held in the stomach by the closure of the muscular valves (**sphincters**) at either end of the stomach.

Since this vigorous activity of the stomach wall mixes the food with the digestive juices from the stomach wall, it helps in the chemical breakdown of food.

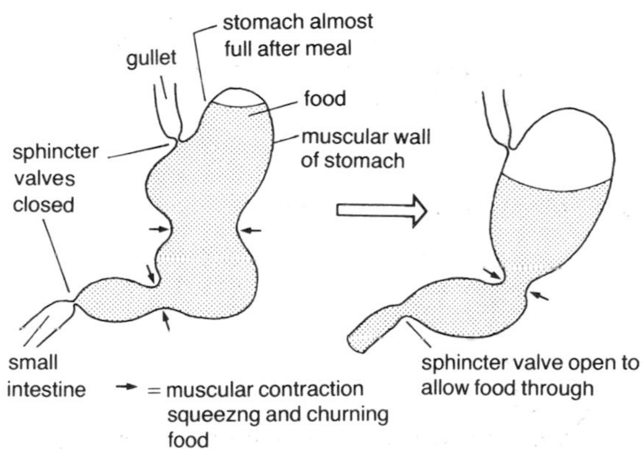

Figure 12.13 Muscular contractions of stomach

KEY QUESTIONS

1 Name the FOUR types of teeth present in a mammal's dentition.
2 In general what are teeth used for?
3 Describe the role of (a) a carnivore's carnassial teeth and (b) a herbivore's molar teeth.
4 Arrange the following parts of the alimentary canal in the correct order, beginning with the mouth: anus, gullet, large intestine, mouth, rectum, small intestine, stomach.
5 Describe the effect of peristalsis on a mouthful of swallowed food.
6 Explain how the stomach helps in the chemical breakdown of food.

Extra Questions

7 a) Explain the mechanism of peristalsis.
b) Name THREE regions of the alimentary canal where peristalsis occurs.
c) Explain how peristalsis works.

Digestive enzymes

Enzymes are substances which speed up the rate of biochemical reactions (see page 51). **Digestive enzymes** promote the breakdown (digestion) of foodstuffs during their passage through the digestive system. Different enzymes are responsible for the breakdown of carbohydrates, proteins and fats.

Action of amylase on starch (carbohydrate)

Amylase is a digestive enzyme present in saliva and pancreatic juice.

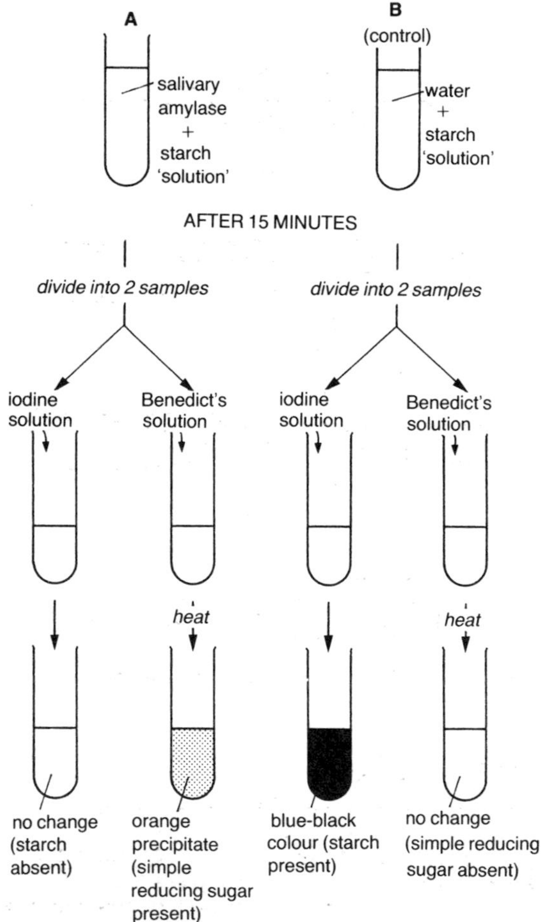

Figure 12.14 Action of salivary amylase

Look at the experiment shown in figure 12.14. From the results it is concluded that in tube A the enzyme amylase has digested starch to simple sugar. In control tube B which does not contain enzyme, the starch has remained undigested.

Importance of control

If tube B had not been set up, it would be valid to suggest that starch would have changed to sugar whether amylase had been present or not.

Action of pepsin on protein

Pepsin is a digestive enzyme present in the human stomach.

The protein used in the experiment shown in figure 12.15 is egg white (albumen). From the results it is concluded that in tube A the enzyme pepsin has digested insoluble particles of egg white protein (which began as a cloudy suspension) to smaller soluble molecules (leaving a clear solution). In B no digestion occurred because no enzyme was present.

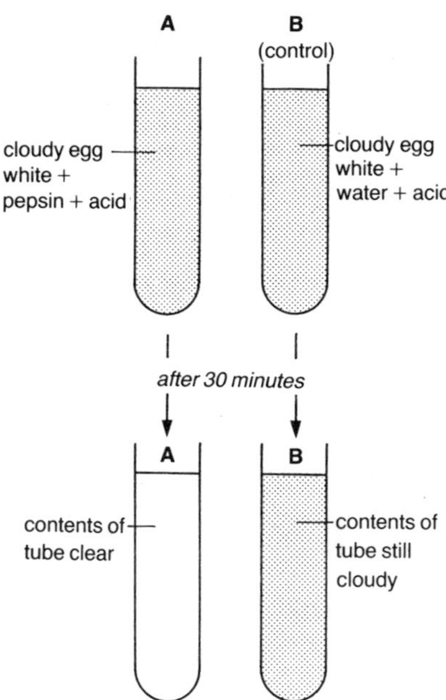

Figure 12.15 Action of pepsin

Action of lipase on fat

Lipase is a digestive enzyme made by the pancreas and active in the small intestine. The fat used in the experiment shown in figure 12.16 is the fat present in the cream in milk.

A little dilute alkali is added to each so that the contents begin at pH 8–9 (green). After an hour the contents of tube A have turned orange (pH 5), whereas tube B remains unchanged. It is concluded therefore that lipase digests fat to acids (fatty acids).

Bile is not an enzyme. It converts large drops of fat into tiny droplets, thus increasing the surface area of fat upon which lipase can act.

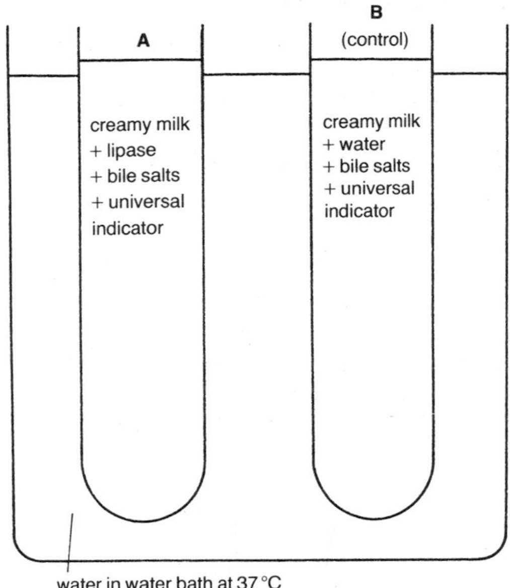

Figure 12.16 Action of lipase

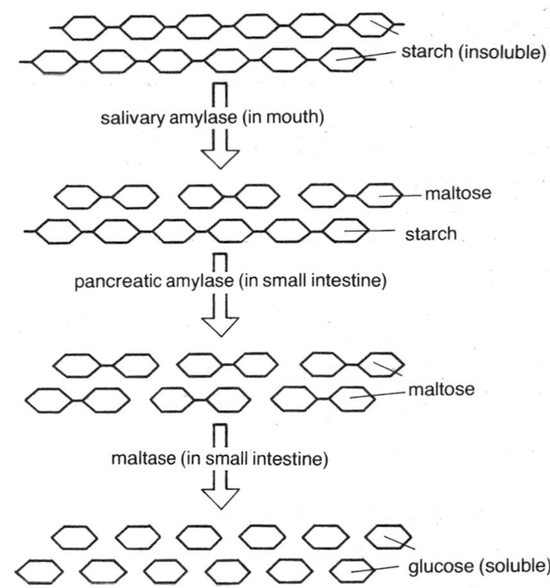

Figure 12.17 Role of enzymes in carbohydrate digestion

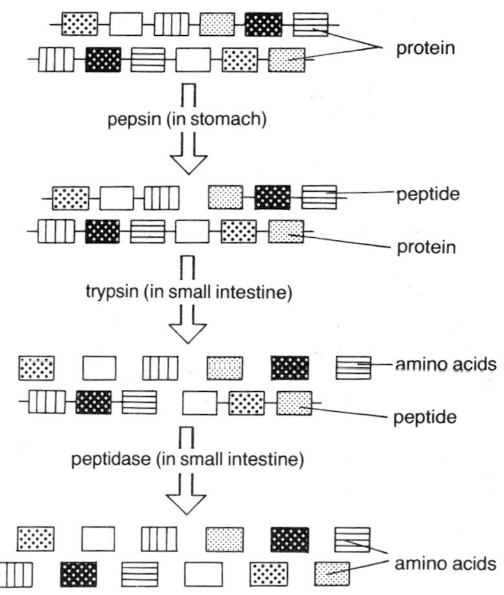

Figure 12.18 Role of enzymes in protein digestion

More to do

Types of digestive enzyme

There are several different types of digestive enzyme. Each works on a particular type of substrate as shown in table 12.2

Complete digestion

The complete digestion of food often involves the activities of several digestive enzymes as shown in figures 12.17 and 18.

KEY QUESTIONS

1 What job is performed by digestive enzymes in the human gut?
2 Name THREE enzymes and for each one, state the type of food upon which it acts.
3 Why is control tube B included in the experiment shown in figure 12.15?

Extra Questions

4 a) Give ONE example of a protease, a lipase and an amylase.
b) For each, state the enzyme's substrate and the products of digestion formed.

type of enzyme	example(s)	type of food digested (i.e. substrate)	end products of digestion
amylase	salivary amylase pancreatic amylase	starch starch	maltose maltose
protease	pepsin trypsin peptidase	protein protein peptides	peptides peptides and amino acids amino acids
lipase	pancreatic lipase	fat	fatty acids and glycerol

Table 12.2 Types of digestive enzyme

Absorption in the small intestine

One of the main functions of the small intestine is to absorb the end products of digestion through its wall and then pass them into the bloodstream. The small intestine is very efficient at this job because of its structure (see figure 12.19). It is very **long** and its internal surface is **folded** and bears thousands of tiny **finger-like projections**. As a result the small intestine presents a **large absorbing surface area** to the digested food.

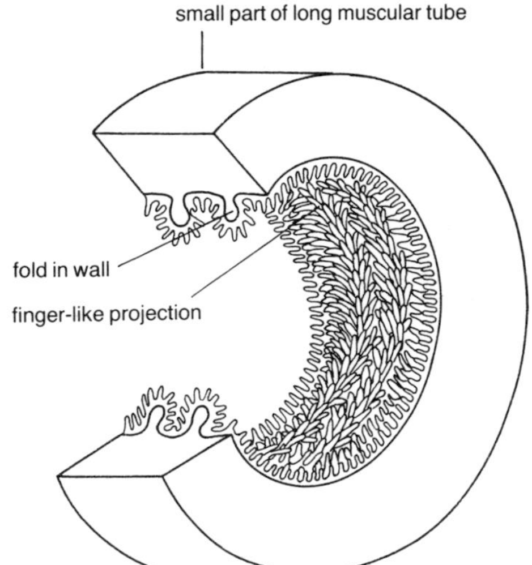

Figure 12.19 Structure of small intestine

More to do

Variety of end products

Imagine the small intestine of a person who has eaten an egg sandwich. As a result of complete digestion, the starch in the bread has been broken down to glucose, the protein in the egg to amino acids and the fat in the butter or margarine to fatty acids and glycerol. All of these different end products of digestion must now be absorbed and transported round the body.

Role of villi

The finger-like projections present on the surface of the small intestine are called **villi**. The internal structure of a villus is shown in figure 12.20. In addition to presenting a large surface area, villi are ideally suited to the jobs of absorption and transport of digested food for the following reasons.

Each villus is covered by a cellular lining (epithelium) which is only **one cell thick**. The end products of digestion are therefore able to pass through rapidly.

Each villus contains a dense network of **blood capillaries** into which glucose and amino acids pass, ready to be carried into the bloodstream leading to the liver and then to all parts of the bodyl

Each villus also contains a tiny **lymphatic vessel** (lacteal) which collects the products of fat digestion ready for distribution round the body.

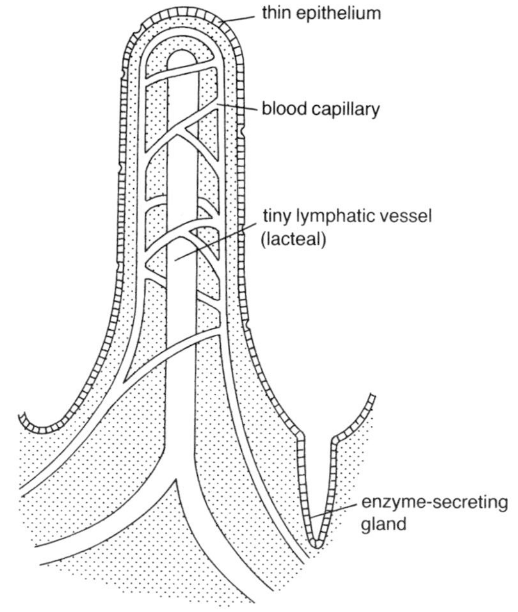

Figure 12.20 Structure of a villus

Large intestine

Material passing into the large intestine consists of undigested matter, bacteria and dead cells. The large intestine absorbs **water** from this unwanted material leaving **faeces**. Faeces are eliminated by being passed into the rectum from where they are later expelled through the anus.

KEY QUESTIONS

1 a) State THREE ways in which the structure of the small intestine is suited to its function
b) Explain why these features enable digested food to be easily absorbed.
2 State ONE function of the large intestine.

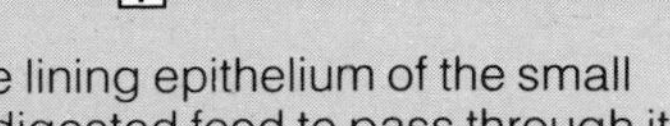

3 What feature of the lining epithelium of the small intestine enables digested food to pass through it easily?
4 Explain how the internal structure of a villus is related to the transport of digested food.

13 Reproduction

Survival of the species

Reproduction is the production of new members of a species. For a species to survive, it must produce sufficient young to replace those lost through old age, disease and other causes of death.

Sex cells

Most animals reproduce sexually. Both sexes make **sex cells** (gametes). The male produces a large number of **sperm** and the female produces a smaller number of **eggs**.

A sperm (see figure 13.1) consists of a head region (mainly a nucleus containing genetic material) and a tail which enables it to move.

An egg is larger than a sperm because in addition to its nucleus it has a store of food in its cytoplasm. It lacks a tail and cannot move of its own accord.

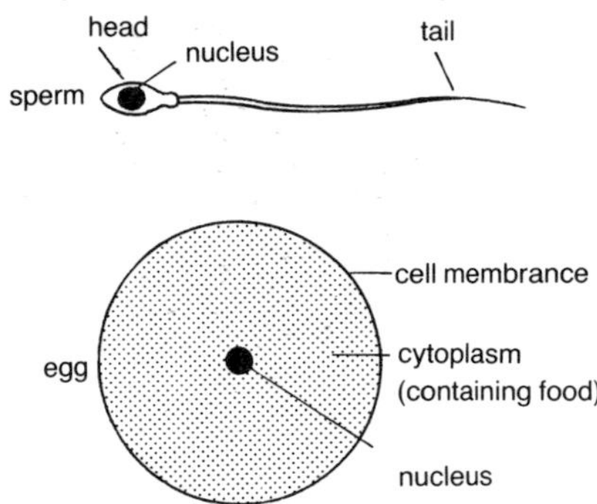

Figure 13.1 Sex cells

Fish (e.g. stickleback)

During the breeding season, the male builds a nest of waterweeds and develops a red belly which attracts the female. The two fish perform **courting** movements (see figure 13.2). Finally the female lays her eggs in the nest and swims away. The male then enters the nest and deposits his sperm in the water adjacent to the eggs. The sperm are attracted to the eggs and swim towards and fertilise many of them.

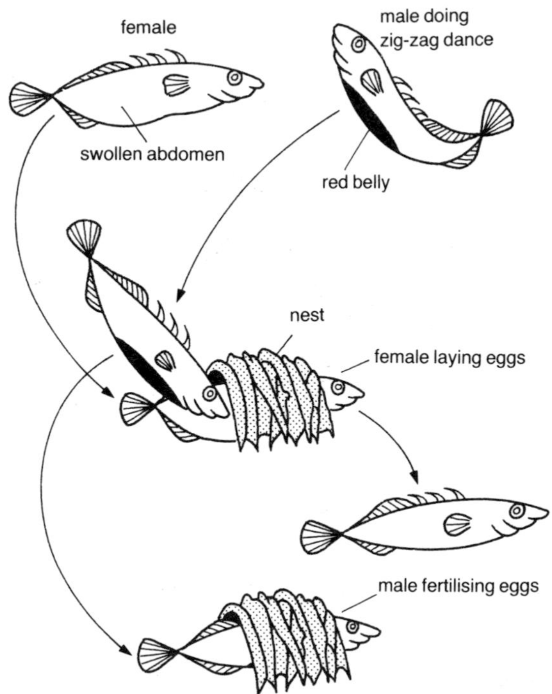

Figure 13.2 Mating in sticklebacks

Fertilisation

For a new individual to be formed by sexual reproduction, **fertilisation** must occur. This process involves a sperm reaching an egg and then the sperm's nucleus entering the egg and fusing with the egg's nucleus to form a single cell called a **zygote**. The zygote is the first cell of the new individual.

Achieving fertilisation

In many animals **mating** occurs. This brings the sperm close to the eggs and increases the chance of fertilisation occurring.

Mammal (e.g. human)

Land animals do not live in a watery environment into which sperm and eggs can be released. Instead, a fluid containing sperm is produced by the male and deposited in the female's body during copulation.

Reproductive organs of a mammal

The human reproductive organs are shown in figure 13.3. Sperm cells are produced in the **testes** of the male mammal. Eggs are produced in the **ovary** of the female and released at regular intervals.

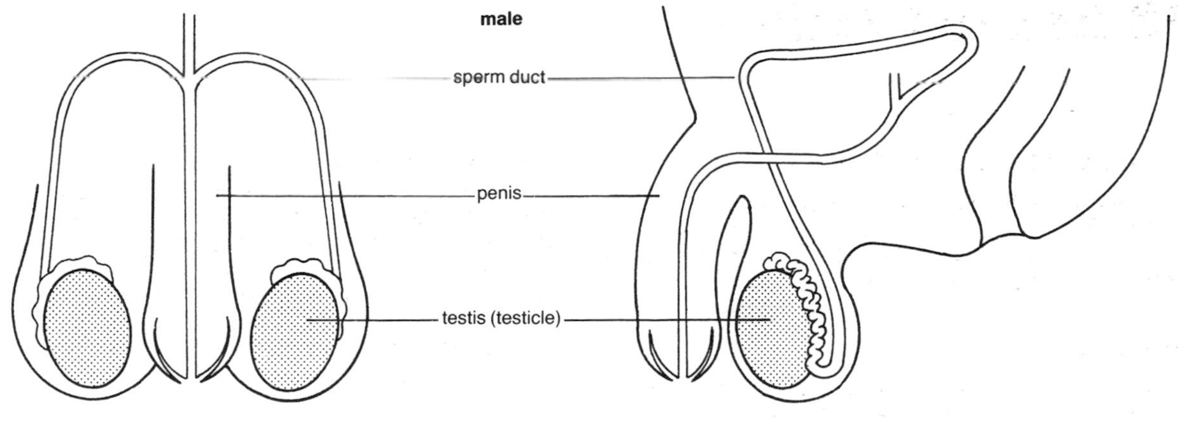

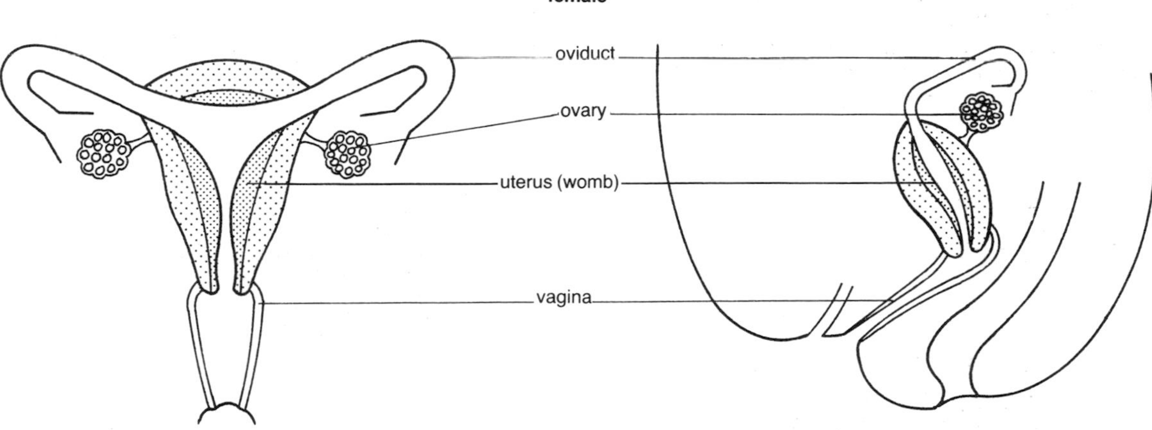

Figure 13.3 Human reproductive organs

Since an egg is unable to move of its own accord, it is the job of the oviduct to transport the egg towards the uterus. One end of the oviduct is funnel-shaped and fringed with hair-like cilia. It moves towards the ovary and picks up the egg as shown in figure 13.4. The inner lining of the oviduct has more cilia whose movements carry the egg along towards the uterus.

Copulation

A male mammal has an organ called the penis for depositing sperm into the female. During **copulation** (sexual intercourse in humans) the penis (stiff and erect, since it has received an extra supply of blood) is inserted into the vagina. Muscles around the testes and sperm ducts finally force millions of sperm up the ducts and out into the upper end of the vagina near the uterus. The sperm swim up the uterus and into the oviducts. It is here that one sperm fertilises one egg.

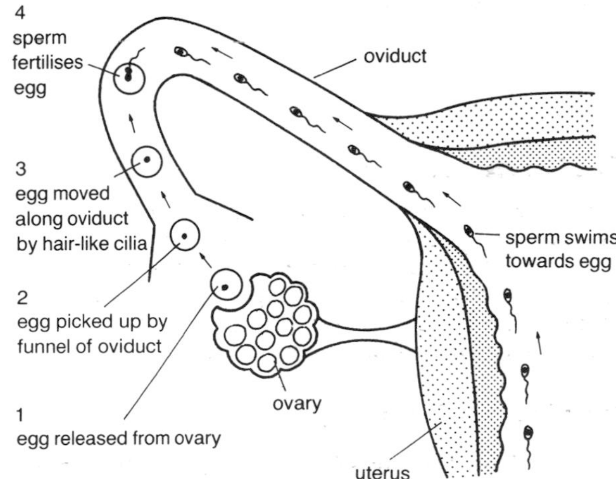

Figure 13.4 Fertilisation in oviduct

More to do

Internal and external fertilisation

In fish, fertilisation is said to be **external** because it occurs outside the parents' bodies in the surrounding water and not inside the female.

In mammals, the sperm cells do meet the eggs inside the female parent's body and fertilisation is therefore said to be **internal**.

Internal fertilisation is essential amongst land animals because there is no water in the animal's immediate environment to carry the sperm cells to the eggs.

Fertilisation and cell division

Although many sperm may meet an egg, only one sperm fertilises the egg. The zygote formed becomes an **embryo** by undergoing cell division as shown in figure 13.5.

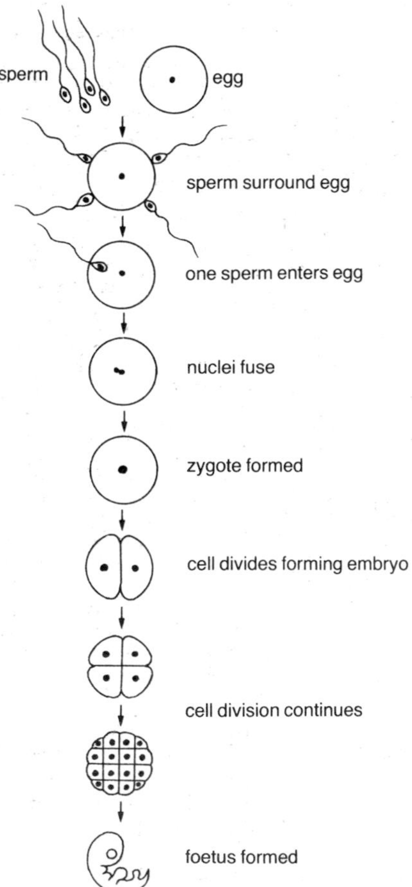

Figure 13.5 Fertilisation and cell division

KEY QUESTIONS

1 Give TWO differences in structure between an egg and a sperm.
2 a) What is fertilisation?
  b) What is formed as a result of fertilisation?
3 Rewrite the following sentences, choosing the correct word from each pair.
  a) During mating a male fish deposits sperm/eggs in the water inside/outside the female's body.
  b) During copulation a male/female mammal deposits sperm/eggs inside/outside the female's body.
4 Name the organ that produces (a) eggs in a female mammal and (b) sperm in a male mammal.
5 In which region of a female mammal's reproductive system does fertilisation normally take place?

6 What is meant by (a) internal and (b) external fertilisation?
7 Why is internal fertilisation essential to land animals?
8 a) What name is given to the first cell of a new individual formed as a result of fertilisation?
  b) Briefly describe the first steps in the development of this cell into a multicellular organism.

Protection of developing embryo

Fish

Very few species of fish protect their fertilised eggs. Many of the embryos formed are therefore eaten by predators.

Some do survive thanks largely to their **protective covering** which takes the form of a fairly tough yet flexible membrane (see figure 13.6).

Those fish embryos that do survive obtain food from the **yolk** enclosed inside their own egg.

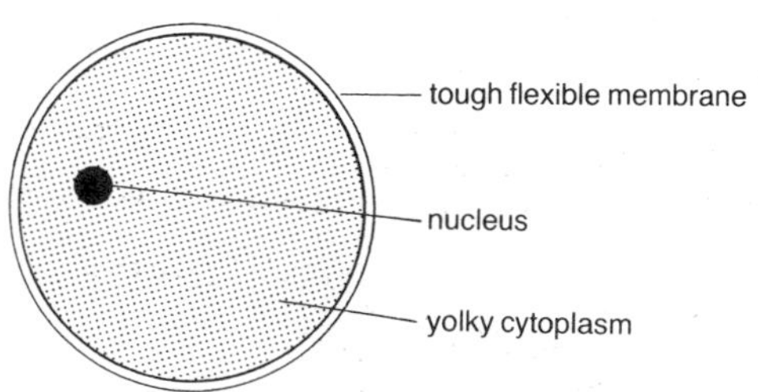

Figure 13.6 Fertilised fish egg

Mammal

As a result of cell division, the fertilised egg becomes an **embryo** while it is being transported along the oviduct (see figure 13.7). When the embryo passes into the **uterus** it becomes attached to (implanted into) the uterus wall which has developed a spongy lining (rich in blood) ready to receive it.

The mammalian embryo remains within the female parent's body during its period of development (**gestation**). It is attached by its **umbilical cord** to the **placenta** and bathed in and cushioned by fluid in the **water sac** (amnion) as shown in figure 13.8. Its mouth and nose are not used for feeding or breathing during this time. Instead, it obtains its food and dissolved oxygen from the mother's blood circulation across the placenta.

The placenta is an organ which allows the blood supplies of the mother and the embryo to come in close contact. Oxygen and food pass from the mother's blood into the embryo's blood which then flows along the umbilical cord to the developing baby.

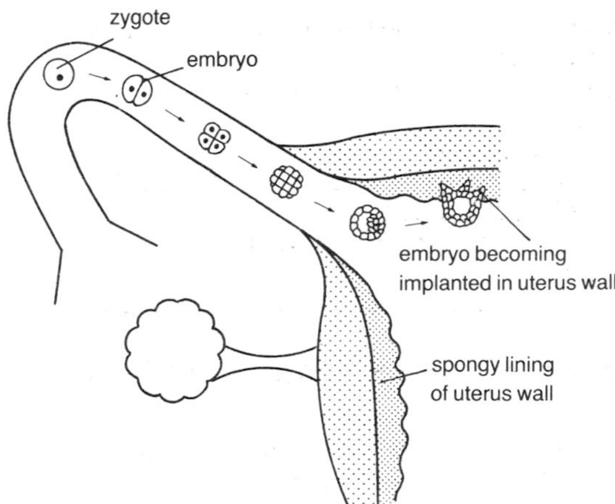

Figure 13.7 Formation and implantation of embryo

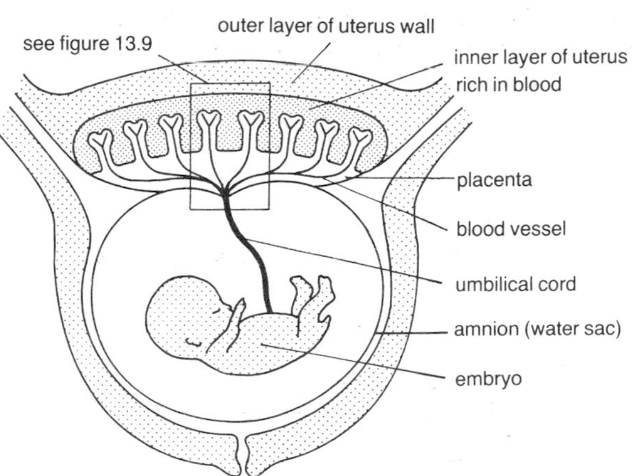

Figure 13.8 Development of human embryo

Carbon dioxide and other wastes are returned to the mother in the opposite direction.

The embryo increases in size and complexity and after two or three months when its species can be recognised from its appearance, it is called a **foetus**.

More to do

Placenta

During gestation the placenta develops into a large disc bearing many finger-like villi (see figure 13.9). These project into a region of the uterus wall which is richly supplied with maternal blood. Each villus contains blood vessels continuous with foetal circulation. Since maternal and foetal blood are now only separated by a thin barrier of membranes, oxygen and dissolved food diffuse into the baby's bloodstream from the mother and carbon dioxide and wastes pass into the mother's blood from the baby.

The placenta is therefore the foetus' organ of **nourishment**, **gaseous exchange** and **excretion**.

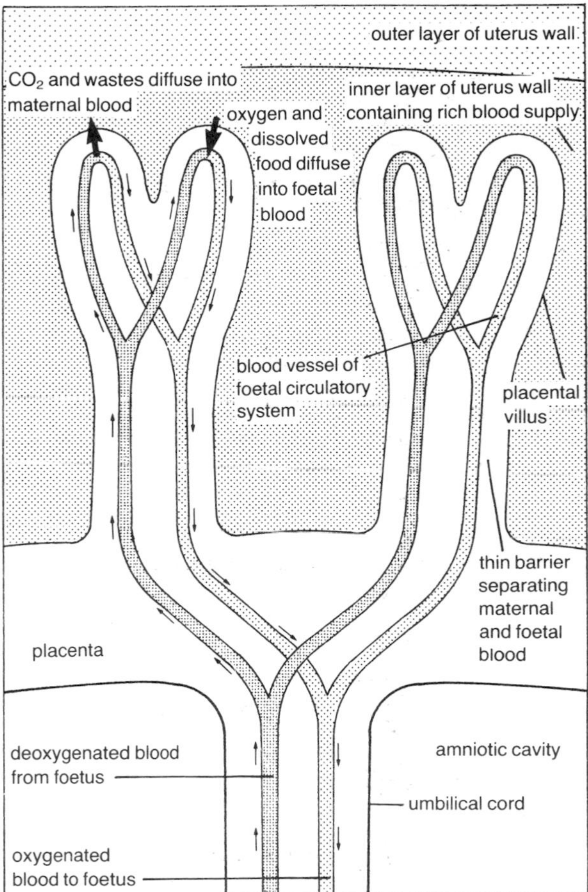

Figure 13.9 Placenta

Number of eggs produced

Fish produce an enormous number of eggs for the following reasons. External fertilisation is inefficient and wasteful. The chance of a sperm cell meeting and fertilising an egg is relatively low. In addition, those eggs that are fertilised receive little or no protection from the parent animals during development.

Mammals produce far fewer eggs because internal fertilisation is more efficient and less wasteful. There is a higher probability of sperm cells fertilising eggs. In addition, the fertilised eggs are well protected inside the female's body during development.

Fish must therefore produce many more eggs than mammals to ensure that a sufficient number are fertilised and survive to maintain the species.

KEY QUESTIONS

1 Give ONE reason why a newly fertilised fish egg may fail to survive.
2 Complete the blanks in the following sentence. In a female mammal, a fertilised ________ passes along a tube called the ________ and into the uterus where it becomes attached to the ________.
3 a) What substance surrounds a developing mammalian embryo during gestation?
b) What is the function of this substance?
c) From where does the developing mammalian embryo obtain food?

Extra Questions

4 Briefly describe the structure and function of the mammalian placenta.
5 Explain why an inverse relationship exists between the number of eggs produced by an animal and the degree of protection given to them by the parent animals.

Care of young

Fish

Young fish emerge from the eggs able to look after themselves without help from their parents. A young fish feeds at first on the remains of its **yolk sac** (see figure 13.10). When this starts to run out, it begins to catch its own food (e.g. pond insects) and gradually it develops into an adult fish.

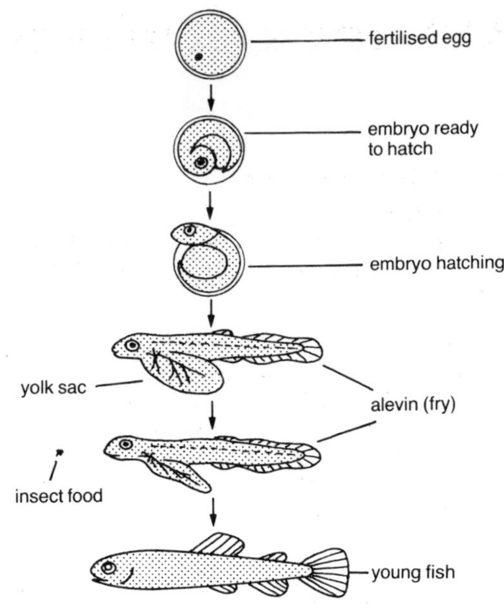

Figure 13.10 Development of young fish

Mammals

Young mammals are unable to look after themselves without parental care. Following birth, the young mammal obtains food by suckling **milk** from the mother's mammary glands. Adult mammals build a **nest** which prevents the young straying into the clutches of predators. The parents' **body heat** protects the young from low temperatures.

Unlike fish, young mammals (especially humans) are dependent on adults for care and protection for a lengthy period of time.

KEY QUESTIONS

1 a) Name a type of vertebrate animal that emerges from its egg able to support itself without help from its parents.
b) Which type of animal is unable to look after itself at birth?
2 Briefly describe how the young of (a) fish and (b) mammals obtain food on hatching or being born.
3 Name TWO possible causes of death from which an adult mammal protects its newly born offspring.

14 Water and waste

Water balance – gain

Water from drink

Much water is gained from liquids (such as milk, tea, coffee, lemonade and water itself) which we drink every day.

Water from food

Investigating the water content of foods

100 g of a fresh food is dried in an oven at 90 °C for two days and its new mass noted. This process is repeated until the sample reaches **constant mass**, showing that all of its water has been removed. The percentage mass of water originally present in the food is calculated by using the formula:

$$\frac{\text{mass of water lost}}{\text{original mass of fresh food}} \times \frac{100}{1} = \text{\% water content of food}$$

Table 14.1 shows the water content of some common foods. All foods contain water, even 'dry' breakfast cereals.

food	% water content
lettuce	95
potato	78
cheese	40
bread	35
cornflakes	5

Table 14.1 Water content of foods

Water from chemical reactions

Look at the experiment shown in figure 14.1. Although the glucose powder is completely dry at the start of the experiment, drops of **moisture** are found near the mouth of the test tube after burning of the food. It is concluded therefore that this water has been formed during burning.

Similarly, water is formed in the cells of the human body during chemical reactions such as aerobic respiration:

glucose + oxygen → **water** + carbon dioxide + energy

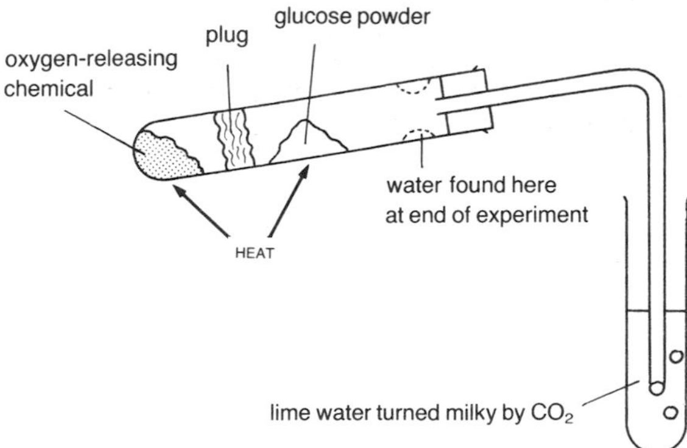

Figure 14.1 Water from a chemical reactoin

Water balance – loss

Water in sweat

Water-sensitive paper is blue when dry and pink when wet. When a dry piece of this paper is taped to the palm of the hand (figure 14.2) for a few minutes, it is found to turn pink. It is therefore concluded that water is lost through the skin (as **sweat**).

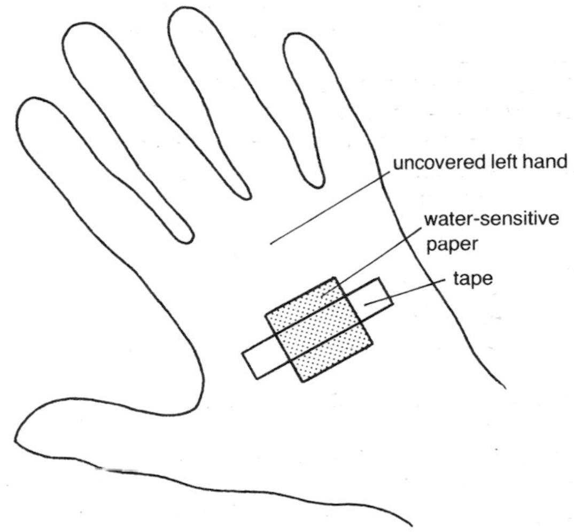

Figure 14.2 Water in sweat

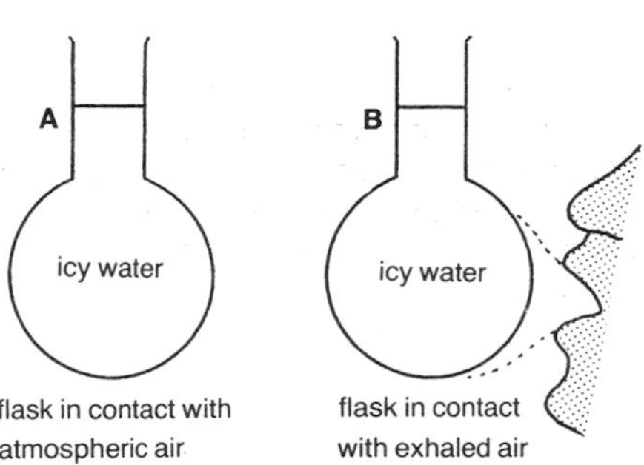

Figure 14.3 Water in breath

Water in breath
Look at the experiment shown in figure 14.3. Much more water is found to condense on the outside of flask B than on flask A. It is therefore concluded that exhaled air contains more water vapour than inhaled air and that some water is lost in **breath**.

Water in urine
Excess water is removed from the body by the kidneys (see page 111).

Water in faeces
Although the large intestine absorbs most of the water from unwanted solid waste, a certain amount of water remains in faeces so that they can be easily expelled from the body.

Water balance – summary

Figure 14.4 shows the daily **water balance** for an average person. In order to maintain water balance, water gain must be equal to water loss.

Maintenance of internal water balance
About 70 per cent of human body weight consists of water. This remains fairly constant from day to day because the **kidneys** regulate the water content of the body. If a person consumes hardly any or a huge amount of water, then the kidneys produce a small or a large volume of **urine** as required thus keeping the body's internal water content constant. Only the kidneys are able to do this. The amount of water lost in sweat, faeces and exhaled air cannot be altered in order to maintain water balance.

More to do

Role of ADH

The kidneys are controlled by a chemical called **anti-diuretic hormone** (ADH) which is released by the pituitary gland (attached to the brain). ADH (one of a group of chemical messengers called hormones) controls the amount of water reabsorbed and returned to the bloodstream by the kidneys (see page 113). Water which is not reabsorbed passes to the bladder as urine to be expelled. This regulation of water balance is summarised in figure 14.5.

KEY QUESTIONS

1 Name THREE ways by which the human body gains water.
2 State FOUR ways in which the body loses water.
3 Which organs are the main regulators of the body's water content?

4 Explain the role of ADH in regulating water balance when
 a) drinking water is in short supply;
 b) a huge volume of water has been drunk.

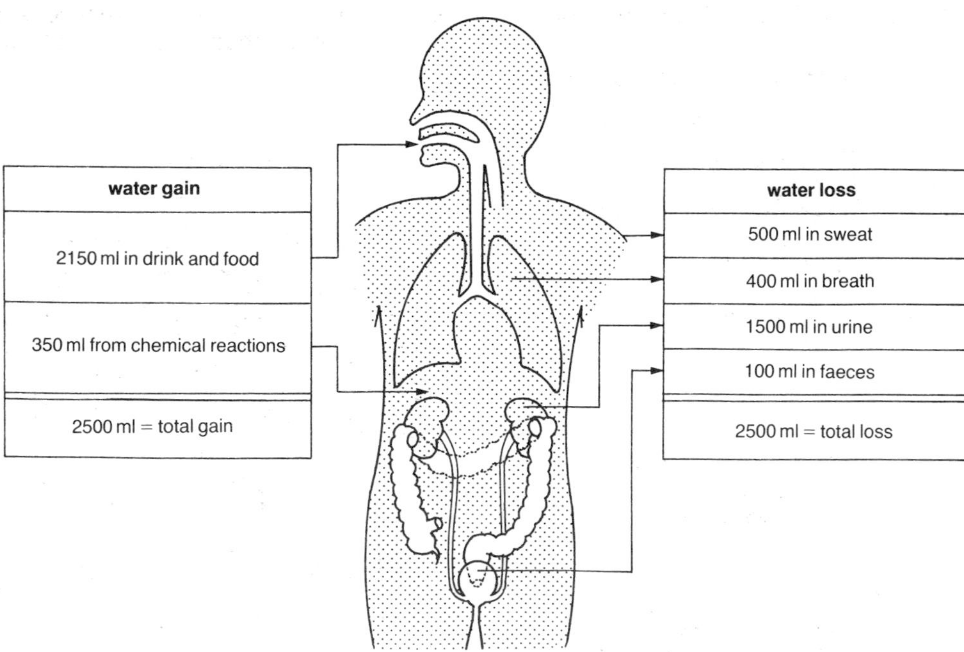

Figure 14.4 Water balance

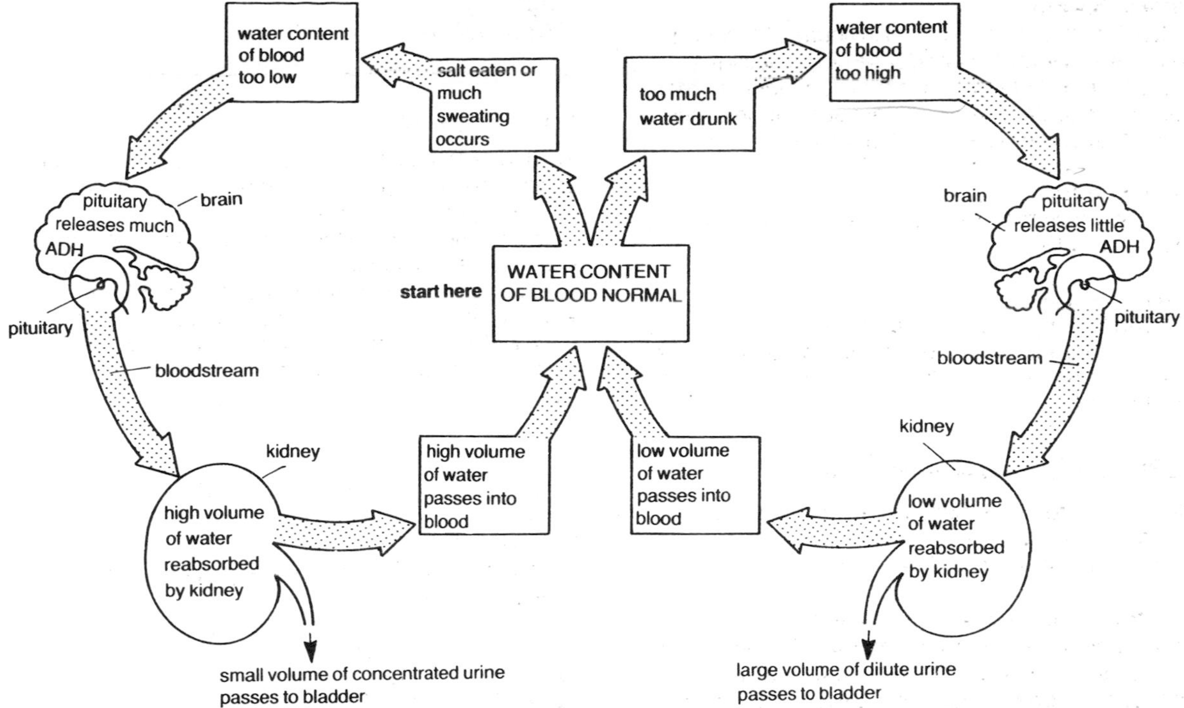

Figure 14.5 Role of ADH

Waste disposal

A great number of chemical reactions occur in the body. Some of these produce harmful wastes such as **urea**. The kidneys remove urea from the blood and pass it (along with excess water as **urine**) to the bladder. The human urinary system is shown in figure 14.6.

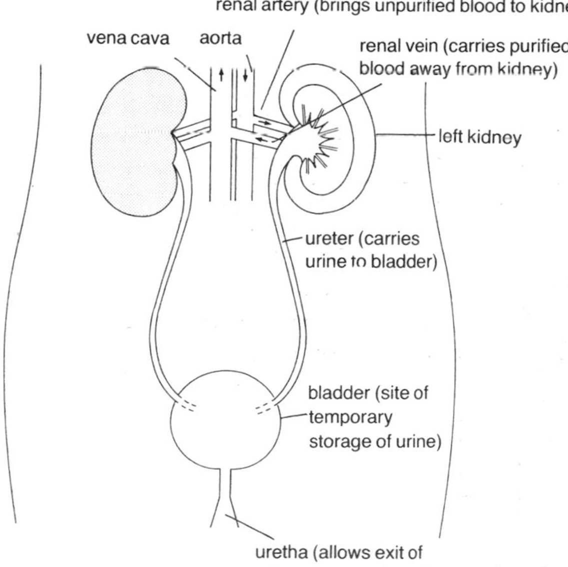

Figure 14.6 Urinary system

Filtration and absorption

The renal artery containing unpurified blood enters a kidney and divides into many tiny branches (see figure 14.7). The blood in each tiny branch is **filtered** by a special filtering unit (a simplified version of which is shown in figure 14.8). Useful substances such as glucose are **reabsorbed** back into the blood vessels

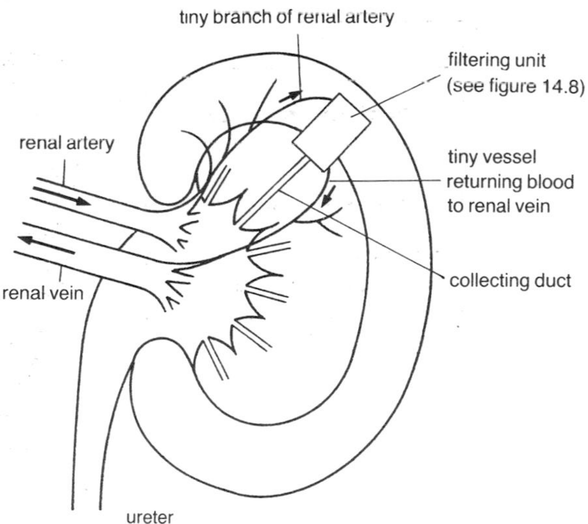

Figure 14.7 Blood supply to kidney

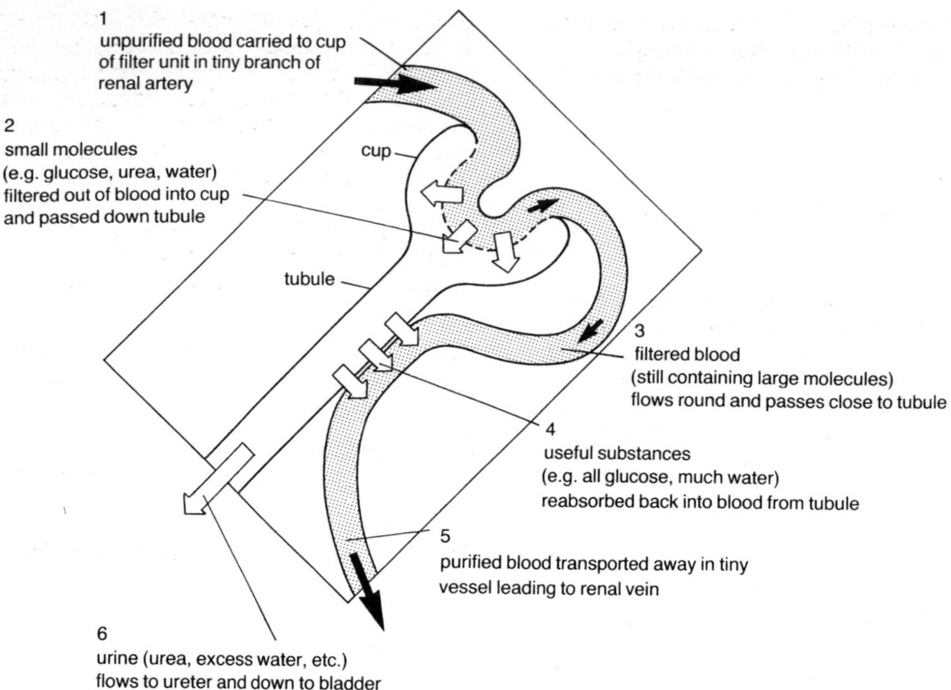

Figure 14.8 Simplified version of filtering unit

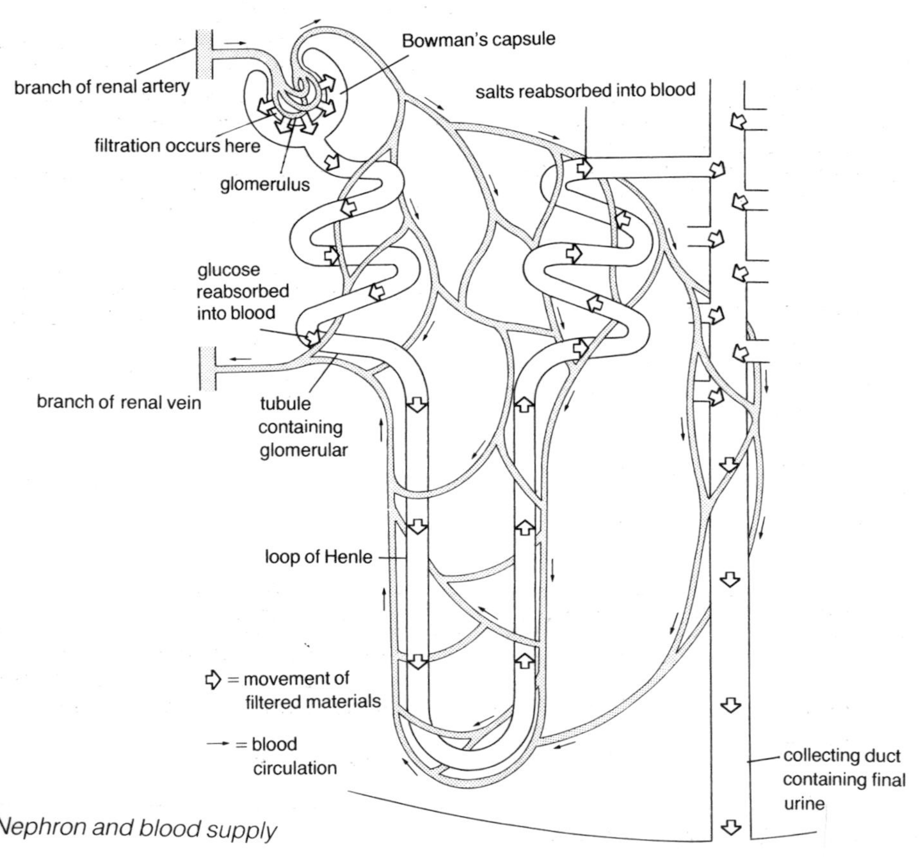

Figure 14.9 Nephron and blood supply

which then unite to leave the kidney as the renal vein which contains purified blood. Thus a kidney's essential functions are filtration and reabsorption.

More to do

Excretion

Excretion is the extraction and elimination from the body of the waste products of metabolism. **Urea** is a poisonous nitrogenous waste made in the liver by the breakdown of extra unwanted **amino acids** (formed during protein digestion). Urea passes from the liver into the blood and is transported round the body to the kidney where it is removed during filtration.

Nephron

Each filtering unit in a kidney is called a **nephron**. Its detailed structure is shown in figure 14.9. It consists of a cup-shaped Bowman's capsule leading into a long tubule.

The renal artery supplying each kidney with blood divides into about a million tiny branches, each of which leads to a **glomerulus**, a tiny knot of blood capillaries surrounded by a Bowman's capsule.

Urine production

Since the blood vessel entering a glomerulus is wider than the vessel leaving it, the blood in a glomerulus is under pressure. As a result, plasma fluid filters out through pores in the capillary walls and collects in the Bowman's capsule. This **glomerular filtrate** contains glucose, salts, urea and water but not plasma proteins or blood cells which are too large to pass through the capillary wall.

As glomerular filtrate passes through a kidney tubule, useful substances (all glucose, some salt and much water) are reabsorbed into the branching network of capillaries surrounding the tubule. Further water is reabsorbed from the liquid flowing down the collecting duct.

The whole process is so effective that about 99 per cent of the water originally present in the glomerular filtrate is reabsorbed. Therefore, of the 120 cm^3 of glomerular filtrate produced every minute, only 1 cm^3 leaves in urine. Final urine also contains all the urea originally present in the filtrate and excess salts.

Table 14.2 compares the composition of glomerular filtrate with urine. To bring about such **selective reabsorption**, kidney cells need energy. This is generated during tissue respiration in kidney cells using oxygen. Blood leaving each kidney is therefore deoxygenated. Table 14.3 compares the composition of blood in the renal artery and vein.

substance	glomerular filtrate (%)	final urine (%)
glucose	0.1	0
salts (sodium, calcium, etc.)	1.0	1.8
urea	0.02	2.0
water	98.5	96.0

Table 14.2 Comparison of glomerular filtrate and urine

KEY QUESTIONS

1 Redraw the following table to match each structure correctly with its function.

structure	function
renal artery	carries urine from kidney to bladder
kidney	transports unpurified blood to kidney
ureter	stores urine ready for expulsion
bladder	carries urine from bladder to external environment
urethra	transports purified blood away from kidney
renal vein	contains special units which filter blood

2 Briefly explain how a kidney purifies blood by referring to the two processes involved.

3 a) What name is given to the poisonous waste that is removed by the kidneys?
b) Name the liquid in which this waste is expelled from the body.

substance	blood in renal artery	blood in renal vein	reason for difference
oxygen	more	less	some oxygen used in cell respiration
carbon dioxide	less	more	some CO_2 produced during cell respiration
glucose	more	less	some glucose used in cell respiration
urea	some	none	excess removed in urine
salts	more	less	excess removed in urine
water	more	less	excess removed in urine

Table 14.3 Comparison of blood in renal artery and vein

Extra Questions

4 a) Where is urea made?
   b) From what substance is urea made?
   c) By what means is urea transported to the kidneys?
5 Explain how urine is produced by referring to the various processes that occur in a nephron.
6 180 litres of glomerular filtrate but only 1.5 litres of urine are produced on average by man per day. What happens to most of the water in the glomerular filtrate?

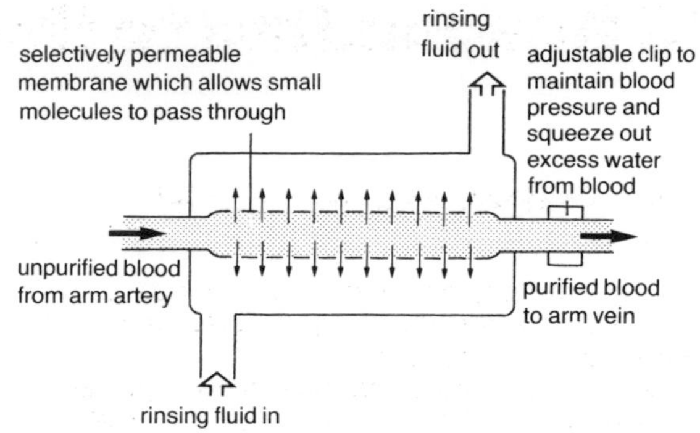

Figure 14.10 Kidney machine

Kidney failure

A person can live with only one kidney but if both kidneys become damaged or diseased, the concentration of poisonous wastes such as urea builds up in the bloodstream. The body's internal balance becomes so disturbed that death eventually results. This condition is called total **kidney failure**. Nowadays death can be prevented by the following treatments.

Kidney transplant

The person receives a healthy replacement kidney from a donor.

Kidney machine

The sufferer has his blood purified every few days by a kidney (dialysis) machine. Figure 14.10 shows a simplified version of such an artificial kidney.

The rinsing fluid contains the same concentration of useful substances (e.g. sugar, salt, etc.) as that of normal blood plasma. There is therefore only an overall (net) loss from the blood of molecules of substances (e.g. urea, excess salt) that are in a higher concentration in the blood than in the rinsing solution. These pass through the membrane and are removed in the rinsing solution.

Comparison

Whereas a natural kidney filters the blood and then reabsorbs the useful materials, a kidney machine only filters out the waste materials from the blood.

Table 14.4 compares the benefits and limitations of the two treatments for kidney failure.

KEY QUESTIONS

1 Under what circumstances does a person need a replacement kidney?
2 a) If no kidneys are available, what alternative method is used to purify the person's blood?
   b) Describe the limitations of this method compared to a kidney transplant.

3 Explain why kidney failure can be fatal.
4 Briefly compare the action of the kidneys with the action of a kidney machine.

	artificial kidney (kidney machine)	replacement kidney (kidney transplant)
benefits	It effectively purifies the blood and therefore saves the life of a person with diseased kidneys. There is no possibility of rejection by the body since it is not inserted into the body.	If successful the person can lead a normal life and does not have to spend two periods of about 12 hours every week connected to a machine.
limitations	It is too large a machine to carry around and it is expensive to run. As a result the sufferer cannot live a totally 'free' life but must spend two periods of about 12 hours every week connected to a machine, usually in a hospital.	The body may reject the replacement kidney. Special drugs used to prevent rejection reduce the body's ability to fight infection. There is also a moral problem. Healthy kidneys are in short supply so who decides which patient should get one when available?

Table 14.4 Comparison of kidney failure treatments

15 Responding to the environment

Effect of light on behaviour of woodlice

A **choice chamber** (see figure 15.1) is a piece of apparatus which allows one difference (variable factor) to be studied at a time.

When several woodlice are inserted through the hole in the choice chamber, they move around at random. Those animals that happen to be in the light side are found to respond by moving quickly. This increases their chance of moving out of the light and into the dark side. Those animals that happen to be in the dark side are found to respond by moving slowly or coming to a standstill. This behaviour increases their chance of staying in the dark side.

After a few minutes most of the animals are found to have gathered in the dark side.

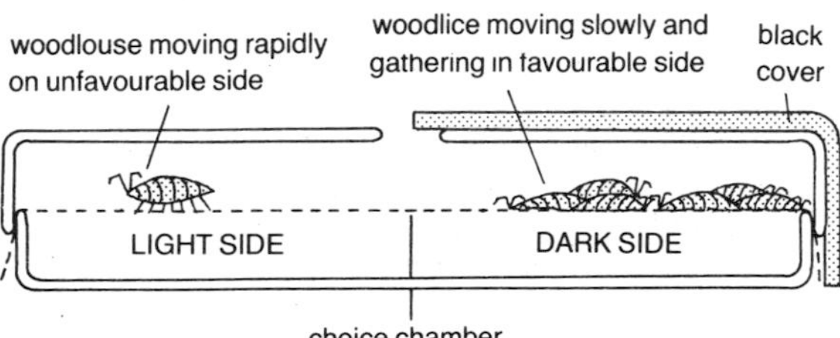

Figure 15.1 Choice chamber

Effect of light on behaviour of blowfly maggots

Blowfly maggots have no eyes yet they are also sensitive to light. They are found to move directly away from light and gather in dark places.

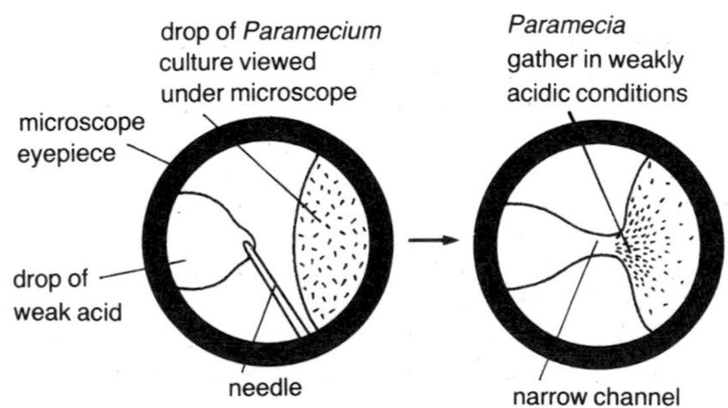

Figure 15.2 Response of Paramecium *to weak acid*

Effect of weak acid on *Paramecium*

Paramecium is a tiny unicellular animal which lives in pond water. Look at the experiment shown in figure 15.2. When the needle is used to connect the drop of vinegar (a weak acid) to the drop of *Paramecium* culture, the tiny animals are found to move towards and gather in the weakly acidic solution.

More to do

Significance of responses to environmental stimuli

In addition to slowing down and stopping in dark places, woodlice also congregate in damp places. Their behaviour is of survival value because when in light they are easily spotted and eaten by predators and when in dry conditions they lose water through their permeable skins and die.

Blowfly maggots hatch from eggs laid in manure heaps and in animal corpses. The maggots must remain in these dark places if they are to receive food, moisture and shelter so it is of survival value if they move away from light.

Paramecium feeds on bacteria which cause decay and live in water with a slightly acidic pH. So if these tiny animals move towards such water there is a good chance that they will find food and survive.

KEY QUESTIONS

1 a) What is a choice chamber?
  b) What ONE variable factor is investigated in the experiment shown in figure 15.1?
  c) Why must only one variable factor be studied at a time?
  d) Why are several woodlice used in the experiment rather than just one?
2 Describe how a woodlouse's behaviour changes when it moves from (a) dark to light conditions and (b) light to dark.
3 Give TWO further examples of a situation where an environmental factor has an effect on a named animal's behaviour.
4 For both blowfly maggot and woodlouse, explain in turn why the animal's response to the stimulus light is of survival value.

Rhythmical behaviour

This is behaviour which is repeated at definite intervals. The time interval may be quite short or as long as a year. Although such behaviour is controlled internally by the animal's '**biological clock**', the clock is normally triggered by an external stimulus. Table 15.1 gives some examples of rhythmical behaviour.

More to do

Disturbance of man's natural rhythms

Heart rate, oxygen consumption, carbon dioxide production, sensitivity to pain and ability to perform physical and mental tasks are just a few of the many human processes that show **diurnal** rhythms. They all show most activity during daytime and least at night.

When a person quickly crosses several time zones in a jet plane, he ends up being ahead or behind the time to which his body was accustomed at the start of the journey. His body feels confused and upset on discovering that it is now, say, early evening when it feels as though it should be, say, the middle of the night. Such disturbance of the body's natural diurnal rhythms is called jet lag.

KEY QUESTIONS

1 Describe TWO features common to all forms of rhythmical behaviour.
2 Describe the rhythmical behaviour shown by (**a**) fiddler crab, (**b**) cockroach and (**c**) swallow. In each case identify the external trigger stimulus involved.

Extra Questions

3 For each of the animals named in question 2, explain why the rhythmical behaviour shown by the animal helps it to survive.
4 Explain what is meant by the term **jet lag**.

type of rhythm	animal	description of rhythmical behaviour	external trigger stimulus	significance
tidal (occurs at each turn of tide)	green flatworm	animal rises to surface of sand when it is covered with water but buries itself when the sand dries out	rhythmical movement of tidal water under influence of moon	behaviour prevents animal's slimy skin from drying out, which would be fatal
	fiddler crab	animal emerges from burrow at low tide and retreats into burrow when tide comes in again	rhythmical movement of tidal water under influence of moon	behaviour enables animal to find food fragments left on tidal mud
circadian (occurs once every 24 hours)	cockroach	animal shows peak of activity at night	onset of darkness	behaviour enables animal to feed unobserved by predators
	man	animal produces less urine during the night	onset of darkness and sleep	animal's sleep is not disturbed by sensation of full bladder
annual (occurs once a year)	rabbit	animal courts mate and breeds during spring	onset of longer daylengths	new born are allowed several months to grow before harsh winter conditions return
	swallow	animal migrates to warmer climate in autumn	onset of shorter daylengths	behaviour enables animal to escape cold temperature and shortage of food in winter
	squirrel	animal hibernates in late autumn	onset of shorter daylengths	behaviour enables animal to survive extreme conditions of winter

Table 15.1 Rhythmical behaviour

PROBLEM SOLVING

1 The following pie chart show the sources of fat eaten by a person consuming an unhealthy diet.
a) What percentage of the person's fat intake comes from cakes?
b) The total mass of fat consumed by the person in one day was 240 g. What mass of fat was contained in the meat that the person ate?

2 Compare the following diagrams carefully. In each case the arrows indicate the direction in which the teeth act when the jaws move.

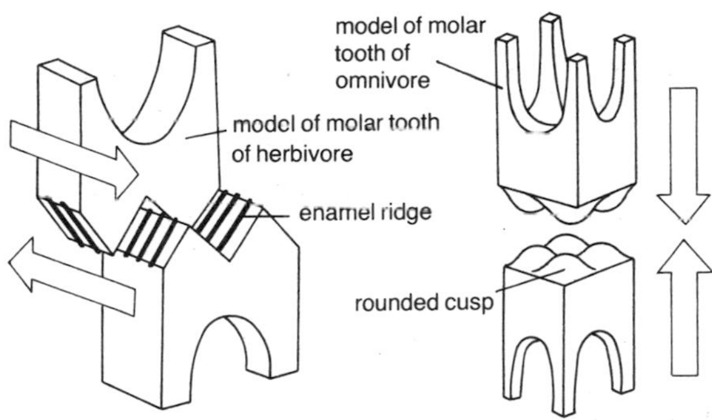

Using all the words and phrases given in the accompanying word bank, write a short paragraph to compare the action of the molar teeth in a herbivore with that in an omnivore.

word bank

between enamel ridges
omnivore's molar teeth
up and down movement
between rounded cusps
crushed
herbivore's molar teeth
side to side movement
ground up

3 A few drops of iodine solution were added to 5 cm^3 of starch 'solution' in a test tube at 37 °C. The contents of the test tube turned blue-black. 1 cm^3 of 1 per cent pancreatic amylase was then added to the test tube. After five minutes the blue-black colour had disappeared. It was therefore concluded that the enzyme pancreatic amylase is able to digest starch.
a) In what way would the time required for the blue-black colour to disappear be affected in a repeat experiment by
(i) keeping the test tube at 10 °C?
(ii) using 5 per cent pancreatic amylase?
b) Name ONE factor that must be kept the same in each repeat of the experiment to make it fair.

4 The gestation period for a ferret is six weeks, whereas a female rat's pregnancy lasts for only half of this time. Calculate how long a dog's gestation period lasts if it is three times longer than that of a rat.

5 The following table refers to the gain in the mass of a human foetus (developing embryo) during pregnancy.

age of foetus in weeks	12	16	20	24	28	32	36
mass of foetus in grams	50	150	300	600	1150	1700	2250

a) Using similar graph paper and the axes shown below, draw a line graph of the information in the table.

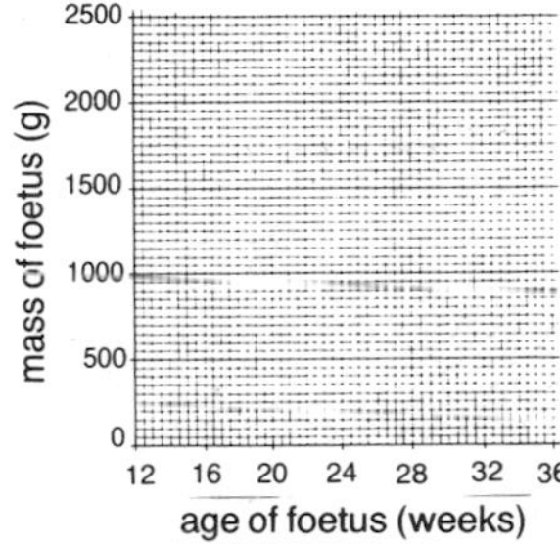

b) State the mass of the foetus at 20 weeks.
c) How many weeks did it take for the 20-week foetus to double its mass?
d) How many grams did the foetus gain between weeks 24 and 36?

Extra Questions

e) State the mass of the foetus at 27 weeks.
f) How many grams did the foetus gain between weeks 31 and 35?

6 The shaded part of the graph below shows the limits within which a healthy baby's mass should be as it grows.

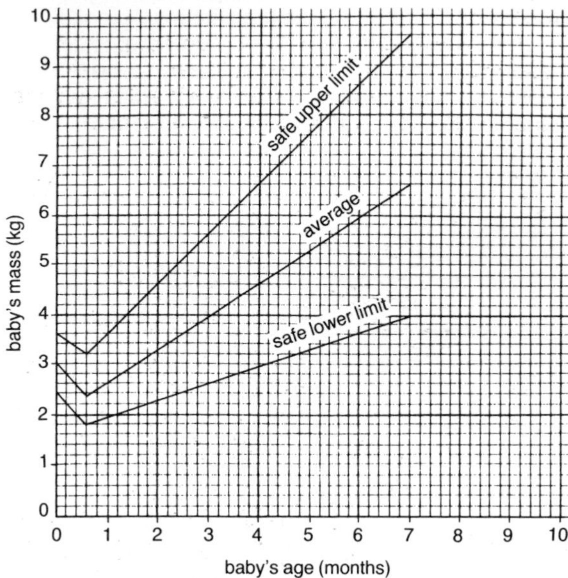

a) What should be the mass at 4 months of a baby that weighed 3 kg at birth?
b) What should be the mass at 3 months of a baby that weighed 3.6 kg at birth?
c) If a baby weighs 4 kg at 7 months, what was its mass at birth according to the graph?

Extra Questions

d) Extend the graph to find out the expected mass at 10 months of
(i) a baby that weighed 3 kg at birth,
(ii) a baby that weighed 2.4 kg at birth.

7 Read the passage and answer the questions based on it.

Insulin is a chemical which controls the amount of glucose in the bloodstream by regulating the amount that is taken up from the blood and stored as glycogen in the liver. When insufficient insulin is made by the body, a disease known as *diabetes mellitus* results. Too much glucose is now present in the person's blood. The glomerular filtrate contains so much glucose that it is not all reabsorbed into the bloodstream. Instead some sugar is excreted in urine.

Three symptoms of *diabetes mellitus* are hunger, thirst and excessive urination accompanied by loss of weight and strength. The condition is effectively treated by injections of insulin and a strict diet.

Diabetics are five times more likely than other members of the population to develop kidney trouble. One example is damage to the kidney's filtration units. When this happens, a blood protein called albumin leaks out into the glomerular filtrate. This protein can only be detected in urine using a dipstick test once a fairly high concentration is being excreted. However, by then the person may have already suffered considerable kidney damage. A new test has been developed which detects tiny amounts of protein in urine. This early detection of the problem will allow future sufferers to apply preventative measures. By sticking to a strict diet and keeping blood pressure down, they will be able to reduce strain on the kidneys and therefore prevent damage to the tiny filtration units.

a) What causes *diabetes mellitus*?
b) State ONE way in which the composition of the blood of an untreated sufferer of *diabetes mellitus* would differ from that of normal blood.
c) What symptoms would make a doctor suspect that a person was suffering from *diabetes mellitus*?
d) Describe how the doctor could test for *diabetes mellitus*.
e) Give TWO ways in which *diabetes mellitus* is treated.
f) Under what conditions is protein excreted in urine?
g) Explain why the newly developed test for the presence of protein in urine is preferable to the traditional dipstick method.

8 Graph A below shows the response of a man to drinking 1 litre of tap water; graph B shows the response of his younger brother to drinking 1 litre of 0.96 per cent salt (sodium chloride) solution. In each case urine was collected at half-hourly intervals.

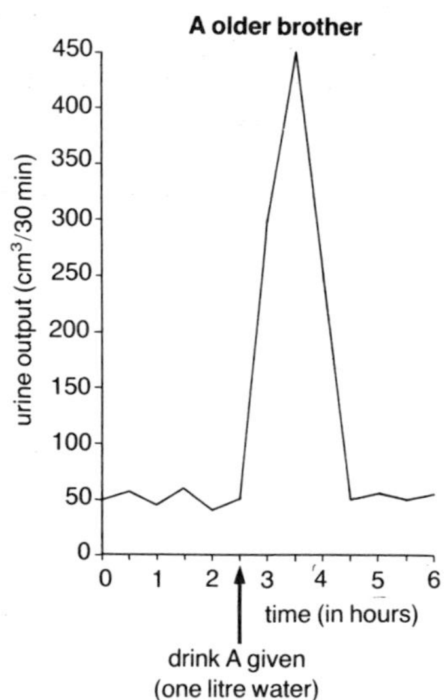

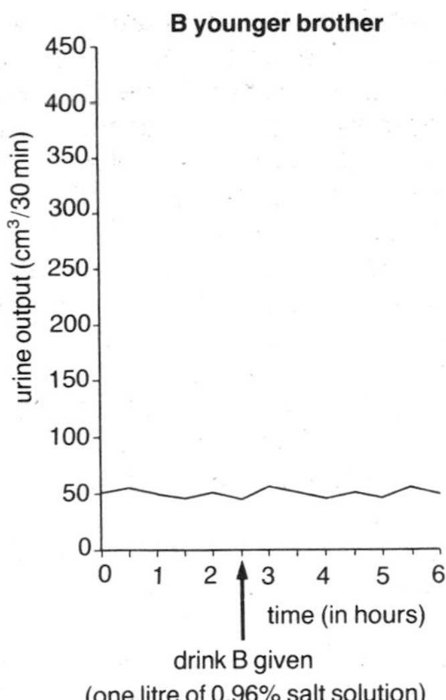

a) What was the older brother's average urine production before drinking the water?
b) By how much had the older brother's urine production increased one hour after drinking the water?
c) After drinking the water, how long did it take for the older brother's urine output to return to normal?

Extra Questions

d) The aim of this investigation was to compare the effect on urine output of drinking 0.96 per cent salt solution with that of drinking tap water. Identify a source of error in the above procedure and state how it could be overcome.
e) Assume that the source of error referred to in (d) has been overcome and that a repeat experiment gives results similar to those graphed above. What conclusion can be drawn about the effect on urine output of 0.96 per cent salt solution compared with tap water?

9 The day and night rhythm that occurs during every 24 hour period in Britain divides living things into two distinct groups: the diurnal (active by day, at rest by night) and the nocturnal (active at night, at rest by day).
The following diagram shows how six animals respond to both day and night. In each case the outer circle shows when the animal is active and the inner circle shows when the animal is at rest.

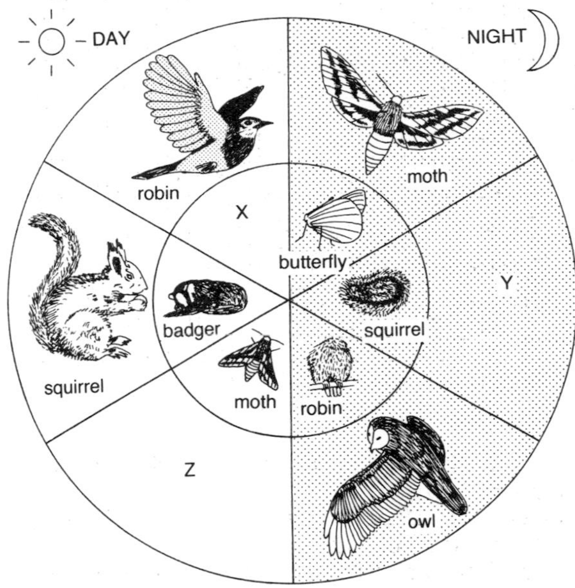

a) Study the diagram carefully and state which of the six animals should have been drawn in spaces X, Y and Z.
b) Classify the six animals according to their type of rhythmical behaviour.

10 The experiment below was set up to investigate the effect of light and dark on blowfly maggots.
a) Name TWO factors that could affect the fairness of the test.
b) Suggest suitable ways of correcting these two factors so that they do not make the experiment unfair.

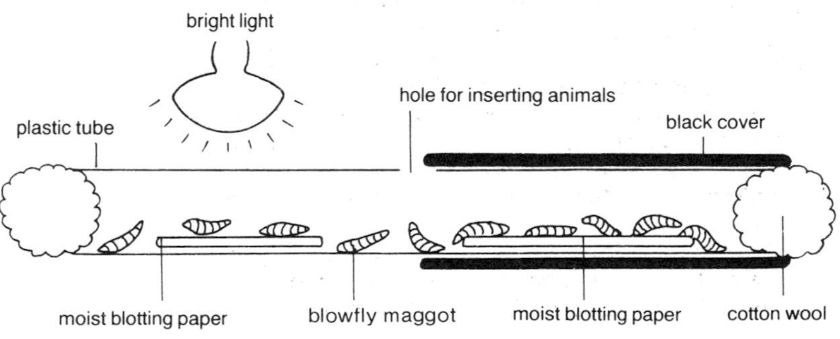

Section 5 The Body in Action

16 Movement

Physical activity

All physical activities (e.g. sport, dance, play and work) require parts of the body to move. These movements are brought about by the action of **muscles** attached to the skeleton. In addition to muscles, many other parts of the body are needed to play their part as shown in figure 16.1.

Skeleton

Man's bony skeleton performs several important functions. It provides the human body with a strong and reliable means of **support**. It acts as a framework for **muscle attachment**. It **protects** vital body parts such as heart, lungs, brain and spinal cord (see figure 16.2).

Composition of bone

Look at the two experiments shown in figure 16.3. A bone immersed in **acid** becomes rubbery and **flexible**. A bone **heated** strongly in an oven becomes **brittle**. It is therefore concluded that bone consists of both flexible material (which is removed by heat but not by acid) and hard mineral material (which is removed by acid but not by heat.

Since the basic shape of the bone is unaltered after each treatment, this shows that the two components of bone must be closely intermixed.

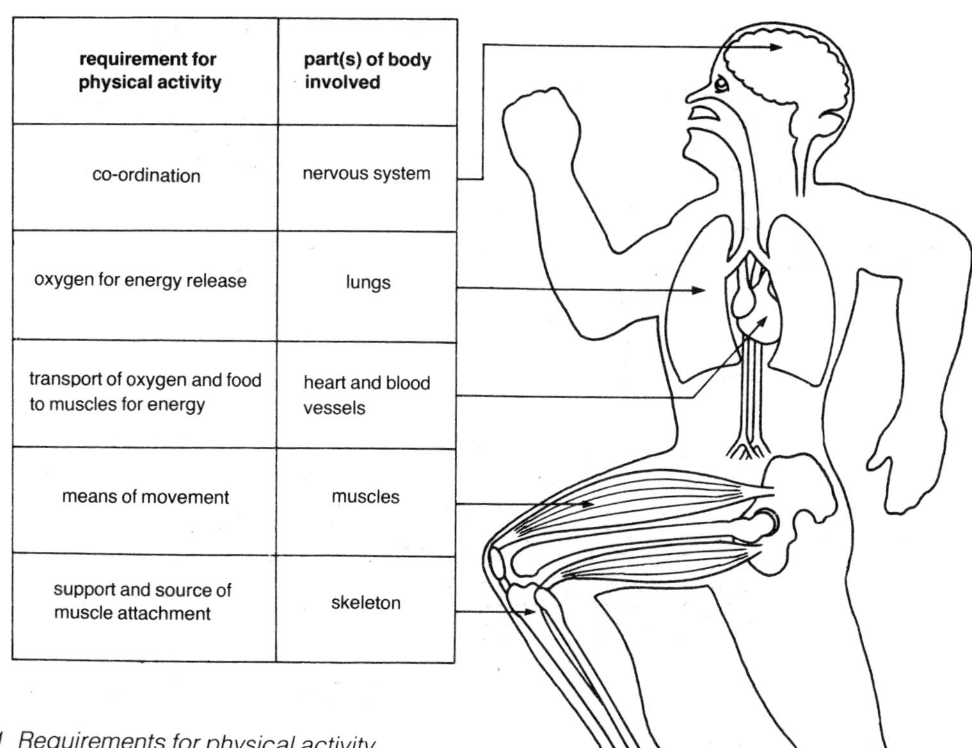

requirement for physical activity	part(s) of body involved
co-ordination	nervous system
oxygen for energy release	lungs
transport of oxygen and food to muscles for energy	heart and blood vessels
means of movement	muscles
support and source of muscle attachment	skeleton

Figure 16.1 Requirements for physical activity

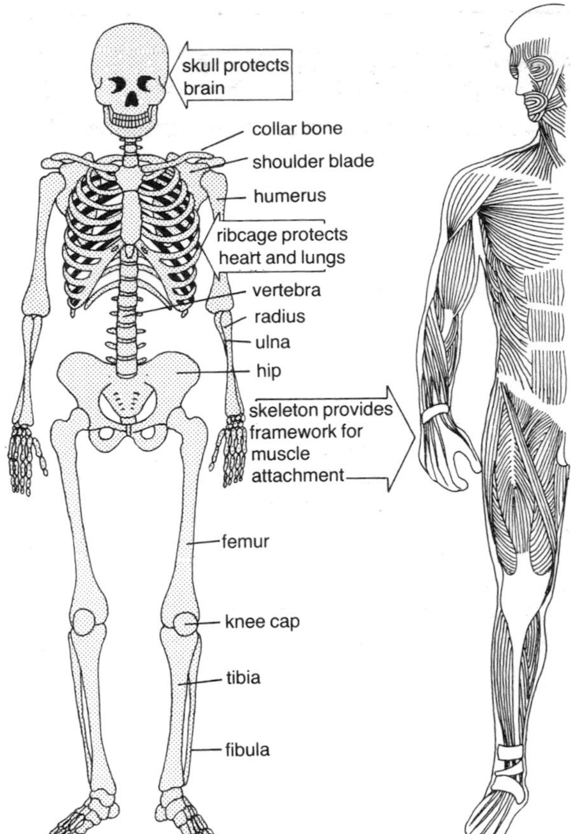

Figure 16.2 *Functions of human skeleton*

More to do

Structure of bone

Although at first glance bone might appear to be dead, it is very much alive. Figure 16.4 shows how it contains living **bone cells**. These extract chemicals such as **calcium** and **phosphate** from the blood in nearby canals and build them into the hard mineral part of the bone (which is largely composed of calcium phosphate).

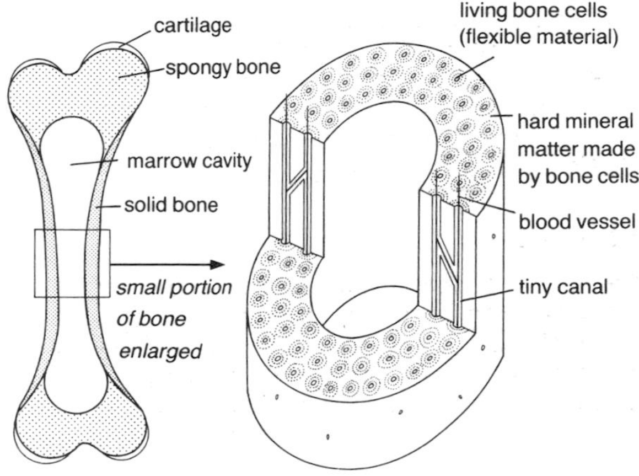

Figure 16.4 *Structure of bone*

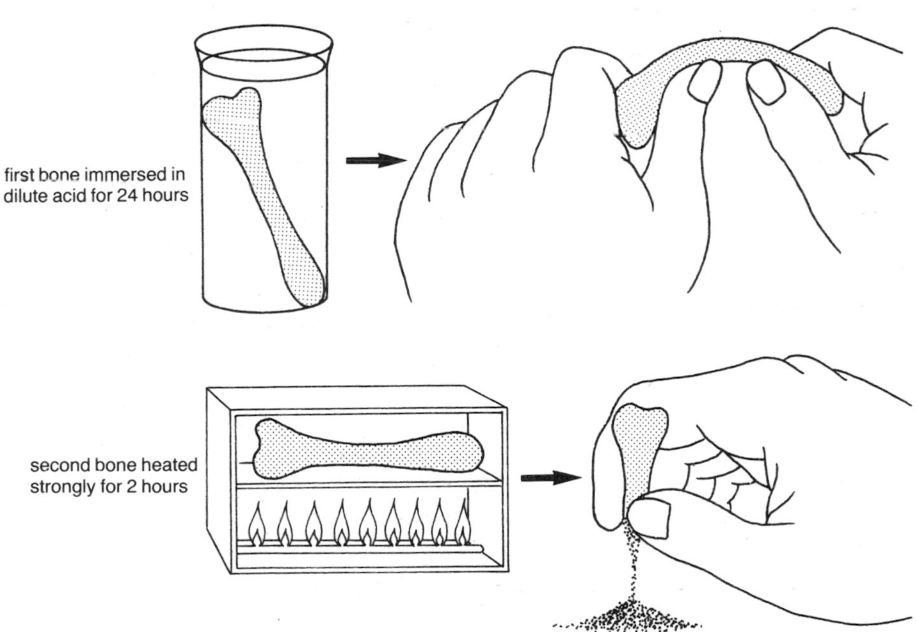

Figure 16.3 *Investigating the composition of bone*

KEY QUESTIONS

1 Give FOUR examples of physical activities.
2 a) In addition to energy, what else does the body require for physical activity?
b) Which parts of the body are involved in providing these requirements?
3 a) Name FOUR vital parts of the human body protected by the skeleton.
b) Give TWO further functions of the human skeleton.
4 a) Of what TWO types of material is bone composed?
b) Describe briefly the two experiments that show this to be the case.

5 a) Is bone dead or alive?
b) A broken bone is able to mend itself. Using only the information given in this chapter, suggest how this is possible.

Joints

The meeting point between two bones is called a **joint**. Although some joints allow little or no movement between bones (e.g. skull), other joints are freely movable.

Ball and socket

At the hip and shoulder, the rounded head (ball) of one bone fits into the socket of another allowing movement in three different planes as shown in figure 16.5.

Hinge

At the knee and elbow, the bones meet together like a hinge and movement is restricted to one plane as shown in figure 16.6.

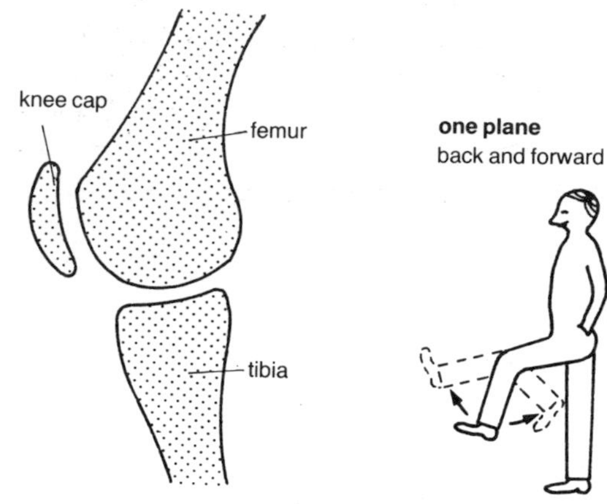

Figure 16.6 Hinge joint (at knee)

Ligaments and cartilage

Ball and socket and hinge joints are held together by strong cords called **ligaments** which attach bone to bone (see figure 16.7). The ends of the bones that meet at a joint are covered with a layer of strong smooth slippery **cartilage** which cushions and protects them during movement.

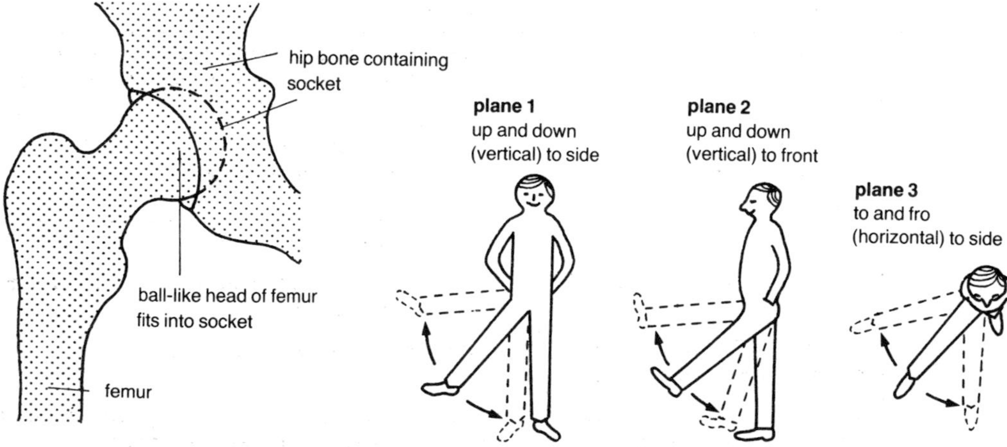

Figure 16.5 Ball and socket joint (at hip)

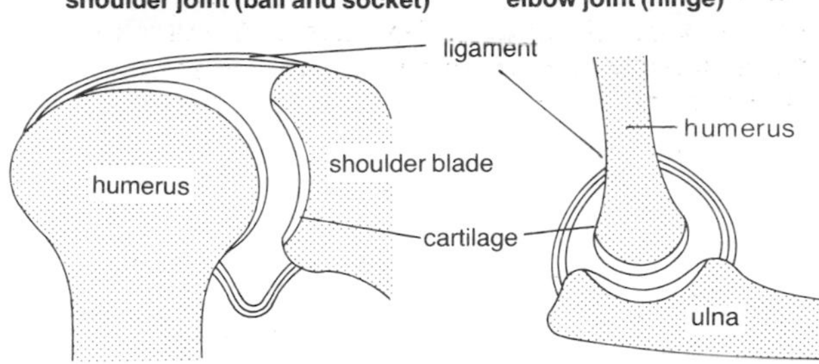

Figure 16.7 Ligaments and cartilage

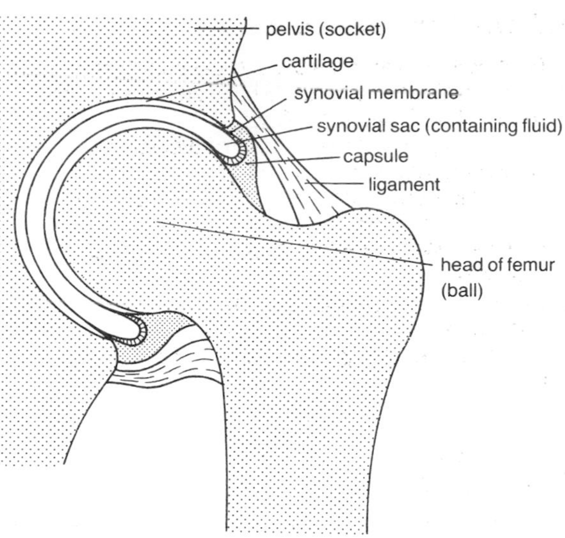

Figure 16.8 Synovial joint

Synovial joint

All ball and socket and hinge joints are called **synovial** joints because each possesses a synovial membrane. Figure 16.8 shows this structure included in a more detailed version of the hip joint. Table 16.1 gives a summary of the structure and functions of the parts of a synovial joint.

KEY QUESTIONS

1 a) Name TWO types of movable joint and give two examples of each.
b) For each type of joint, state the range of movement(s) that it allows.
2 Name the structures that hold joints together.
3 What function is performed by the cartilage present in a joint?

Extra Questions

4 Why is a synovial joint so-called?
5 Describe the structure and function of the following parts of a synovial joint: synovial sac, capsule, synovial membrane.

Muscles

Muscles are attached to bones by structures called **tendons** (see figure 16.9). When a muscle contracts it becomes shorter. Its two ends are therefore brought more closely together. Since these are attached to two

part	structure	function
bones	hard supporting material	provide source of attachment for muscles which move joint; serve as a set of levers which greatly magnify movements of muscles acting on them
cartilage	smooth and slippery yet tough and hard-wearing material	cushions and protects ends of bones; absorbs shock and helps to allow free movement of joint
synovial membrane	moist slippery membrane	secretes synovial fluid which lubricates inner joint surfaces, allowing friction-free movement
synovial cavity	enclosed space full of synovial fluid	acts as cushion, preventing friction between cartilage-covered ends of bones
capsule	tough sleeve-like extension of membrane on outside of bone	completely encases and protects joint
ligaments	strong cords of dense fibrous tissue with small degree of elasticity	lash bones firmly together while being able to withstand sudden stresses during movement of joint

Table 16.1 Parts of a synovial joint

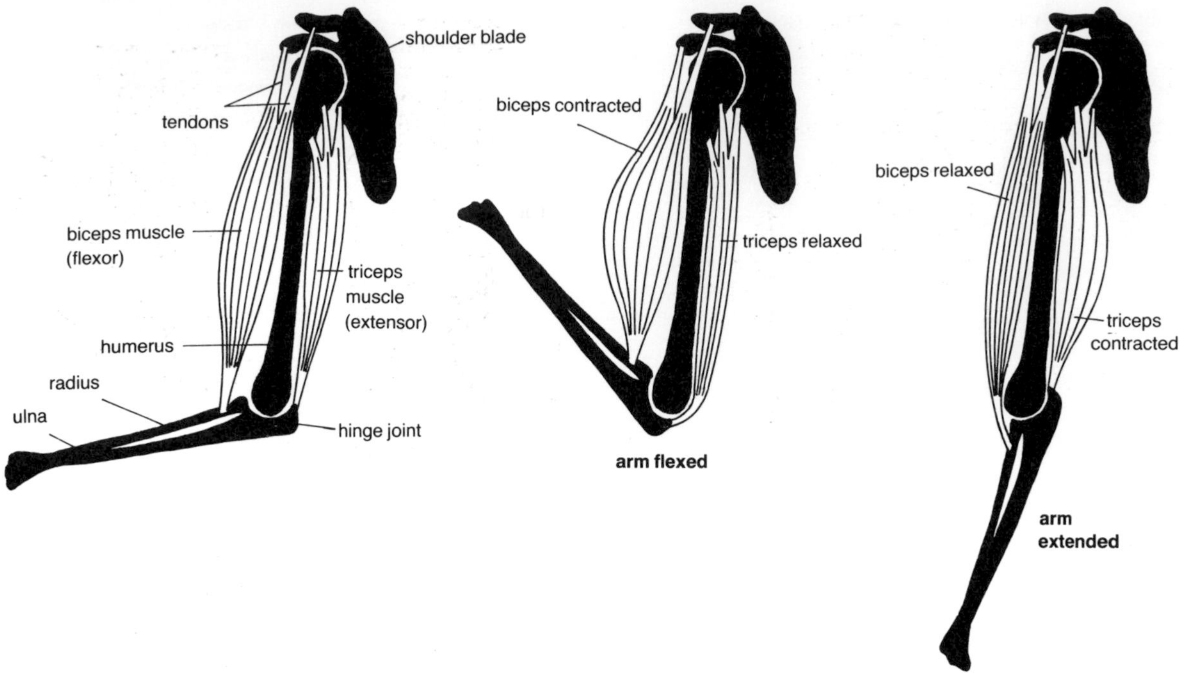

Figure 16.9 Movement of arm

different bones, one of the bones is made to move.

When the biceps muscle (figure 16.9) contracts, the arm bends (i.e. flexes). When the triceps muscle contracts, the arm straightens (i.e. extends). Thus, during these movements of the arm (at the hinge joint), one muscle contracts and the other becomes relaxed.

Two such muscles which produce movement of a limb in opposite directions make up an opposing (**antagonistic**) pair.

Need for pair of opposing muscles

A muscle is able to contract (and then later become relaxed) but it is unable to actively lengthen of its own accord. Thus, following contraction, a muscle depends on the action of its antagonistic partner to restore it to its original length. Following the extension of the arm, for example, the biceps (flexor muscle) is needed to contract and pull the relaxed triceps (extensor muscle) back to its elongated shape. Following flexion, the triceps contracts and pulls the biceps back to its original shape.

Such an arrangement of paired muscles working antagonistically by pulling against one another in a controlled way, allows limbs to flex and extend in a smooth co-ordinated way at hinge joints.

More to do

Tendons

Tendons are made of tough inelastic tissue and attach muscles to bones. When a muscle contracts, the force generated is transferred via the tendon to the bone. If free to do so, the bone will move. This is only possible because the tendon is **inelastic**. If tendons were elastic then they would stretch when pulled on by muscles and the bones would fail to move properly.

Injury

Strenuous activity can create extra strains on joints and muscles. In some cases this can cause injuries.

Joint damage

Dislocation

This is caused by a sudden blow which alters the position of one or more of the bones at a joint. A dislocated shoulder (figure 16.10) may result, for example, from a rugby tackle or a fall. Although the bone can be readily returned to its proper position, often the joint remains weakened for some time.

Cartilage displacement

As a result of intense activity, one of the cartilage pads in the knee may become loose and move out of

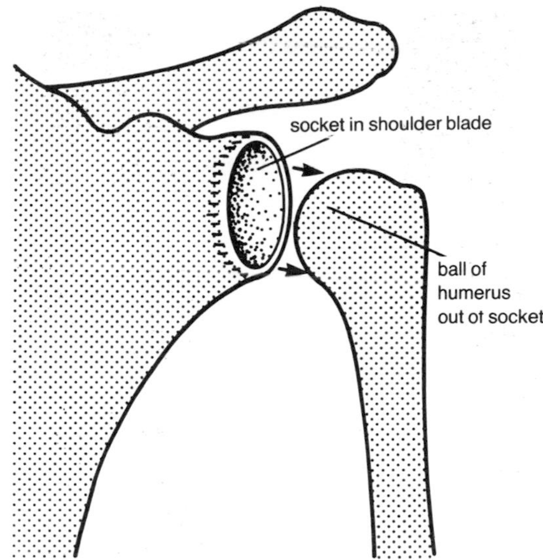

Figure 16.10 Dislocated shoulder

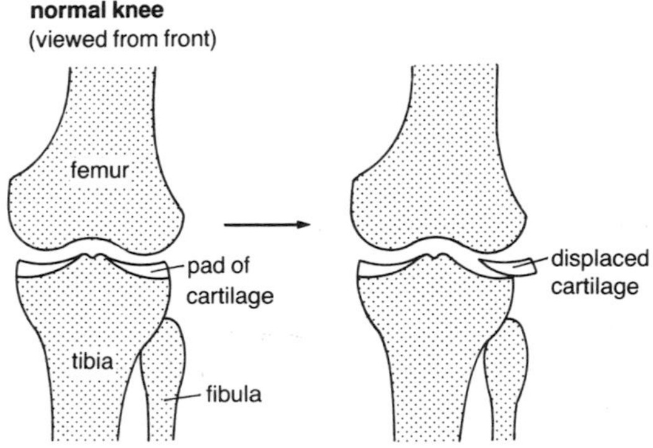

Figure 16.11 Displaced cartilage

position (see figure 16.11). If it locks the knee, preventing movement, then it may have to be removed by surgery.

Torn ligament
A ligament may be torn if the ankle becomes wrenched (sprained) during vigorous activity (see figure 16.12). The ankle bones become displaced since they are no longer being held together by the ligament. This type of injury normally needs a long period of support and rest to allow recovery.

Muscle and tendon damage
Often an awkward or too vigorous movement causes pulling of a muscle or stretching or even tearing of a tendon. A common site of damage is the tendon (hamstring) at the back of the thigh (figure 16.13). Such injuries require a long period of rest.

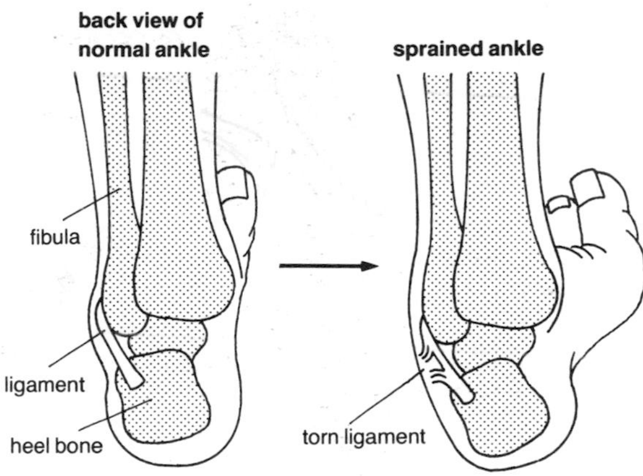

Figure 16.12 Torn ligament

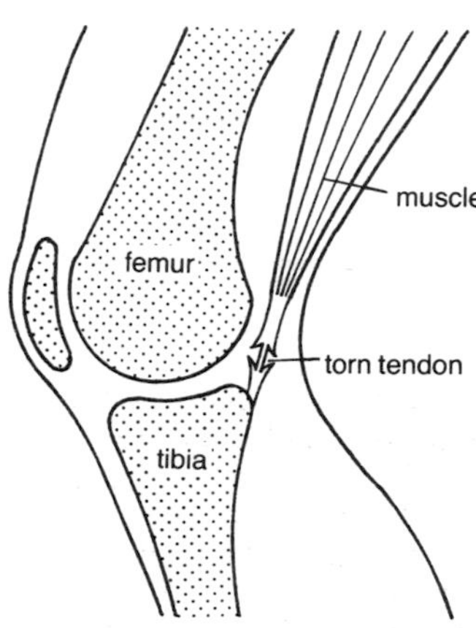

Figure 16.13 Torn tendon

KEY QUESTIONS

1 What is the function of a tendon?
2 Briefly describe how movement of a named limb is brought about by muscular contraction.
3 a) Give an example of a pair of opposing (antagonistic) muscles.
b) Explain why a pair of opposing muscles is needed at a joint.
4 a) Name TWO joints which may become injured during strenuous activity.
b) For each of these describe briefly the nature of the injury.

5 a) Of what type of tissue are tendons composed?
b) Explain why the structure of a tendon is suited to its function.

17 The need for energy

When asleep, human beings need energy for keeping warm, for breathing, for heartbeat and for many other workings of the body such as digestion and excretion. When awake, extra energy is needed for movement. The food that we eat provides all of this energy.

Energy content of foods

Energy is measured in units called **kilojoules** (kJ). Different foods have different energy values, as shown in table 17.1. The method used to find a food's calorific value is described on page 58.

foodstuff	calorific value (kJ)
beans, baked	3.9
biscuits, sweet	23.5
bread, brown	10.1
bread, white	10.1
butter	33.5
cabbage	1.2
cheese, cheddar	17.8
cocoa powder	19.0
coffee powder	0.0
cornflakes	15.2
chicken	5.8
eggs, fresh	6.8
fish, white fried	8.4
fish, white steamed	2.9
kidney, fresh	4.2
margarine	33.5
meat, fresh beef	13.6
milk	2.8
peanuts	25.5
potatoes, boiled	3.7
potatoes, deep fried	10.0
tea, dry	0.0

Table 17.1 Energy values of foods

Energy requirements

The total amount of energy required per day depends on the **body size**, **age**, **sex** and **occupation** of the person (see table 17.2).

Imbalance

When energy input (energy in food eaten) is greater than energy output (energy used up by body), then the body suffers an energy **imbalance** and stores the extra energy as fat (see figure 17.1). To use up this fat and so lose weight, the slimmer must eat a diet in which the total energy content of the food is below his or her daily energy requirement.

An energy imbalance also occurs if the person's energy input over a long period is far below the energy output. The person becomes severely underweight and unhealthy until he or she reverses the imbalance by eating more.

person	daily energy requirement (kJ)
2-year-old child	5000
6-year-old child	6500
12–15-year-old girl	9600
12–15-year-old boy	11700
woman (light work)	9500
woman (pregnant)	10000
woman (heavy work)	12500
man (light work)	11500
man (moderate work)	13000
man (very heavy work)	15500

Table 17.2 Daily energy requirements

Figure 17.1 Energy input and output

KEY QUESTIONS

1 a) State THREE ways in which humans use up energy whilst asleep.
  b) Why is extra energy needed when awake?
  c) From where do humans obtain all of this energy?
2 State the effect on the body of energy input being far below energy output.

3 Sort out the following pairs of foodstuffs into those (a) suitable and (b) unsuitable for inclusion in a slimmer's diet: chicken and beef; cabbage and baked beans; boiled potatoes and chips; cocoa and black coffee.

Comparison of inhaled and exhaled air

Oxygen content

The apparatus shown in figure 17.2 is used to collect a sample of exhaled air. A candle flame is found to burn for a much shorter time in this exhaled air than in inhaled (atmospheric) air. This shows that exhaled air contains less oxygen than inhaled air. Some oxygen must therefore be absorbed by the body during breathing.

This oxygen is used to release energy from food during aerobic respiration in living cells (see page 58).

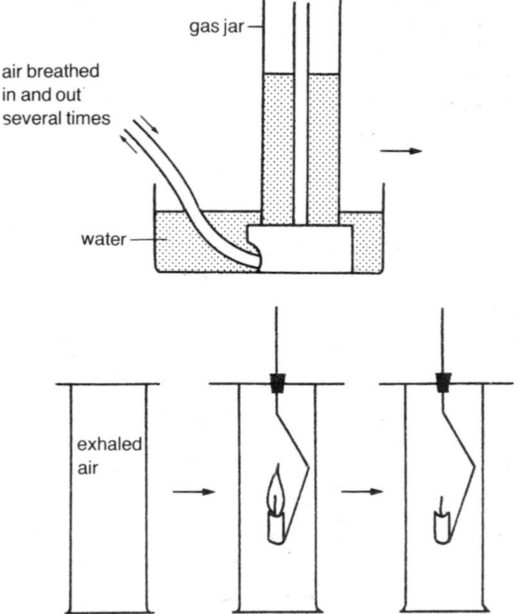

Figure 17.2 Oxgyen content of exhaled air

Carbon dioxide content

When air is inhaled through tube X (figure 17.3), it bubbles through the lime water in A. When air is exhaled through tube X, it passes through the lime water in B before escaping. Lime water B turns much more milky than lime water A, showing that exhaled air contains more carbon dioxide (CO_2) than inhaled air. Some carbon dioxide must therefore be released by the body during breathing.

This carbon dioxide is formed during aerobic respiration in living cells.

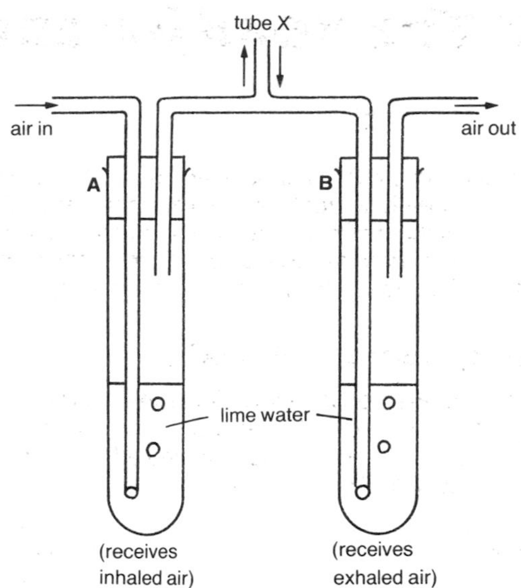

Figure 17.3 CO_2 content of inhaled and exhaled air

More to do

Analysis of inhaled and exhaled air

An air bubble is drawn into the **J-tube** (figure 17.4) by turning the screw and then the apparatus is left in a trough of water for two minutes. The length of the air bubble (now at the temperature of the water) is measured and then concentrated **potassium hydroxide solution** (which absorbs CO_2) is drawn in. The bubble is remeasured after a further two minutes in the water trough. Any change in length must be due to the removal of carbon dioxide by the potassium hydroxide solution.

The oxygen in the bubble is removed next by introducing **alkaline pyrogallol** solution (which absorbs oxygen). Any change in length of the bubble is noted following another two minutes in the water. (All measurements must be made at the same temperature because any variation in temperature would alter the bubble's length and invalidate the results.)

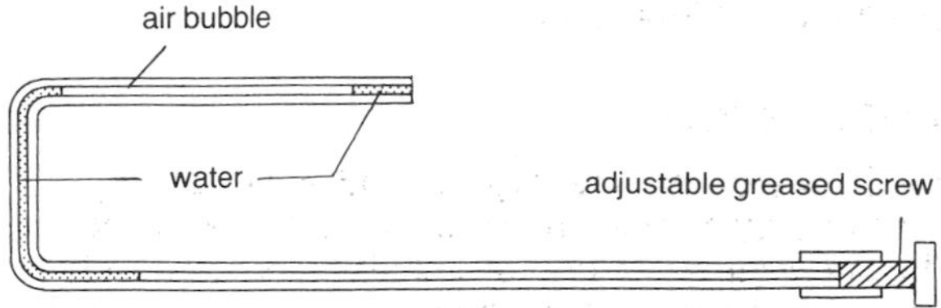

Figure 17.4 J-tube

Table 17.3 shows a typical set of results. The carbon dioxide content of inhaled air is so small that it cannot be measured accurately by this method.

	inhaled air	exhaled air
length of original bubble	100 mm	100 mm
length of bubble after using potassium hydroxide (CO_2 now removed)	100 mm	96 mm
percentage of CO_2	0% (0.03% = correct value)	4% $\left(\frac{100-96}{100} \times 100\right)$
length of bubble after using alkaline pyrogaliol (O_2 now removed)	80 mm	80 mm
percentage of O_2	20% $\left(\frac{100-80}{100} \times 100\right)$	16% $\left(\frac{96-80}{100} \times 100\right)$

Table 17.3 J-tube results

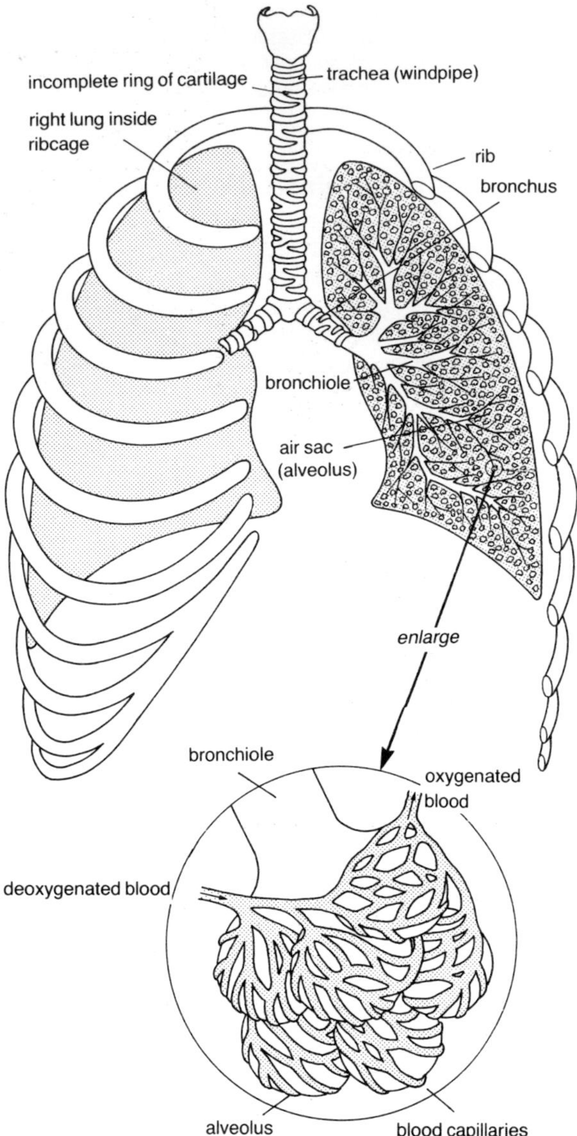

Figure 17.5 Human respiratory system

Human respiratory system

The **lungs** (figure 17.5) are a mammal's organs of gaseous exchange. Air entering by the nose or mouth passes via **larynx**, **trachea**, **bronchus** and **bronchioles** which end in tiny air sacs called **alveoli**.

The alveoli are so numerous that they provide a large surface area for gas exchange and give the lungs a sponge-like texture.

The trachea and bronchi are held permanently open by incomplete rings of **cartilage**. Otherwise they would collapse and close, leading to suffocation.

Cilia and mucus

The trachea and bronchi are lined with tiny hair-like **cilia** (figure 17.6) and glandular cells which secrete sticky **mucus**. Rhythmic beating of the cilia sweeps mucus containing trapped dust and germs upwards to the larynx from where it passes into the gullet. Germs are normally killed by acid in the stomach.

Associated structures

The two lungs are contained in and completely fill the chest cavity, an airtight space bounded by the chest wall (containing **ribs** and **intercostal muscles**) and the **diaphragm** (a dome-shaped muscular sheet stretched across the lowest ribs).

Each lung is surrounded by two thin sheets of tissue called the **pleural membranes** (see figure 17.6). The inner membrane completely covers the lungs while the outer membrane lines the inside of the chest wall. The pleural membranes are held together by a very thin layer of fluid between them. This pleural fluid also acts as a lubricant and allows the lungs to slide smoothly over the chest wall during breathing.

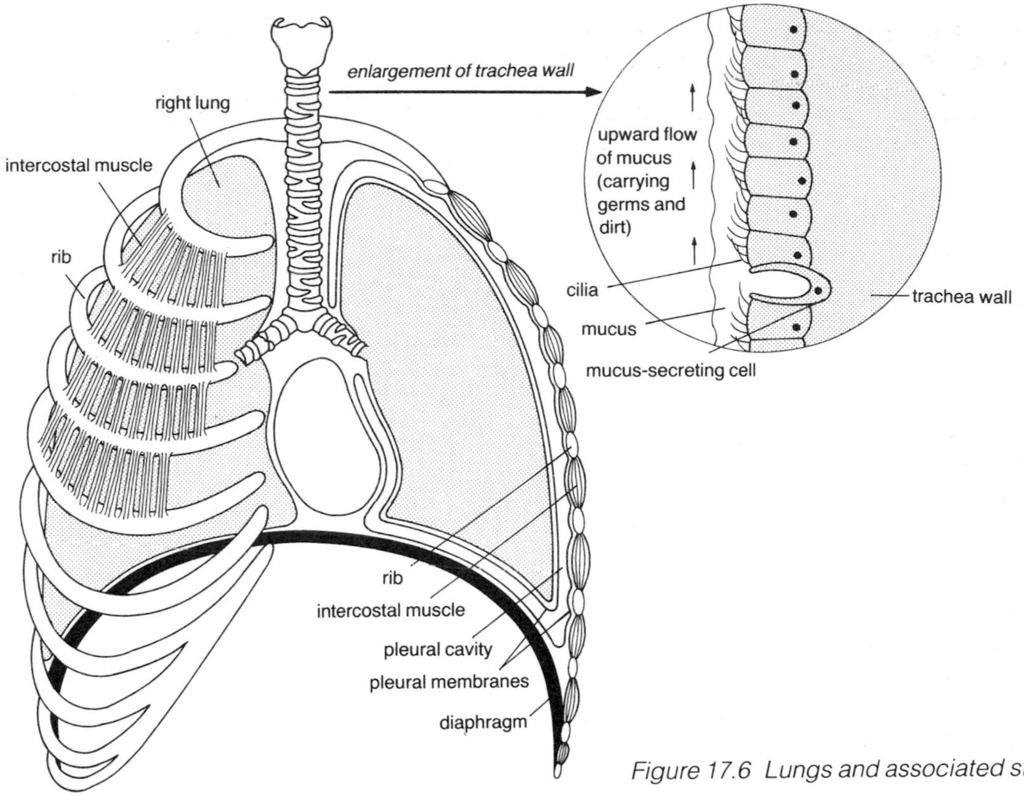

enlargement of trachea wall
right lung
intercostal muscle
rib
upward flow of mucus (carrying germs and dirt)
cilia
mucus
mucus-secreting cell
trachea wall
rib
intercostal muscle
pleural cavity
pleural membranes
diaphragm

Figure 17.6 Lungs and associated structures

Breathing movements

Inspiration

Contraction of the intercostal muscles pulls the rib cage out and up as shown in figure 17.7. At the same time, contraction of the diaphragm lowers (flattens) the floor of the chest cavity. The volume of the chest cavity is therefore increased (and so pressure is decreased) causing air to be inhaled.

Expiration

On relaxation of the intercostal muscles, the rib cage moves down and in. Relaxation of the diaphragm (back to its dome shape) causes a reduction in volume (and increase in pressure) of the chest cavity. Air is therefore exhaled.

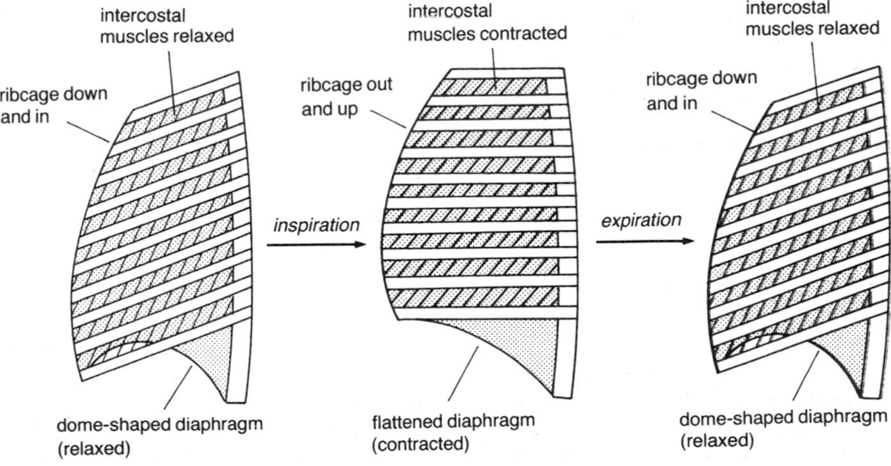

Figure 17.7 Breathing movements

Gas exchange

Blood arriving in a lung is said to be **deoxygenated** because it contains a low concentration of oxygen. Since air breathed into an alveolus contains a higher concentration of oxygen, diffusion occurs.

Oxygen first dissolves in the moisture on the surface of the thin lining of an alveolus (see figure 17.8) and then diffuses into the blood in the surrounding blood vessels (capillaries). The blood therefore becomes **oxygenated** (rich in oxygen) before leaving the lung and passing to all parts of the body.

Since deoxygenated blood contains a higher concentration of carbon dioxide than the air breathed into an alveolus, carbon dioxide diffuses from the blood into the alveolus from where it is exhaled.

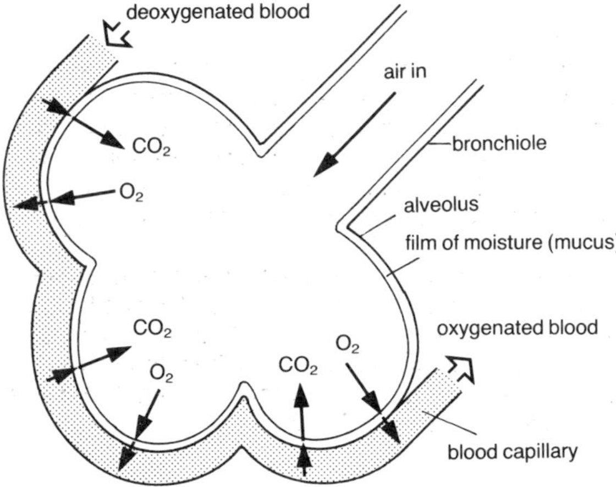

Figure 17.8 Gas exchange in an alveolus

More to do

Important features of a respiratory surface

The alveoli and associated blood vessels make the lungs efficient gas exchange structures for the reasons given in table 17.4.

KEY QUESTIONS

1 Name the gas that is used to release energy from food.
2 Rewrite the following sentence to include only the correct words.
During breathing oxygen/carbon dioxide is absorbed into the body and oxygen/carbon dioxide is released.
3 Arrange the following words in the order in which inhaled air would pass through them: bronchus, air sac, trachea, bronchiole, larynx.
4 At the ends of which tubes are alveoli found?
5 Briefly describe the process of gas exchange between an air sac and a surrounding blood vessel.
6 Why are the trachea and bronchi surrounded by rings of cartilage?

Extra Questions

7 Explain why the trachea is lined with cilia.
8 Describe the processes of inspiration and expiration in humans.
9 What features make the lungs efficient gas exchange structures?

feature	function
alveoli present large surface area	to absorb oxygen
alveolar surface is moist	to allow oxygen to dissolve
alveolar surface is thin	to allow oxygen to diffuse through into blood easily
network of tiny blood vessels surrounds alveoli	to pick up and transport oxygen

Table 17.4 Features of human respiratory surface

Blood and circulation

The circulatory system consists of the **heart** (a muscular pump) and the **blood vessels** (a system of tubes) which carry blood to all parts of the body.

Mammalian heart

The heart is divided into two separate sides as shown in figure 17.9. Each side has two hollow chambers: an **atrium** and a **ventricle**. The diagram shows the four heart chambers viewed from the front of the person. The right hand side of the heart is therefore on the left side of the diagram and vice versa. The wall of the heart is made of muscle.

Blood flow

The path taken by blood as it flows through the heart and its associated blood vessels is shown in figure 17.10. Blood from the body enters the right atrium (sometimes known as auricle) by the main vein (**vena cava**). The right atrium contracts, pumping blood into the right ventricle. Contraction of the muscular ventricle wall forces blood into the **pulmonary artery**. This vessel carries blood to the lungs where it takes up oxygen (see page 133).

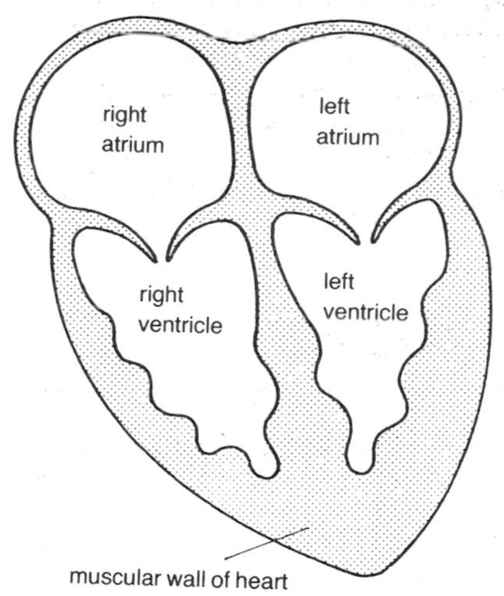

Figure 17.9 Chambers of the heart

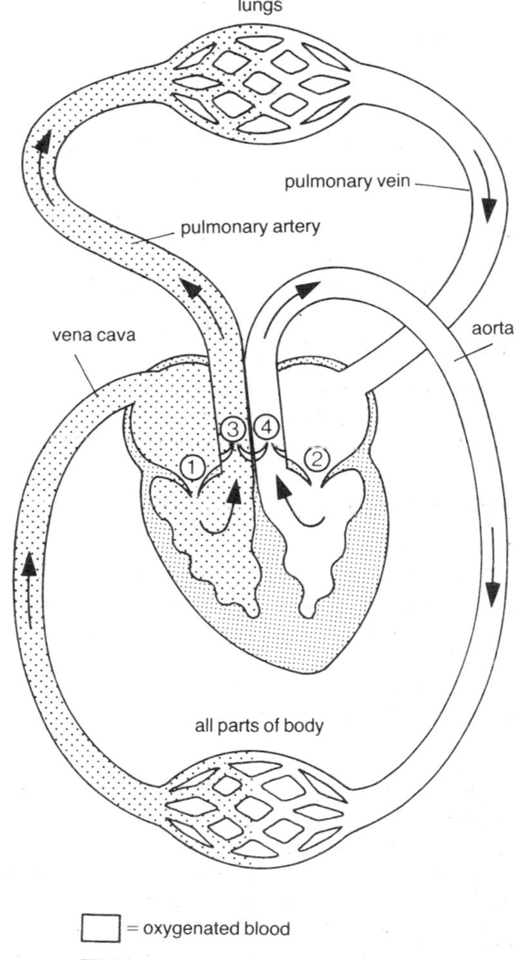

Figure 17.10 Heart and associated blood vessels

The **pulmonary vein** returns the oxygenated blood to the left atrium of the heart. The blood is then pumped into the left ventricle where strong contraction of the muscular wall forces blood into the main artery (**aorta**).

The aorta takes blood to all parts of the body, which use up the oxygen as the blood passes through, making it become deoxygenated. Deoxygenated blood returns by the main vein (vena cava) to the right atrium of the heart to repeat the cycle.

Thickness of ventricle walls

The wall of the left ventricle is particularly thick and muscular since it is required to pump blood all round the body. The wall of the right ventricle is less thick since it only pumps blood to the lungs.

Valves

Figure 17.10 also shows the positions of the four heart **valves**. Valves 1 and 2 are situated between the atria and the ventricles. When they are open, blood passes from the atria into the ventricles. When the ventricles contract, the blood, under pressure, closes valves 1 and 2. This prevents blood from flowing back into the atria.

Valves 3 and 4 are situated between the ventricles and the two arteries that leave the heart. Once blood has been pumped through valves 3 and 4 they close, preventing backflow of blood from the arteries down into the ventricles. Blood is therefore only able to travel in one direction through the heart.

Blood vessels

An **artery** is a vessel which carries blood away from the heart. It has a thick muscular wall (see figure 17.11) to withstand the **high pressure** of oxygenated blood spurting along it. (The pulmonary artery is exceptional in carrying deoxygenated blood.) Each time the heart beats, blood is forced along the arteries at high pressure and this can be felt as a **pulse** beat.

An artery divides into smaller vessels and finally into a dense network of tiny thin-walled **capillaries** (figure 17.11) which are in close contact with living cells.

Capillaries unite to form larger vessels which in turn form **veins**. A vein (figure 17.11) is a vessel which carries blood back to the heart. Although muscular, its wall is thinner than that of an artery since deoxygenated blood at **low pressure** oozes along it. Valves are present in veins to prevent backflow. (The pulmonary vein is exceptional in carrying oxygenated blood.)

A simplified version of the human circulatory system is shown in figure 17.12.

Coronary artery

The first branch of the main artery (aorta) leaving the heart is called the **coronary** artery (see figure 17.13). This artery supplies the muscular wall of the heart itself with oxygenated blood. If this vessel becomes blocked, blood flow to the heart wall is prevented and the person suffers a heart attack.

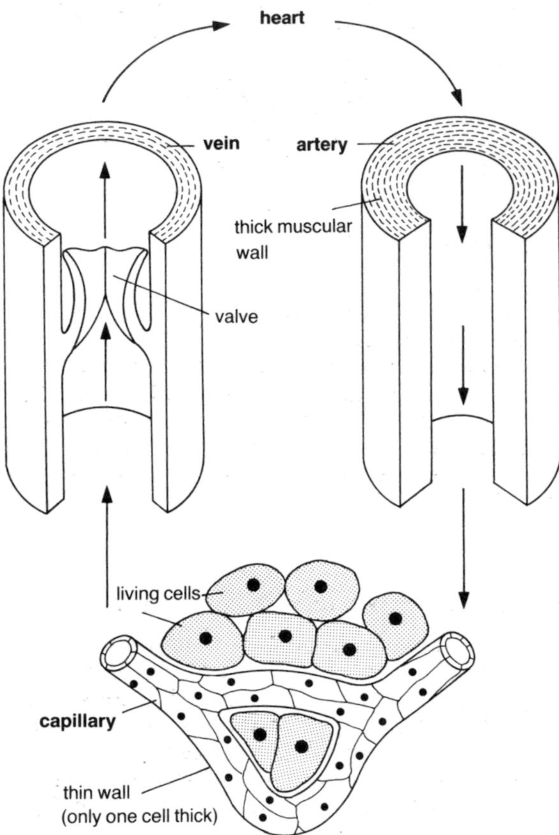

Figure 17.11 Types of blood vessel

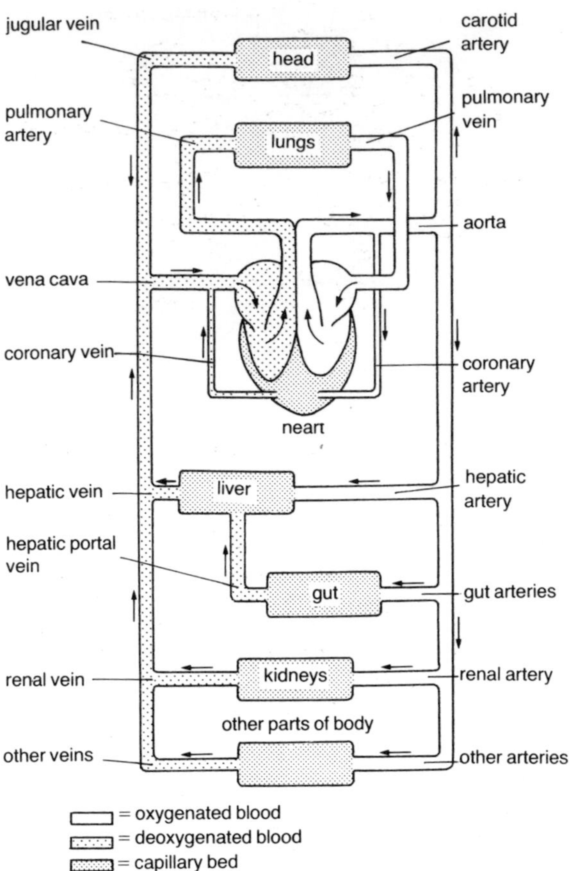

Figure 17.12 Blood transport system in humans

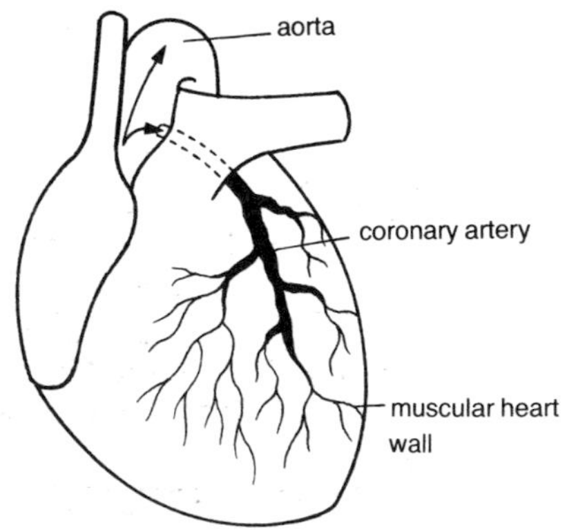

Figure 17.13 Coronary artery

Composition of blood

Figure 17.14 shows a view of blood highly magnified under a microscope. Blood consists of a watery yellow fluid called **plasma** containing many dissolved substances such as glucose, amino acids and respiratory gases (oxygen and carbon dioxide). Plasma also contains blood cells.

Red blood cells

These are very small and extremely numerous (approximately 5 million per mm^3). Their biconcave disc shape offers maximum surface area for oxygen uptake. As blood passes through the lung capillaries, oxygen diffuses into the plasma and then into the red blood cells which carry it round the body.

White blood cells

Compared with red blood cells, white blood cells are much less numerous (approximately 5 000 per mm^3). Their job is to kill germs.

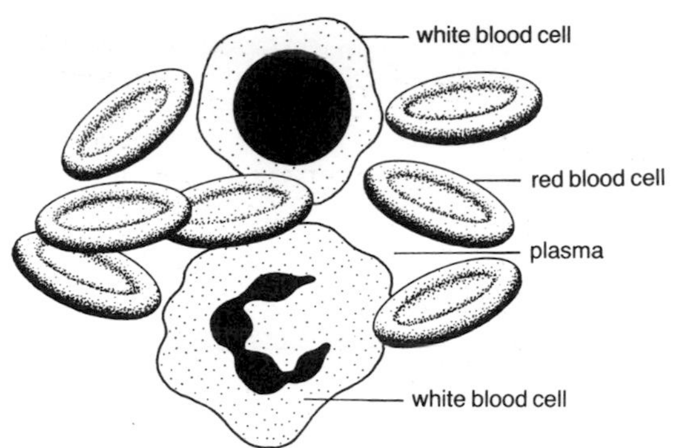

Figure 17.14 Blood

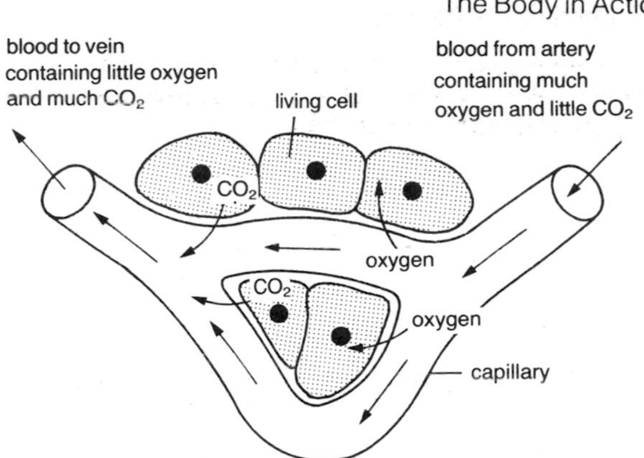

Figure 17.15 Gas exchange in a capillary

Gas exchange between blood and body cells

As oxygenated blood flows through a capillary close to body cells, oxygen diffuses out of the capillary into the surrounding cells (see figure 17.15). Since the cells are respiring, they are constantly using up oxygen and producing carbon dioxide. The carbon dioxide diffuses out of the cells and into the capillary.

The entry of oxygen into body cells and the exit of CO_2 from them is a further example of gas exchange.

Function of haemoglobin

The pigment in red blood cells is called **haemoglobin**. In the presence of a high concentration of oxygen, haemoglobin readily combines with the oxygen to form **oxy-haemoglobin**. When the surrounding concentration of oxygen is low, oxy-haemoglobin readily releases the oxygen again.

As blood passes through lung capillaries, oxy-haemoglobin is formed. When the blood reaches the capillaries beside respiring body cells (whose oxygen concentration is low) oxy-haemoglobin quickly releases the oxygen, which diffuses into the cells (see figure 17.16).

Features of capillary network

Capillaries are very narrow tubes which branch repeatedly, forming such a **dense network** that every living body cell is close to a capillary. The combined **surface area** of the capillary network is enormous and the capillary walls are only **one cell thick**. These properties of the capillary network allow efficient gas exchange to occur between the bloodstream and the body cells.

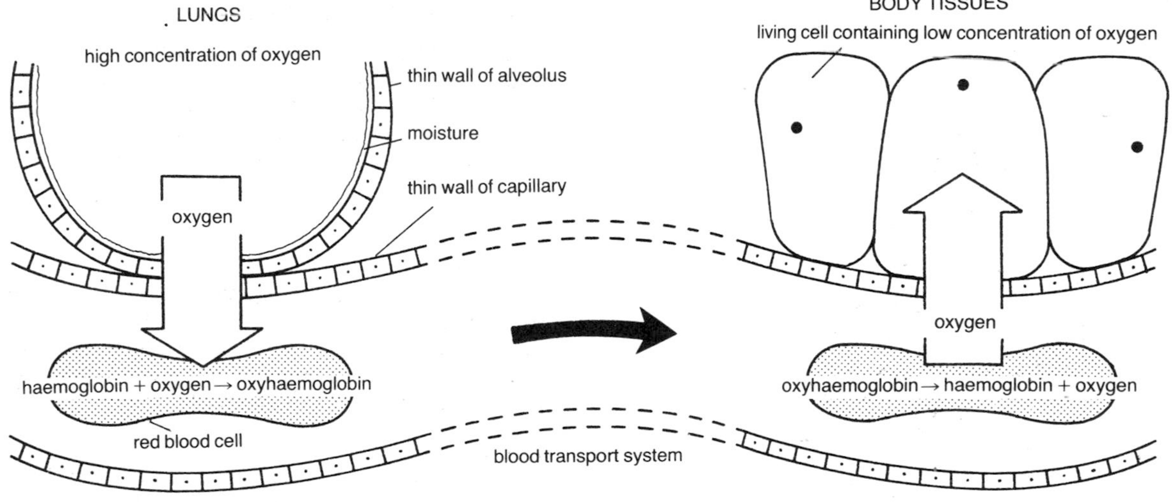

Figure 17.16 Role of haemoglobin

KEY QUESTIONS

1 Name the FOUR chambers of the mammalian heart.
2 Describe the route taken by blood as it flows from the vena cava to the aorta.
3 State the function of the heart valves.
4 Describe the positions of the two types of heart valve.
5 Explain why one ventricle's muscular wall is thicker than the other.
6 Which blood vessel supplies the heart's muscular wall with blood?
7 Rewrite the following sentence by choosing the correct word from each set of brackets.

Blood leaves the heart in {an artery / a capillary / a vein}, flows through a dense network of thin-walled {arteries / capillaries / veins}, and returns to the heart in {an artery / a capillary / a vein}.

8 What is the function of (**a**) red blood cells and (**b**) plasma in the transport of respiratory gases and food?
9 Describe gas exchange between body cells and the blood in the surrounding capillaries.

Extra Questions

10 Explain the function of haemoglobin in the uptake and transport of oxygen.
11 Describe TWO features of a capillary network which allow efficient gas exchange.

18 Co-ordination

Need for co-ordination

All the organs and systems in a healthy human body are **co-ordinated**. This means that they co-operate with one another and work together efficiently to bring about the many functions that are essential to life.

Physical activity is one of these vital functions and it requires co-ordination. Consider the tennis player shown in figure 18.1, about to hit the ball. Arrow 1 represents information (such as the position of the ball) being picked up by the **sense organs** (e.g. eyes). Arrow 2 represents this information being passed to the **central nervous system** (CNS) where it is processed. The CNS then sends out messages (arrows marked 3) to the appropriate **muscles** which respond in the ways required to perform the activity (e.g. right arm operates racquet while left arm provides balance and legs keep body stable). Since there are hundreds of muscles in the body, this is a very complex process.

Physical activities and all other body functions are controlled by the nervous system.

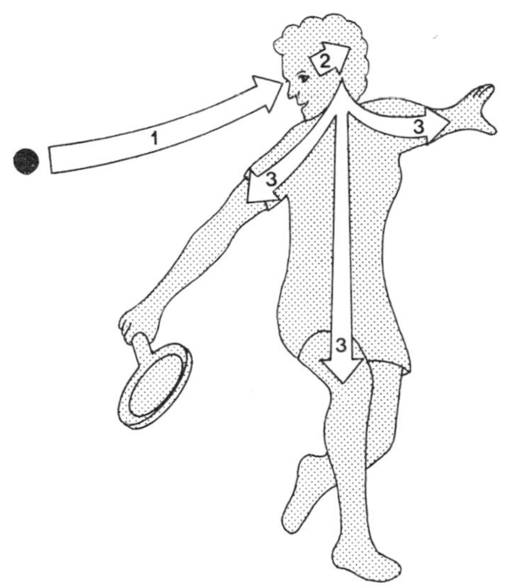

Figure 18.1 Co-ordination of physical activity

Eye

Table 18.1 gives a summary of the functions of the parts of the eye shown in figure 18.2.

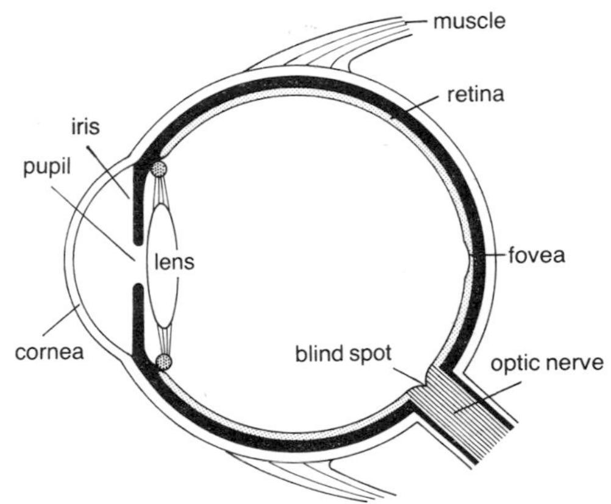

Figure 18.2 Structure of eye

part	function
cornea	tough, transparent layer at front of eye. Allows light to enter and begins to bring it to a focus.
iris	coloured part containing muscles which alter diameter of pupil.
pupil	opening surrounded by iris. Controls amount of light entering eye. Activity of iris muscles enlarge it in dim light to admit maximum light, and reduce it in bright light to admit minimum light, thus preventing damage to retina.
lens	flexible transparent biconvex structure. Focuses light on to retina.
retina	light-sensitive layer containing two types of receptor cell: **rods** (sensitive to dim light) and **cones** (sensitive to bright light and colour). Light converted to nerve impulses.
fovea	small depression in retina containing cones only. Point of most accurate vision.
blind spot	point where receptors are absent and fibres from retina enter optic nerve.
optic nerve	carries nerve impulses from retina to brain.

Table 18.1 Functions of parts of eye

Judgement of distance

Look at the experiment shown in figure 18.3. Several attempts at landing the ring on the upright of the clamp stand are made with one eye covered and then repeated without an eye being covered.

From the typical set of results shown in table 18.2, it can be seen that it is easier to perform the task using both eyes. It is therefore concluded that judgement of distance is more accurate using two eyes rather than one.

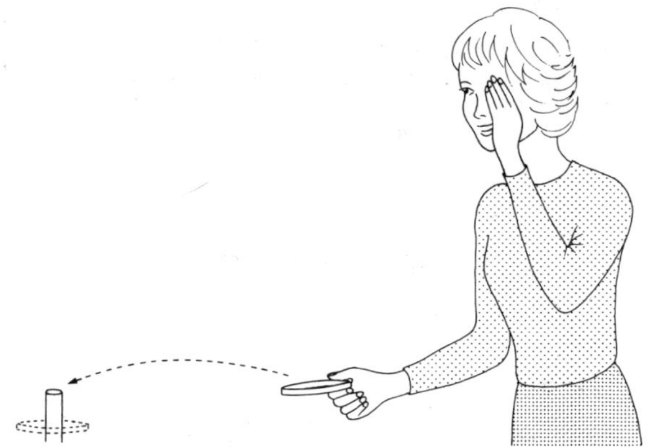

Figure 18.3 Ring-throwing experiment

score out of 25 for one eye covered	score out of 25 for no eyes covered
𝍸 𝍸 𝍸 𝍸	𝍸 𝍸 𝍸 𝍸 III

Table 18.2 Ring-throwing results

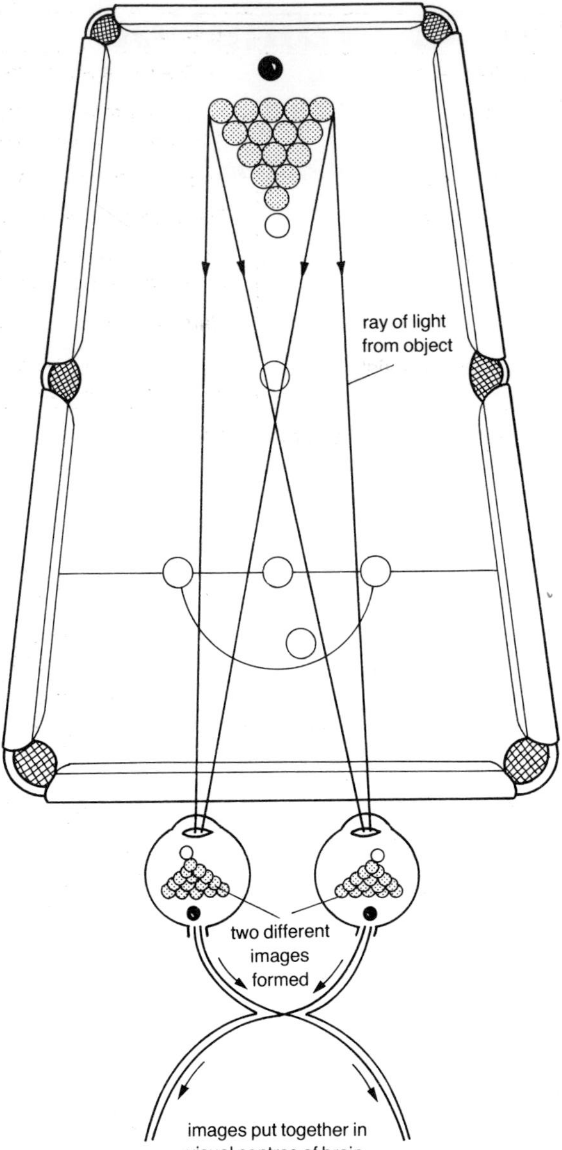

Figure 18.4 Binocular vision

Binocular vision

When an object is viewed, the image formed on the retina of the left eye is slightly different from the one formed on the retina of the right eye, as shown in figure 18.4. The two different images are joined together in the visual centres in the brain, producing a **three-dimensional** picture.

This gives the observer an appreciation of depth (in addition to length and breadth). It is for this reason that the use of two eyes rather than one allows distance to be more accurately judged.

Ear

Table 18.3 gives a summary of the functions of the parts of the ear shown in figure 18.5.

part	function
auditory canal	Air-filled tube which directs sound waves on to eardrum.
eardrum	Thin membrane stretched completely across auditory canal. Set vibrating by sound waves which it passes on to middle ear bones.
middle ear bones	Amplify and transmit sound vibrations from eardrum to oval window.
oval window	Thin membrane which transmits sound vibrations into liquid-filled inner ear.
cochlea	Coiled liquid-filled tube lined with sound receptor cells possessing hairlike endings. Stimulated 'hairs' convert sound vibrations in the liquid to nerve impulses.
auditory nerve	Carries nerve impulses from cochlea to brain.
semi-circular canals	Three tubes containing liquid which moves in response to movements of head. Messages sent to part of brain which controls muscular activity essential for balance.

Table 18.3 Functions of parts of ear

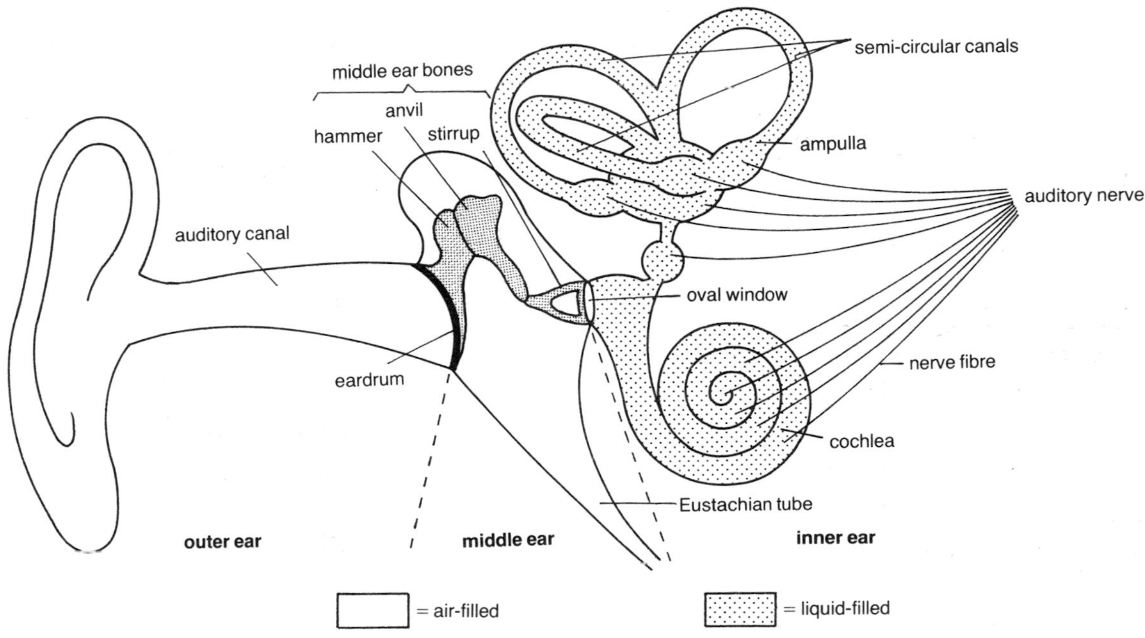

Figure 18.5 Structure of ear

Judgement of direction of sound

Look at the experiment shown in figure 18.6. The subject is blindfolded. The observer moves silently to any one of the positions shown and snaps his fingers. The subject responds by pointing in the direction from which the sound appears to have come. Each position is tested an equal number of times but in a random order. The experiment is then repeated with an efficient earplug fitted into one of the subject's ears.

A typical set of results is shown in table 18.4. From the results it is concluded that judgement of direction of sound is more accurate using two ears rather than one.

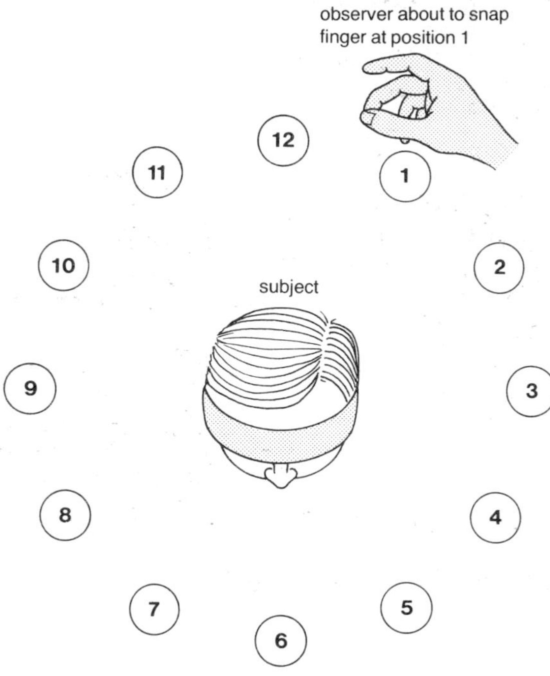

Figure 18.6 Judging direction of sound

two ears		one ear	
correct	wrong	correct	wrong
84	12	65	31

Table 18.4 Sound direction results

More to do

Balance

The three liquid-filled **semi-circular canals** are situated at right angles to one another, two in a vertical plane and one in a horizontal plane as shown in figure 18.7. Each possesses an ampulla containing a receptor which is stimulated by movements of the liquid in the canal. These are caused by the canal being rotated in its respective plane by movements of the head as shown in figure 18.8.

Messages are carried by a nerve to a region of the brain (the cerebellum) which controls the muscular activity essential for balance and posture. For accurate sensations of balance, human beings also rely on information from their eyes and joints.

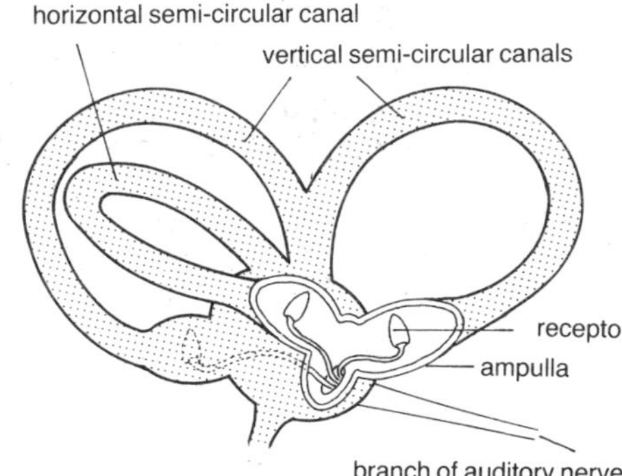

Figure 18.7 Semi-circular canals

KEY QUESTIONS

1 State the function of each of the following parts of the eye: cornea, iris, lens, retina, optic nerve.
2 In what way is the ability to judge distance affected by having one eye covered?
3 State the function of each of the following parts of the ear: ear drum, middle ear bones, cochlea, auditory nerve.
4 In what way is the ability to judge direction of sound affected by the presence of a plug in one ear?
5 State the function of the semi-circular canals.

6 How is the arrangement of the semi-circular canals related to their function?

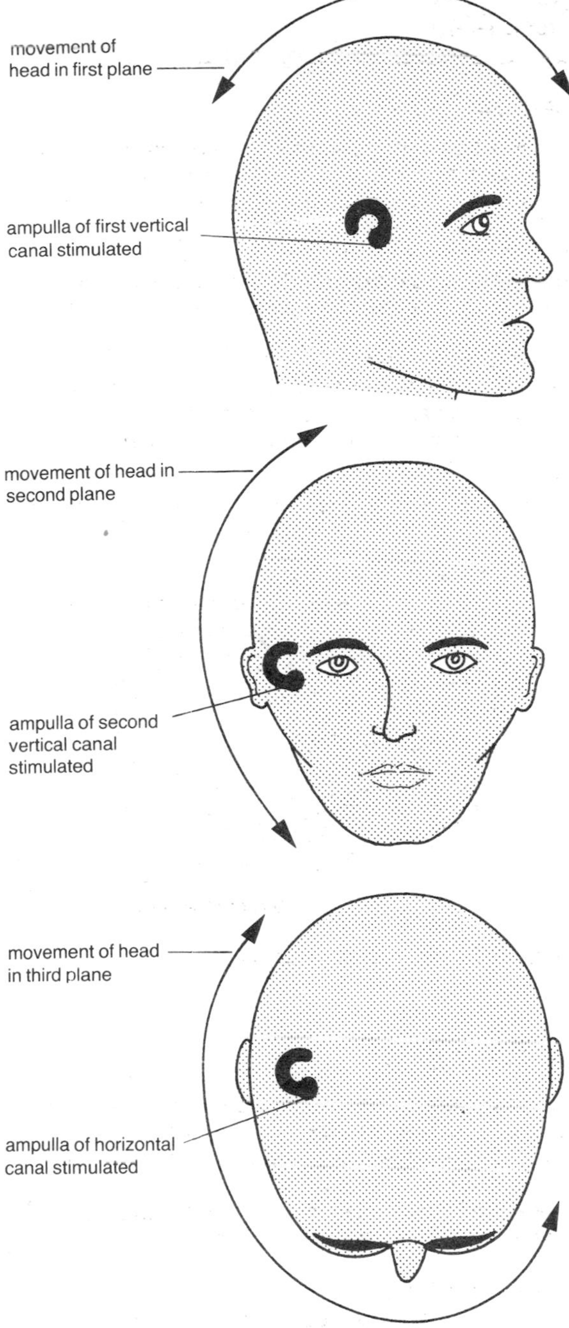

Figure 18.8 Role of semi-circular canals

Human nervous system

The human nervous system (see figure 18.9) is composed of three parts – the **brain**, the **spinal cord** and the **nerves**. The brain and spinal cord make up the **central nervous system (CNS)**. The CNS is connected to all parts of the body by the nerves which lead to and from all organs and systems. This arrangement ensures that all parts work together as a co-ordinated whole with the brain exerting overall control.

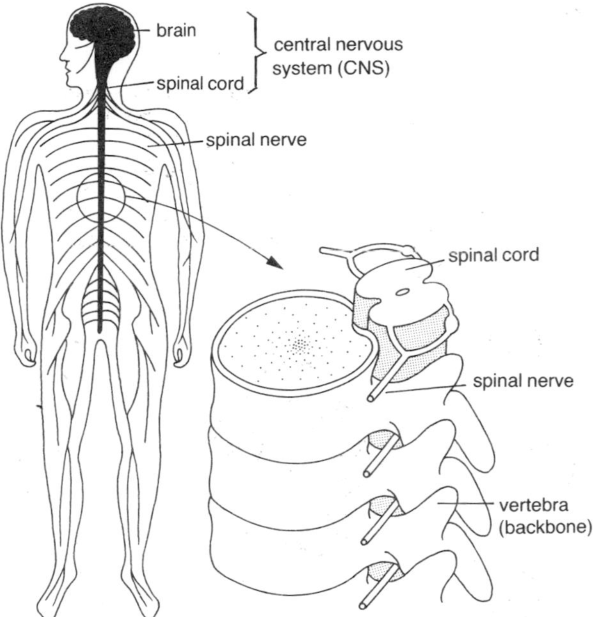

Figure 18.9 Human nervous system

Flow of information

Figure 18.10 shows how one set of nerves (sensory) carry information from the body's sense organs to the CNS and another set of nerves (motor) carry messages from the CNS to other parts of the body such as the muscles.

The messages arriving from the senses keep the CNS informed about all aspects of the body and its surroundings. The CNS sorts out all of this information and stores some of it. When the body is required to perform a particular physical activity, the CNS sends out messages to certain muscles which then make the appropriate response.

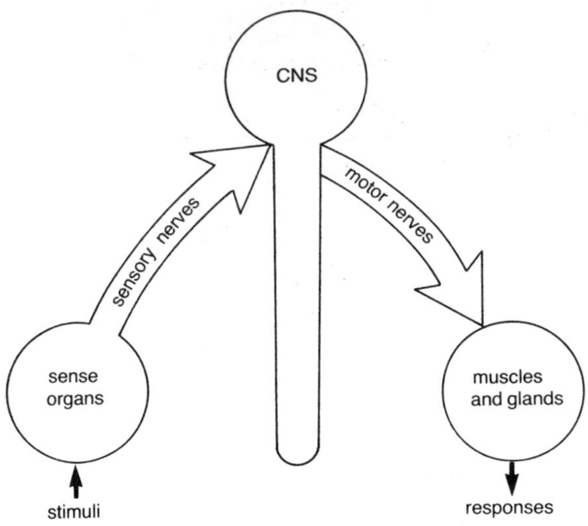
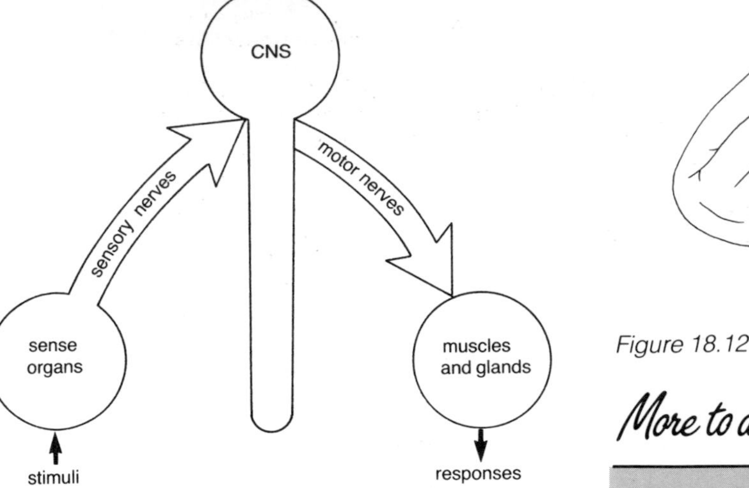

Figure 18.10 Flow of information

Brain

The brain consists of several different regions as shown in figure 18.11. The **medulla** controls the rate of breathing and heartbeat. The **cerebellum** controls balance and muscular co-ordination.

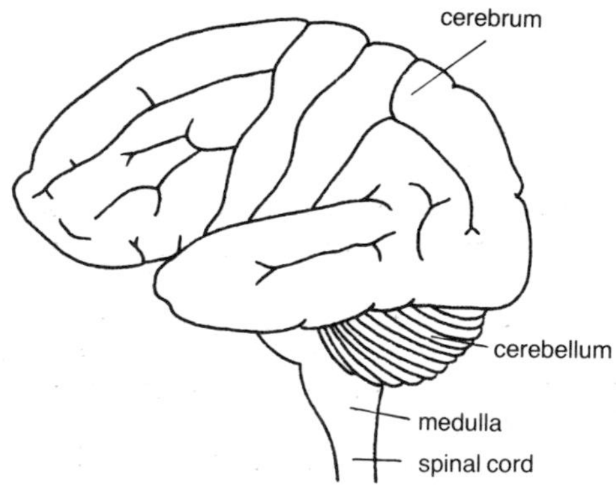

Figure 18.11 Human brain

The largest part of the brain is called the **cerebrum**. It is divided into two halves (hemispheres). Figure 18.12 shows the left cerebral hemisphere. Each region of the cerebrum is concerned with a specific function. The sensory area, for example, receives impulses from the sense organs. In the light of this information, the motor area transmits impulses to all parts of the body, making them work together efficiently. The unlabelled areas in figure 18.12 are concerned with mental processes such as memory, reason, imagination, conscious thought and intelligence.

motor area sensory area

hearing sight

Figure 18.12 Left cerebral hemisphere

More to do

Neurones and reflex arc

The nervous system is made of nerve cells called **neurones**. A neurone consists of a cell body attached to nerve fibres. The simple arrangement of three different types of neurone shown in figure 18.13 is called a **reflex arc**.

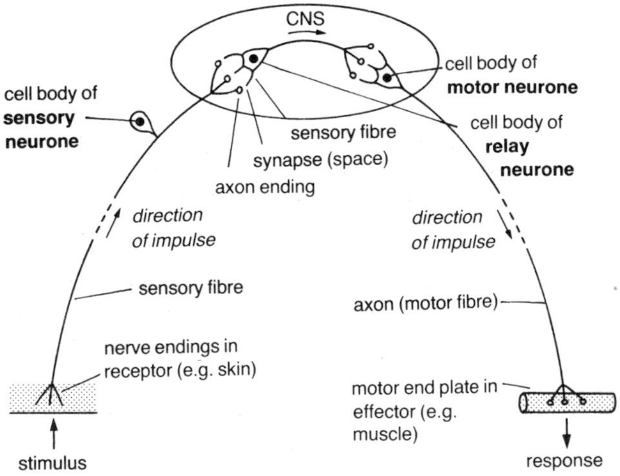

Figure 18.13 Reflex arc

Nerve impulses are carried to the CNS by sensory fibres and away from it by motor fibres (**axons**). A tiny space (**synapse**) occurs between the axon ending of one neurone and the sensory fibre of the next. When a nerve impulse arrives, the tiny knob at the end of the axon branch releases a **chemical** which diffuses across the space and triggers off an impulse in the sensory fibre of the next neurone in the arc.

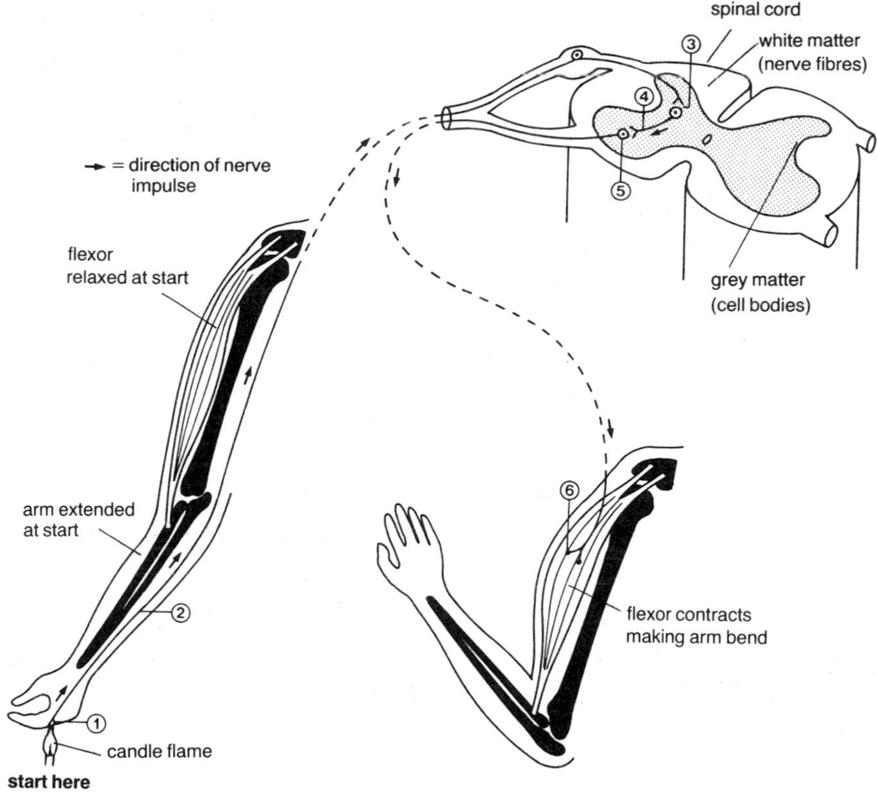

Figure 18.14 Reflex action

Reflex action

The transmission of a nerve impulse through a reflex arc results in a **reflex action**. Figure 18.14 shows an example of a reflex action called limb withdrawal.

When the back of the hand accidentally comes in contact with intense heat, this stimulus is picked up by the pain receptors in the skin (1) and an impulse is immediately sent up the fibre of the sensory neurone (2). In the grey matter of the spinal cord, the impulse crosses the first synapse (3) and passes through the relay neurone (4). Once across the second synapse, the impulse is picked up by the motor neurone (5) and quickly conducted to the axon endings (6) (motor end plates) which are in close contact with the flexor muscle of the arm. Here a chemical is released which brings about muscular contraction (the response) which makes the arm bend, moving it out of harm's way.

A reflex action is a rapid, automatic, involuntary response to a stimulus. Reflex actions protect the body from damage. Since they do not need conscious thought by the brain, many reflex actions may still be performed for a short period by an animal whose brain has been destroyed.

KEY QUESTIONS

1 Name THREE parts of the human nervous system.
2 Describe the important role played by the nervous system.
3 The following structures are involved in the transmission of a nervous impulse. Arrange them in the correct order: motor nerve, receptor, effector, CNS, sensory nerve.
4 a) Which part of the brain is concerned with learning and memory?
b) Name TWO other parts of the human brain and for each state its function.

Extra Questions

5 Make a simple diagram of a reflex arc.
6 What is meant by the term **reflex action**?
7 Blinking in response to a speck of dust landing on the surface of the eye is an example of a reflex action. Explain how it works.

19 Changing levels of performance

Muscle fatigue

In the experiment shown in figure 19.1, a set of bathroom scales is used to measure force. A piece of apparatus called a dynamometer (figure 19.2) can also be employed for this purpose. The apparatus is gripped as tightly as possible and the force (in newtons) recorded. This procedure is repeated until ten trials have been completed.

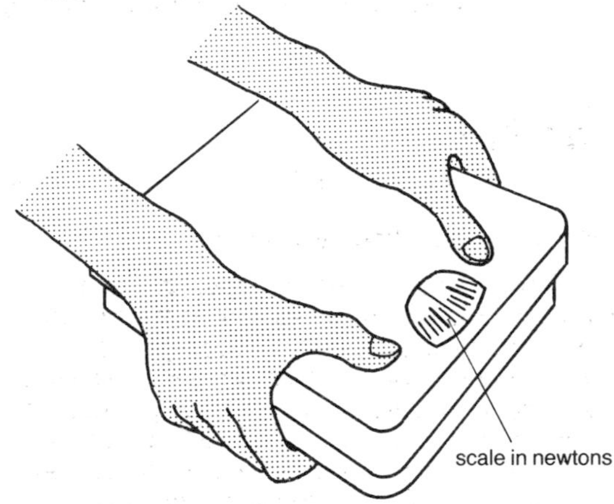

Figure 19.1 Measuring force

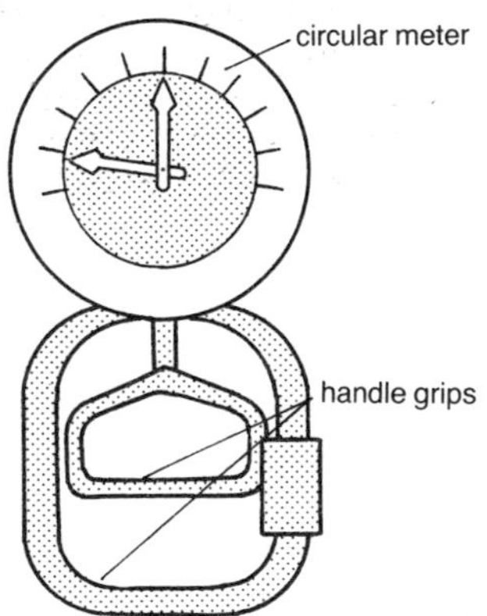

Figure 19.2 Dynamometer

trial number	force (newtons)
1	400
2	380
3	390
4	350
5	330
6	290
7	300
8	290
9	270
10	240

Table 19.1 Hand grip results

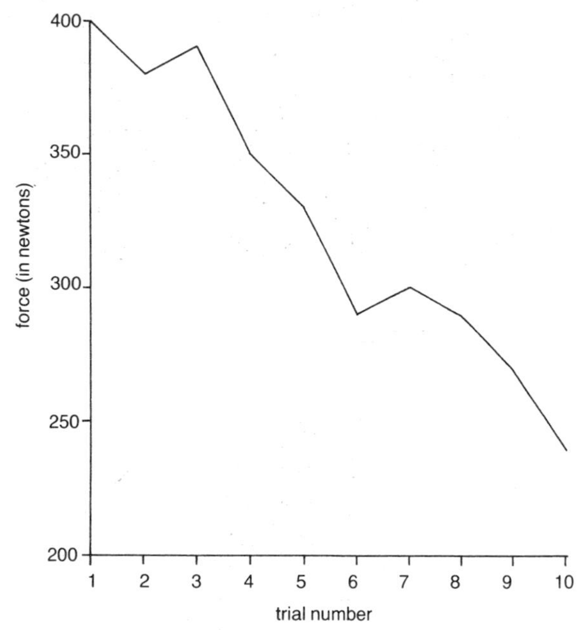

Figure 19.3 Graph of hand grip results

Table 19.1 gives a typical set of results using bathroom scales. These are graphed in figure 19.3. From the results it can be seen that the force that can be applied decreases over the ten trials. It is therefore concluded that rapidly repeated contraction of muscle results in fatigue. (This is also true of continuous muscle contraction, such as when holding up a heavy weight for a period of time.)

In the hand grip experiment, the person feels discomfort and pain gradually building up in the hands. If the experiment is continued beyond ten trials, eventually there comes a point when the person is unable to continue due to **muscle fatigue**. This results from a lack of oxygen in the muscle cells and the build-up of a chemical called **lactic acid** which prevents the muscle from contracting.

Effect of exercise on pulse rate and breathing rate

The boy shown in figure 19.4 begins the experiment by sitting at rest and having his resting pulse rate and breathing rate measured.

He then exercises by stepping up and down on to the box every 2 seconds. After 3 minutes he stops exercising and his pulse rate and breathing rate are measured immediately and then at the end of every minute during his recovery time, until the original resting values are obtained as shown in table 19.2.

From the results it can be seen that both breathing rate and pulse rate increase with exercise. This is because working muscles demand more energy and therefore more oxygen is required in the muscle cells for respiration.

The breathing rate increases in order to get more oxygen into the bloodstream. The pulse rate increases because the heart must beat faster in order to deliver the blood to the muscles.

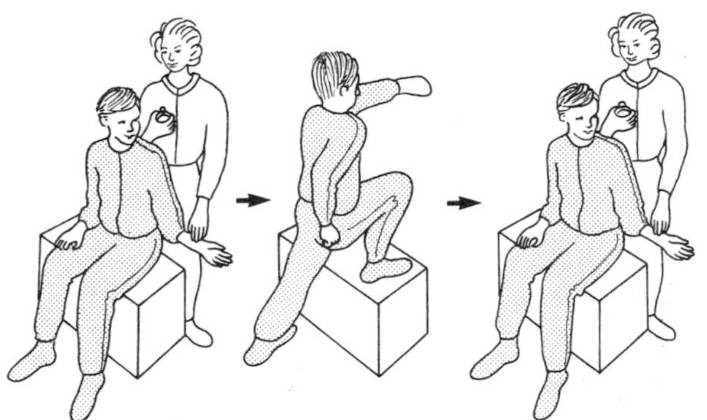

Figure 19.4 Step test

		pulse rate (beats/min)	breathing rate (breaths/min)
at rest		65	12
recovery	1	142	29
time	2	120	24
(minutes)	3	105	20
	4	94	19
	5	88	16
	6	80	14
	7	76	14
	8	70	13
	9	68	12
	10	65	12

Table 19.2 Effect of exercise on pulse and breathing

More to do

Anaerobic respiration

Under normal circumstances the energy needed for contraction of muscles comes from aerobic respiration involving glucose and oxygen (see page 60). However, during vigorous activity when muscles are working continuously, the supply of oxygen cannot meet the demand. Under such conditions of oxygen shortage, the muscles continue to use glucose but respire **anaerobically** as in the word equation:

glucose ⟶ lactic acid + a little energy

Thus, during intense exertion the concentration of lactic acid in the muscle and bloodstream builds up (see figure 19.5). As the level increases it causes discomfort and pain and reduces the efficiency of the muscles, causing them to fatigue rapidly.

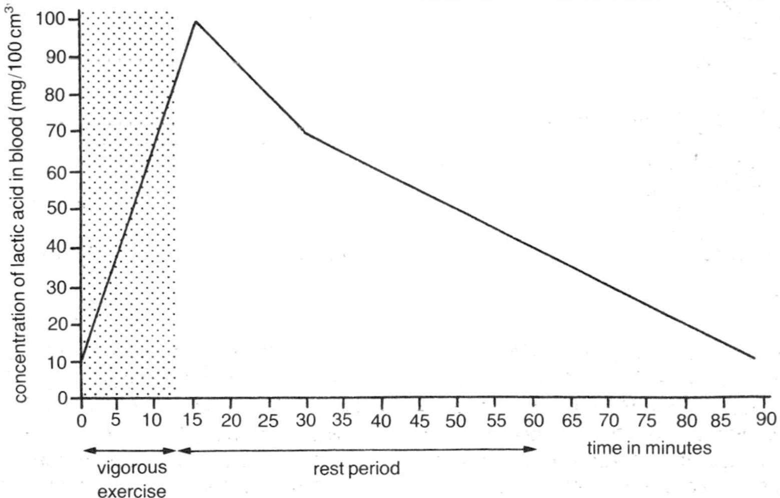

Figure 19.5 Lactic acid graph

Oxygen debt
Oxygen is required to remove the lactic acid so that the muscles can work properly again. The body is therefore said to build up an 'oxygen debt' during anaerobic respiration. In the presence of oxygen this debt is repaid during a rest period and the lactic acid concentration falls.

KEY QUESTIONS

1 a) What name is given to the feeling of discomfort and pain resulting from continuous or rapidly repeated contraction of muscle?
b) What causes this feeling of discomfort and pain?

2 a) In what way do pulse rate and breathing rate change during vigorous exercise?
b) Explain why.

3 Give the word equation for anaerobic respiration in human muscle.

4 Explain muscle fatigue in terms of anaerobic respiration.

5 What is meant by the term 'oxygen debt'?

Effect of exercise on athletes and non-athletes

The data presented in table 19.3 refers to average values obtained from a group of trained, fit athletes and a group of untrained, unfit non-athletes of the same age and sex performing the step test.

From the results it can be seen that during exercise, the pulse rate, breathing rate and lactic acid level rise less in athletes than in untrained people.

Recovery time

This is the time taken for pulse rate, breathing rate and blood lactic acid level to return to normal resting values. The results recorded in table 19.4 show that for each of these measurements, the athletes recovered more quickly than the untrained persons. The recovery time following a set amount of exercise (e.g. 3 minute step test at 30 steps/minute) can therefore be used as an indication of **fitness**.

recovery time (minutes)			
for breathing rate and pulse rate		for blood lactic acid level	
athletes	non athletes	athletes	non athletes
5	12	25	60

Table 19.4 Effect of training on recovery times

More to do

Effect of training

Training (see figure 19.6) involves vigorous exercise over a period of several weeks or months. During this time the heart muscle tissue gradually increases in strength and efficiency and the volume of the heart increases. As a result, the same volume of blood can be supplied to the muscles using fewer heartbeats. Since the heart of a trained athlete beats less often but more strongly than that of a non athlete, the athlete's circulatory system is more efficient.

In a trained athlete, oxygen is absorbed into the bloodstream from the lungs and delivered to the working muscles more quickly than before training. As a result, pulse rate, breathing rate and level of blood lactic acid rise less than for an untrained person doing the same exercise. In addition, the time taken to pay back the 'oxygen debt' and remove excess lactic acid from the bloodstream is reduced in a trained person.

Reaction time

In many sporting activities it is often important to react quickly to a certain signal. The time interval between detecting a stimulus (e.g. hearing the starting gun) and responding (e.g. beginning to run) is called **reaction time**.

	pulse rate (beats/min)		breathing rate (breaths/min)		lactic acid level (mg/100 cm^3 blood)	
	athletes	non athletes	athletes	non athletes	athletes	non athletes
at rest	65	69	14	14	10	10
after 3 minutes' exercise	140	170	28	33	40	65

Table 19.3 Effect of exercise on athletes and non athletes

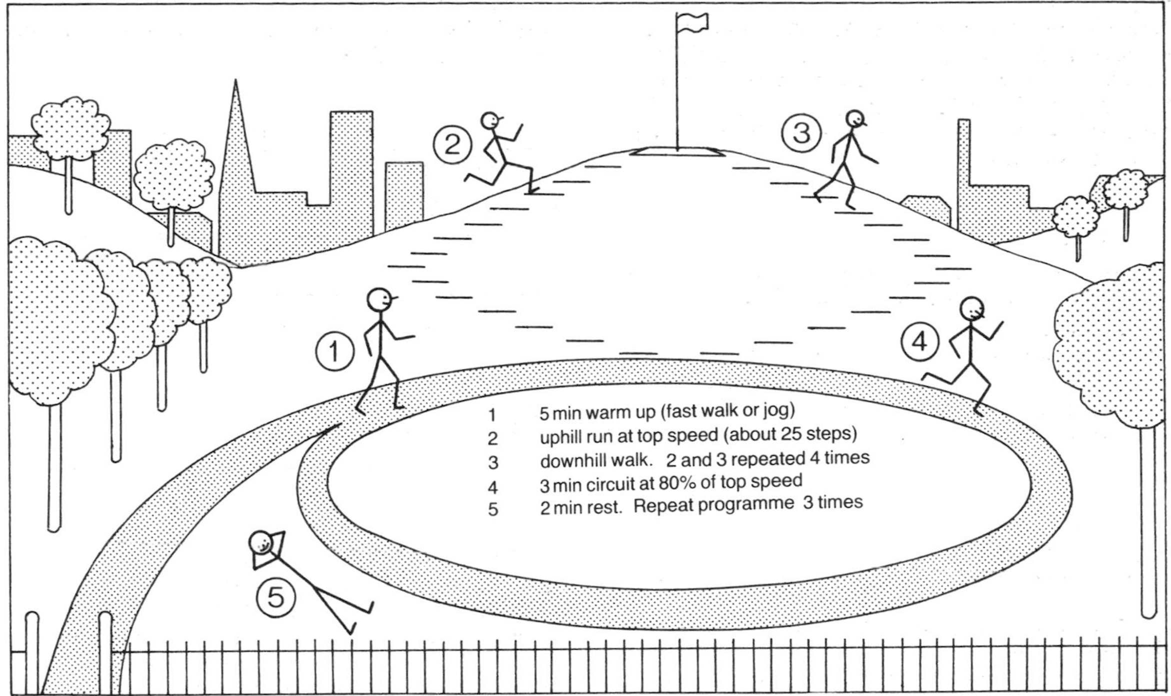

Figure 19.6 Simple training programme

Reaction time can be measured using the apparatus shown in figure 19.7 where the stimulus is the visual signal on the screen and the response is pressing the space bar on the keyboard.

Table 19.5 gives a typical set of results which are graphed in figure 19.8. From the graph it can be seen that reaction time improves with practice. However, a definite limit is eventually reached when reaction time shows no further improvement.

Figure 19.7 Measuring reaction times

trial number	time (seconds)
1	0.041
2	0.027
3	0.020
4	0.018
5	0.019
6	0.017
7	0.017
8	0.018
9	0.019
10	0.018

Table 19.5 Reaction time results

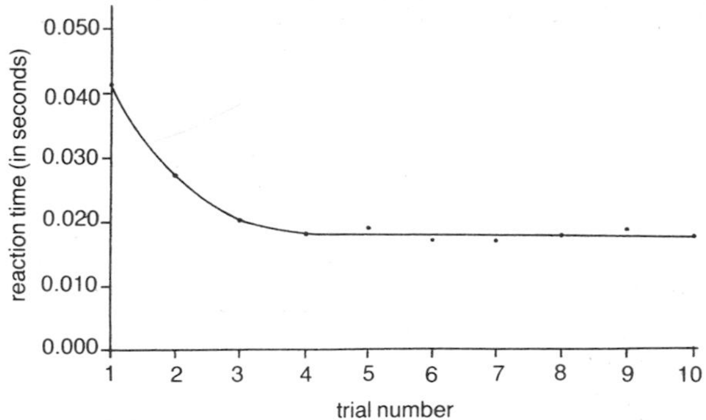

Figure 19.8 Graph of reaction times

KEY QUESTIONS

1 Compare the effect of the same amount of a vigorous exercise on the pulse rate, breathing rate and lactic acid level of an athlete and an untrained person.

2 a) What is meant by the term 'recovery time'?
b) Briefly describe how recovery time can be used as an indication of physical fitness.

3 a) What is meant by the term 'reaction time'?
b) How can a person improve his/her reaction time?

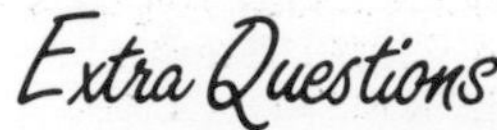

4 What effect does training have on the efficiency of the lungs and circulation?

5 Explain the relationship between the effects of training and recovery time.

PROBLEM SOLVING

1 The diagram shows one leg of a person sitting on the edge of a table.
Which muscle must contract in order to:
a) lift the lower foot and toes but leave the heel on the ground?
b) stand up with the foot flat on the ground?
c) stand up on tiptoe?

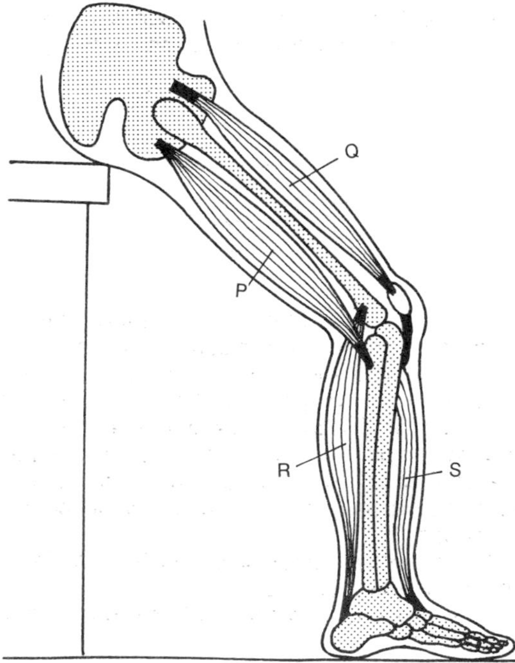

2 Look back to the experiments shown in figure 16.3 on page 121. Predict the effects of heating a bone in the oven and then placing it in the acid for twenty-four hours. Explain your answer.

3 One end of a muscle is attached at the point of insertion to the part of the skeleton to be moved while the other is anchored to the part of the skeleton which remains stationary – the point(s) of origin. The human biceps has two origins (1 and 2 in the accompanying diagram of a boxer's arm) each with its own tendon attaching it to the scapula. The biceps is inserted to the radius (at point 3).

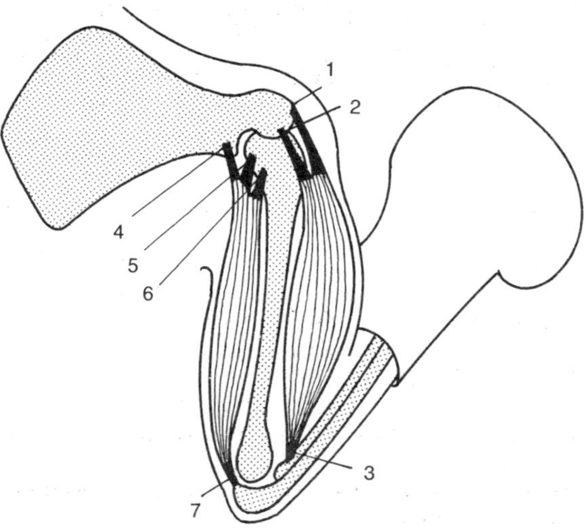

Which number in the diagram indicates:
a) the triceps' point of insertion in the ulna?
b) the triceps' point of origin in the scapula?

4 At high altitudes the air is thinner and less oxygen is gained by the body per breath. Graphs A and B refer to the altitudes reached by and the red blood corpuscle (cell) counts of a group of climbers on an expedition in the Himalaya Mountains.

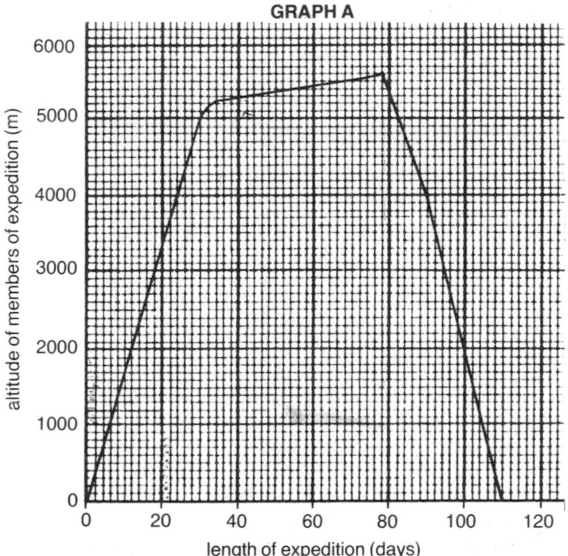

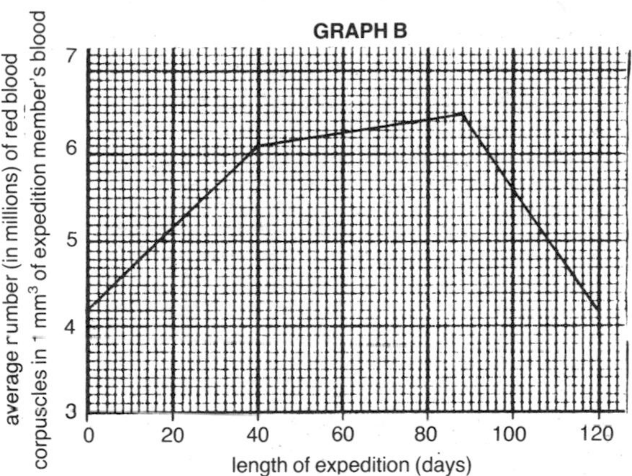

a) State the climbers' (i) starting altitude (ii) altitude after 40 days.
b) How many red corpuscles per mm^3 were, on average, present in a climber's blood (i) at the start of the expedition, (ii) after 40 days?
c) (i) On which day did the climbers reach the highest altitude?
(ii) What was the average number of red corpuscles per mm^3 present in a climber's blood on this day?
d) (i) In general, what relationship exists between the two graphs?
(ii) Explain why such a relationship is necessary for survival. (Use the information at the start of this problem-solving item to help you.)
e) How long did the members of the expedition spend at an altitude of 4000 m or above?
f) Some people live at high altitudes permanently. State one way in which their blood would differ from that of people living at sea level.

Extra Question

g) Which one of the graphs lags slightly behind the other? Suggest why.

5 The following apparatus can be used to compare the amount of carbon dioxide in fresh air and breathed air.

What TWO changes should be made to the apparatus in order to make the experiment a fair test.

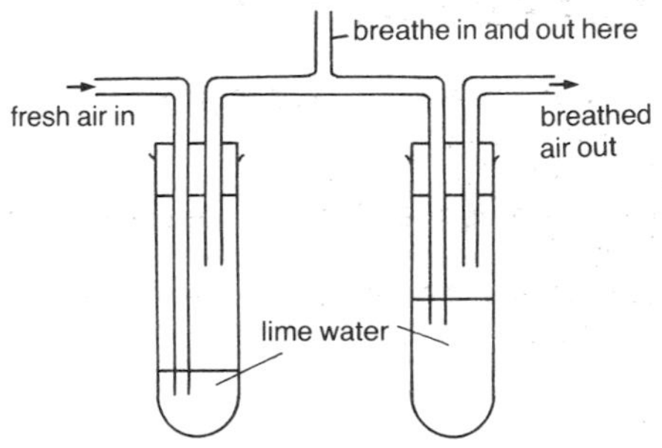

6 Two boys read in a book that it is better to use both eyes rather than just one to judge distance. To put this statement to the test they did the following experiment.

The first boy presented his fist as shown in the diagram. The second boy stood two paces away, covered one eye and made ten attempts to tap the tip of his partner's finger from above with the blunt end of a pencil. He then uncovered his eye, moved forward one pace and with both eyes open, made five further attempts to tap his partner's finger tip. He found that he was more successful with both eyes open.
a) Identify TWO sources of error in this experiment which make the test unfair.
b) State how each of these shortcomings could be corrected.

Extra Question

7 The diagram below shows a reflex action which happens when the foot comes into contact with broken glass on the beach.
Match numbers 1–6 with the following statements:
a) leg muscles contract, removing foot from danger
b) impulse enters spinal cord
c) impulse travels down motor neurone
d) pain receptors in sole of foot stimulated
e) impulse leaves spinal cord
f) nerve impulse travels up sensory neurone

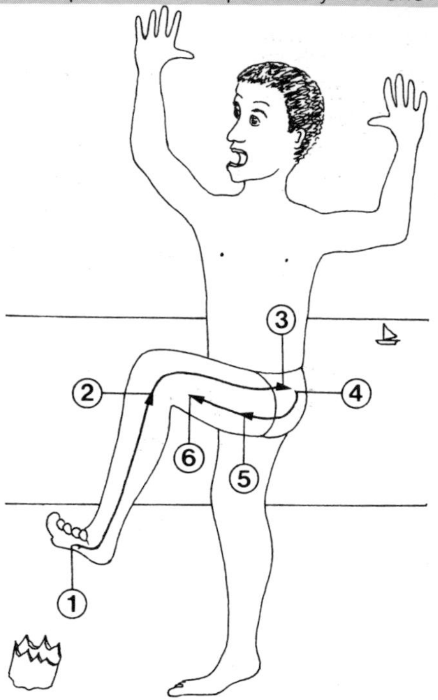

8 Look at the diagram of the ear on page 137 and find the part labelled 'Eustachian tube'. This structure is about 40 mm long and leads from the ear to a region of the throat (behind the nose) where it opens during yawning and swallowing. This allows air to pass into or out of the air-filled middle ear chamber. By this means, equal air pressure is maintained on either side of the ear drum, preventing it from bursting and allowing it to vibrate properly.
a) (i) Under what circumstances does the 'throat' end of the Eustachian tube open?
(ii) Why is this essential for the proper working of the ear?
b) What could happen to the eardrum if air pressure outside underwent a huge change but the Eustachian tube failed to open?
c) People often feel discomfort in their ears when travelling in a plane which is climbing to its cruising altitude. This is because air pressing on the outside of the eardrum is at a different pressure to that pressing on the inside. Why does sucking a sweet help to relieve the discomfort?
d) Suggest why it is possible to suffer temporary deafness during a very heavy cold when much catarrh is present in the nasal passages and throat.

9 The maximum oxygen intake (MOI) which an individual can manage is considered to be a good indication of fitness. Draw TWO conclusions from the information given in the following table, which refers to British men.

age (years)	MOI of athlete (l/min)	MOI of non athlete (l/min)
20–29	4.8	3.2
30–39	4.6	2.9
40–49	4.0	2.7

10 The accompanying table refers to the lactic acid concentration of the blood of an athlete measured over a period of time.

time (in min)	lactic acid concentration of blood (mg/100 cm^3)
0	10
5	10
10	10
15	40
20	70
25	100
30	80
35	60
40	50
45	40
50	37
55	33
60	30

a) Plot a line graph of the information given in the table. Put time on the horizontal axis and extend it to cover a range of 100 minutes.
b) During which of the following times was the athlete exercising vigorously?
(i) 0–10 min, (ii) 10–20 min, (iii) 25–35 min, (iv) 35–45 min.
c) What was the athlete's 'resting level' of lactic acid? (This means the concentration of lactic acid before exercise began.)
d) What was the highest concentration of lactic acid reached?
e) Suggest a reason for the change in lactic acid concentration after minute 25.
f) If the trend shown in your graph between minutes 45 and 60 continues at the same rate, how many more minutes (beyond minute 60) will it take for the blood lactic acid concentration to reach the original level?

Section 6 Inheritance

20 Variation

Species

A **species** is a group of living things which are so similar to one another that they are able to interbreed and produce **fertile offspring**.

The two horses shown in figure 20.1 (and all other horses of every type) belong to the same species (group of interbreeding organisms). The offspring are fertile. This means that they in turn will be able to produce offspring on reaching sexual maturity.

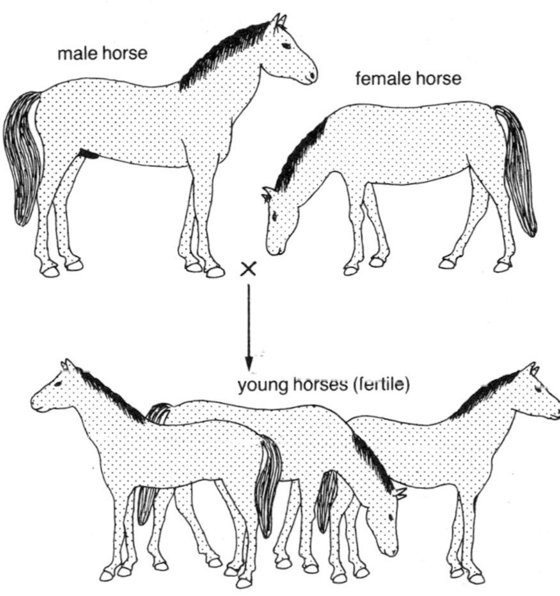

Figure 20.1 Members of one species

Sterile offspring

The members of two very different species (e.g. horses and cattle) cannot interbreed to form offspring. However, sometimes the members of two fairly similar species can interbreed. As the offspring formed in each of the examples shown in figure 20.2 are **sterile** (unable to produce offspring on reaching sexual maturity), this shows that horses, donkeys and zebras belong to three closely related but separate species.

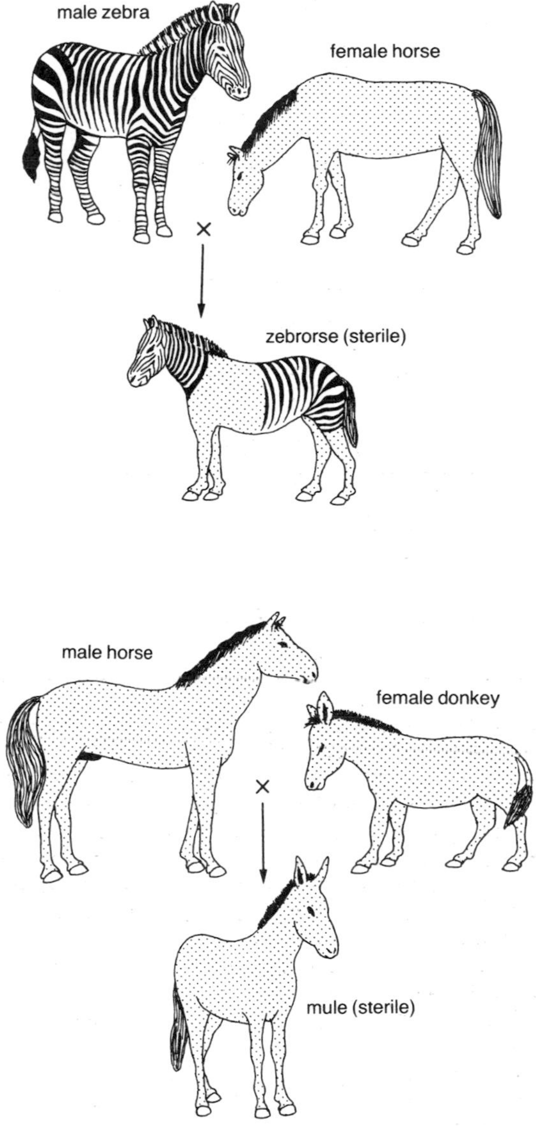

Figure 20.2 Sterile offspring

Variation

Although all the members of a species are very similar to one another (for example, a population of domestic cats all have cat-like features), they are not identical to one another. This is because **variation** occurs within a species. Some cats are bigger, some smaller, some fatter, some thinner and so on as shown in figure 20.3.

Similarly, when a population of limpets (figure 20.4) is examined, the height and diameter of their shells are found to vary. Variation also occurs within a species of plant. Dandelion leaves (figure 20.5) vary in size and shape; daisy ray florets (figure 20.6) vary in number.

Figure 20.3 Size and shape of cats

Types of variation within a species

Discontinuous variation

A characteristic shows **discontinuous variation** if it can be used to divide up the members of a species into two or more **distinct** groups. For example, humans can be divided into two separate groups according to ear lobe type (figure 20.7) and four distinct groups based on fingerprint type (figure 20.8).

The bar graph in figure 20.9 shows an analysis of one hundred forefinger prints with each separate bar representing a distinct group.

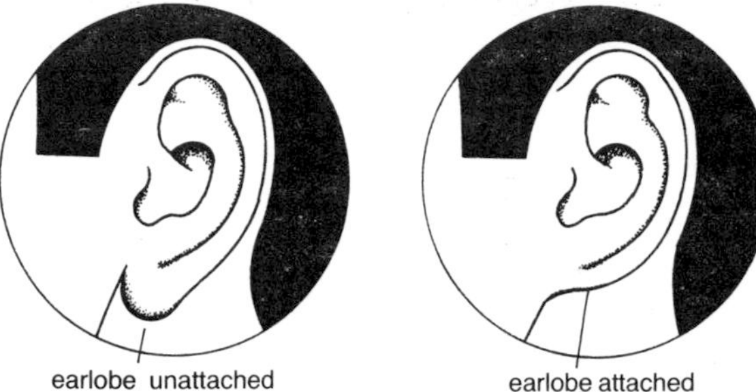

Figure 20.7 Discontinuous variation in ear types

Figure 20.4 Height and diameter of limpets

Figure 20.5 Size and shape of dandelion leaves

Figure 20.6 Number of ray florets in daisy

Figure 20.8 Discontinuous variation in fingerprint types

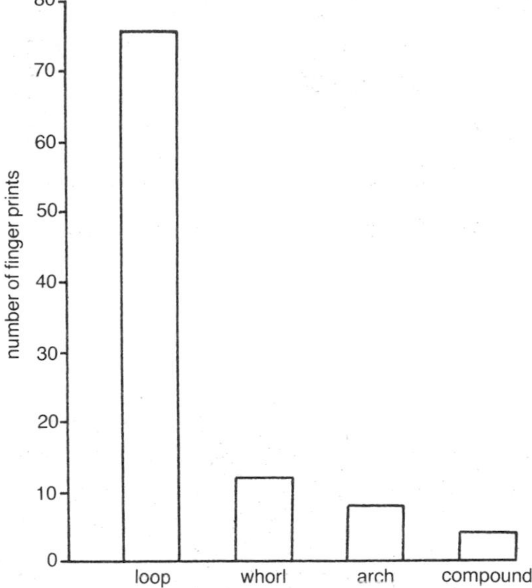

Figure 20.9 Bar graph of discontinuous variation

Continuous variation

A characteristic shows **continuous variation** when it varies (amongst the members of a species) in a smooth continuous way from one extreme to the other (and does not fall naturally into distinct groups). For example, human body height varies continuously from very small to very tall as shown in figure 20.10.

Seed mass is a further example of continuous variation. When sixty castor oil seeds are arranged in order of increasing mass as shown in table 20.1, they do not fall into distinct groups. For convenience, the entire range of the characteristic is divided into small groups (subsets). This allows a **histogram** to be plotted as shown in figure 20.11.

The majority of seeds have a mass that is close to the centre of the range, with fewer at the extremities. When a curve is drawn, a bell-shaped **normal**

seed number	mass (mg)	seed number	mass (mg)	seed number	mass (mg)
1	597	21	751	41	811
2	622	22	755	42	813
3	648	23	758	43	819
4	661	24	762	44	823
5	672	25	765	45	828
6	683	26	768	46	830
7	699	27	769	47	835
8	711	28	770	48	837
9	718	29	770	49	841
10	723	30	774	50	843
11	725	31	777	51	846
12	727	32	782	52	848
13	735	33	785	53	850
14	739	34	787	54	862
15	741	35	791	55	871
16	741	36	793	56	885
17	743	37	799	57	897
18	745	38	800	58	923
19	746	39	801	59	947
20	748	40	806	60	952

Table 20.1 Continuous variation in seed mass

Figure 20.10 Continuous variation in human height

distribution is found to be the result. In this example, the range in mass extends from 550 mg to 999 mg and the most common seed mass is the subset 750 to 799 mg.

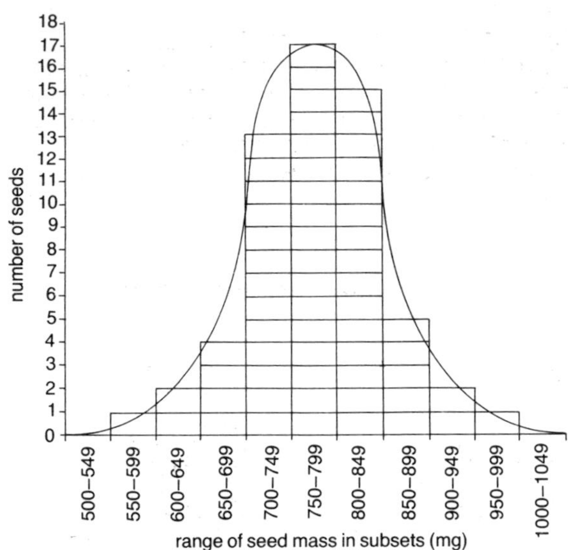

Figure 20.11 Histogram of continuous variation

KEY QUESTIONS

1 What is a species?
2 What is meant by the term **fertile** offspring?
3 State what is meant by continuous and discontinuous variation. Give an example of each to illustrate your answer.

4 Which type of variation is shown by all the examples given in figures 20.3–6?
5 Which of the following is an example of discontinuous variation amongst humans? waist circumference, male or female sex, length of handspan, breadth of foot
6 Under the headings continuous and discontinuous variation, list four ways in which identical twins reared in different environments (**a**) would definitely be exactly alike and (**b**) could differ from one another.

More to do

Variation continued

Further examples of variation are given in table 20.2.

continuous	discontinuous
human body mass	human fingerprint type
human handspan	attached or unattached ear lobes in humans
human heartrate	human blood group type
human neck size	ability or inability to roll human tongue
body length in salmon	human eye colour
number of leaves on lime tree	smooth or wrinkled seed coat in pea plant
number of ray florets in daisy	white or red eye in fruit fly
bean seed mass	green or variegated leaf in spider plant
number of fruits on raspberry bush	white or coloured flower in pea plant

Table 20.2 Examples of continuous and discontinuous variation

21 What is inheritance?

Inherited characteristics

Hair colour, tongue rolling and many other characteristics are determined by **genetic information** which is passed on from one generation to the next. Each person receives two pieces of genetic information about each characteristic, one from each parent. These may both be the same or different.

Family trees

Figure 21.1 shows a **family tree** for the inherited characteristic of hair colour. The genetic information for red hair remains masked if it is paired with that for non red hair. Granddaughter Jan in the family tree only has red hair because both pieces of genetic information that she has inherited are for red hair.

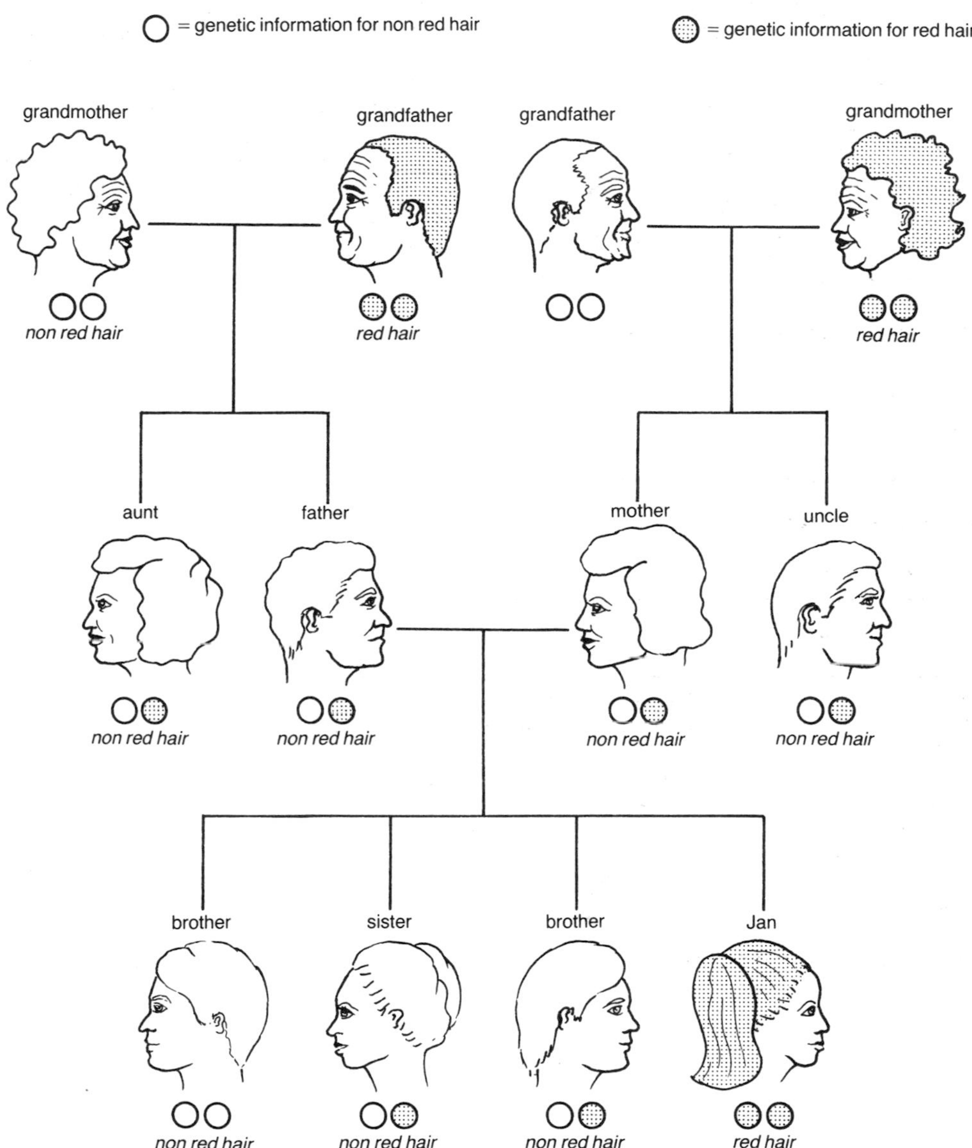

Figure 21.1 Family tree for hair colour

Figure 21.2 shows another way of presenting a family tree. This time the inherited characteristic is ability or inability to roll the tongue. Granddaughter Jan is a non tongue roller because both pieces of information that she has inherited are the inability to roll the tongue.

Her sister is able to roll her tongue because she has inherited the necessary piece of information from her father which masks the piece of non tongue rolling information received from her mother.

In plants, characteristics such as leaf shape and flower colour are similarly determined by genetic information received by offspring from their parents.

Phenotype

Phenotype is an organism's appearance which has resulted from certain genetic information inherited from the parents. For example, a person's phenotype for hair colour is red or non red. Further examples of phenotypes are given in table 21.1.

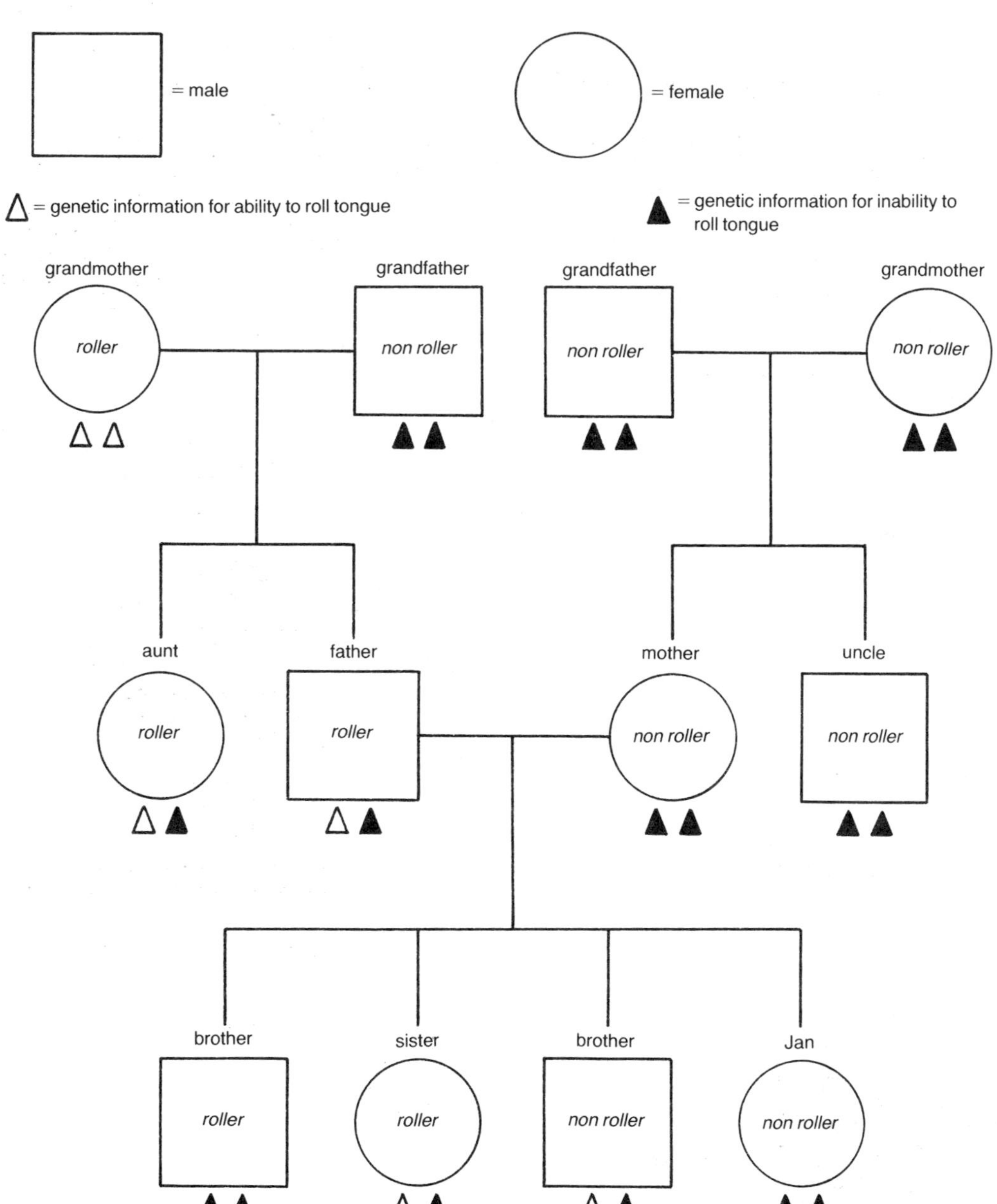

Figure 21.2 Family tree for tongue rolling ability

organism	inherited characteristic	possible phenotypes
man	eye colour tongue rolling ability	blue, brown roller, non roller
fruit fly	wing length eye colour	long, short red, white
pea plant	height seed shape seed colour	tall, dwarf round, wrinkled yellow, green

Table 21.1 Phenotypes

KEY QUESTIONS

1 Rewrite the following sentences and complete the blanks.
Living things possess certain characteristics that are determined by ________ information received from their ________ . These living things pass this genetic ________ on to the next ________ .

2 Copy and complete this table, using all the following words and phrases: attached, ebony, fruit fly body colour, grey, human ear lobe type, long, rat tail length, short, unattached.

inherited characteristic	possible phenotypes

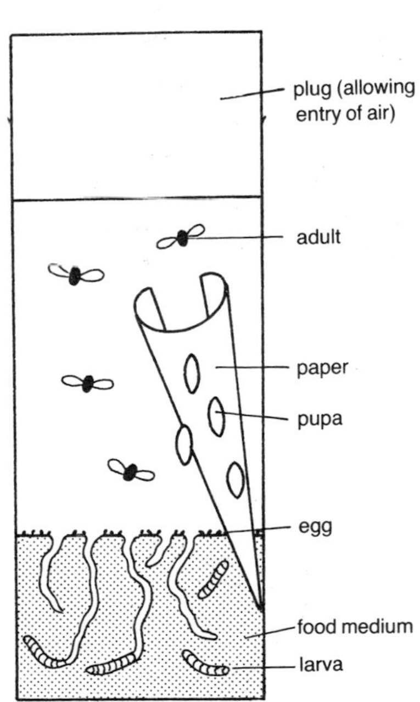

Figure 21.3 Culture tube of fruit flies

Inheritance of wing type in fruit fly (*Drosophila*)

In the cross about to be considered, normal-winged male flies are crossed with curved-winged females. Figure 21.3 shows a culture tube which provides everything that a fruit fly needs during its life cycle. Figure 21.4 shows the method used to set up a cross.

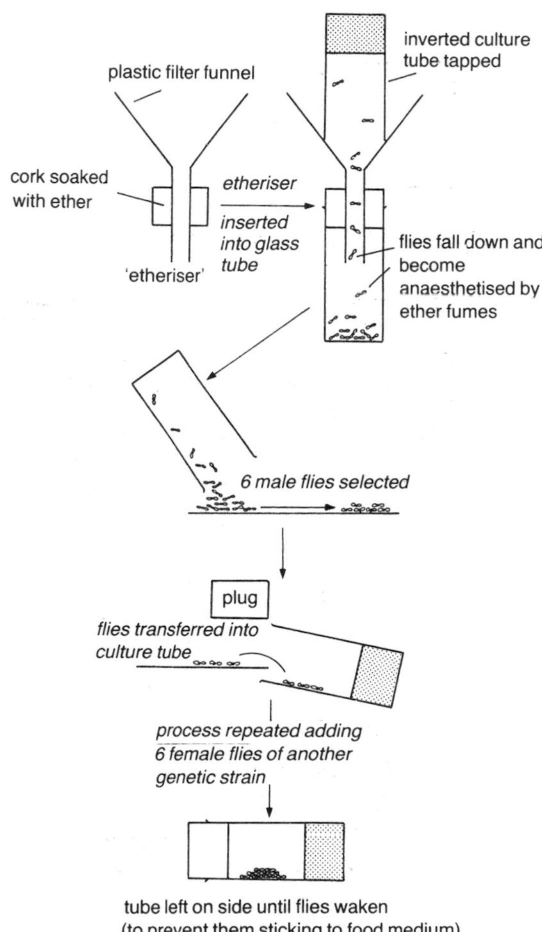

Figure 21.4 Setting up a fruit fly cross

Points to note
1. The female flies must be virgins to ensure that they have not been fertilised in advance by a fly of unknown genetic characteristics.
2. Several of each sex are used to allow for non recovery by some flies from the anaesthetic.
3. After egg-laying, the parents are removed so that they will not be confused with their offspring.

F_1 and F_2
The first generation of offspring produced is called the **first filial generation** (F_1). When they reach adulthood, a few of each sex are transferred to new culture tubes by the same method and left to produce the **second filial generation** (F_2). Figure 21.5 gives a summary of the crosses and the outcome.

The phenotypes of the F_1 generation in such a cross are always found to be uniform (in this case all straight-winged). The curved wing characteristic has disappeared in the F_1 because it has been masked by the straight wing characteristic. Curved wing is therefore said to be the **recessive** characteristic and straight wing the **dominant** one.

The F_2 generation does not contain any new 'in-between' forms of the wing type characteristic. Both straight wing and curved wing appear in the F_2 in their original form unaffected by their union in the F_1 generation. These wing type traits are therefore said to be **true-breeding** characteristics.

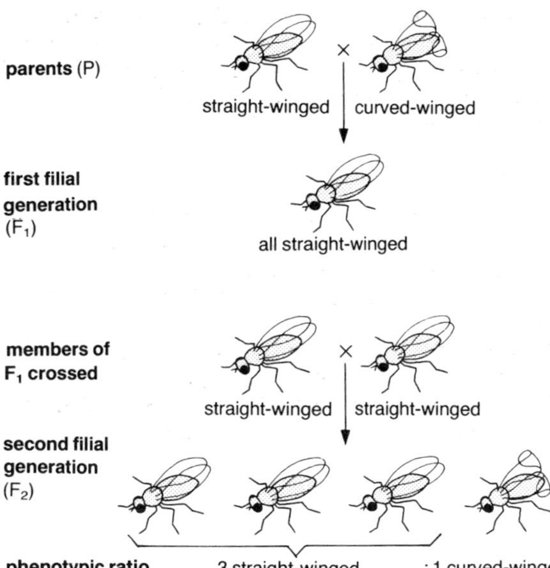

Figure 21.5 Crossing fruit flies

More to do

True-breeding organisms

An organism is said to be true-breeding if on being crossed with a member of the same strain it always produces more organisms of exactly the same kind (e.g. straight-winged crossed with straight-winged are true-breeding if they only produce more straight-winged flies).

An experimental cross such as the one shown in figure 21.5 is described as **monohybrid** because it involves only one difference between the parents. In a cross of this type, each of the original parents is true-breeding for the form of the inherited characteristic that it possesses. Here one parent is true-breeding for the phenotype straight wing and the other true-breeding for curved wing. Each parent shows a different phenotype of the same characteristic (wing type).

All monohybrid crosses of this type produce a **3:1 phenotypic ratio** in the F_2 generation.

KEY QUESTIONS

The following diagram shows the outcome of an early genetics experiment carried out by a famous scientist called Mendel using pea plants.

parents [X]	tall × dwarf
	↓
first filial generation [Y]	all tall
members of F_1 crossed	tall × tall
	↓
second filial generation [Z]	tall and dwarf plants

1 Identify TWO examples of true-breeding characteristics from the crosses.
2 Which of these characteristics is dominant and which is recessive?
3 Which symbols would normally be used at positions X, Y and Z to identify the generations involved?
4 What can always be said about the phenotypes of the F_1 generation resulting from a cross between two true-breeding parents?

Extra Questions

5 Are the parents in the above monohybrid cross true-breeding or non true-breeding?
6 Name TWO different phenotypes of the same inherited characteristic (height) by which the parents differ.
7 Predict the proportions in which the two phenotypes would occur in the F_2 generation.

Chromosomes

Chromosomes are thread-like structures found inside the nucleus of a living cell. When a cell which is undergoing division is stained and squashed, the chromosomes become visible as shown in figure 21.6.

Each normal body cell has two **matching sets** of chromosomes. Man's 46 chromosomes consist of two matching sets of 23 and can be arranged in order of size as 23 pairs as shown in figure 21.7.

The only difference between the two sexes is that a human female has two X chromosomes whereas a male has one X and one Y chromosome (see page 162 for further details).

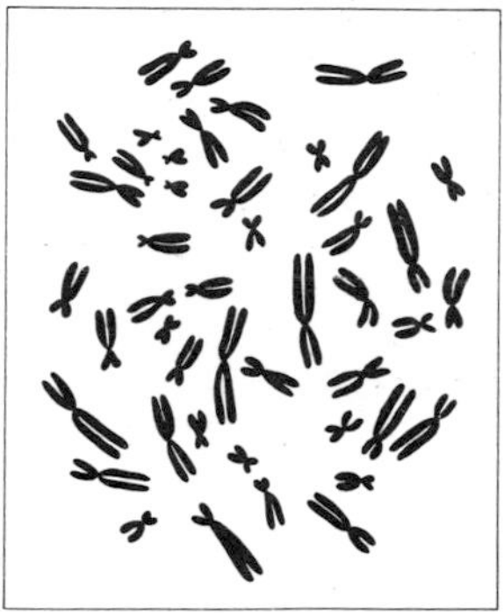

chromosomes in normal body cell of a human female

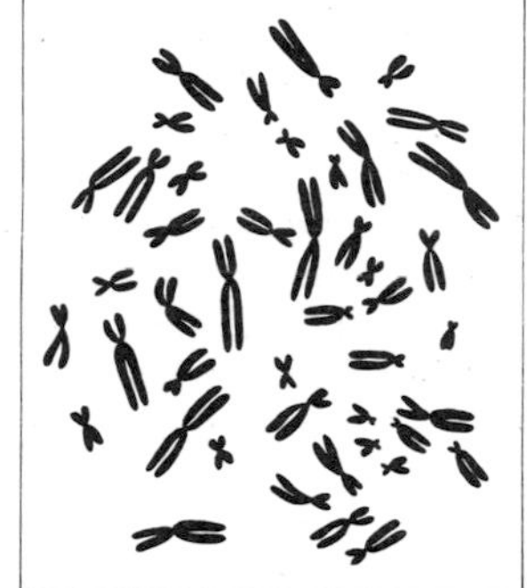

chromosomes in normal body cell of a human male

Figure 21.6 Chromosomes

Figure 21.7 Matching sets of chromosomes

Sex cell production

Sex cells are called **gametes**. The cells that produce gametes are called gamete mother cells. Like other normal body cells, each gamete mother cell contains a double set of chromosomes.

During gamete formation, a gamete mother cell divides, producing four sex cells each carrying a **single set** of chromosomes. Prior to gamete formation, the chromosomes in a gamete mother cell duplicate, producing enough genetic material to make four sex cells. For the sake of simplicity only two sex cells are shown in figure 21.8.

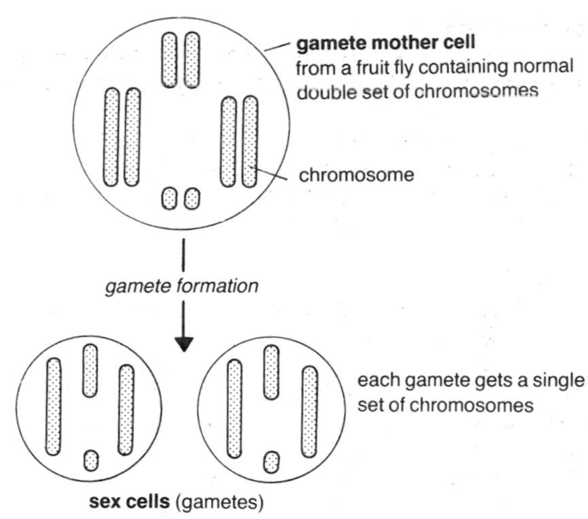

Figure 21.8 Gamete formation

Male animals make gametes called **sperm**. Female animals make sex cells called **eggs** (ova). Flowering plants produce gametes called **pollen** grains (male) and **ovules** (female).

Fertilisation

Fertilisation (figure 21.9) occurs when two different gametes (made by members of one generation) fuse to form a **zygote** (which becomes a member of the next generation). Thus gametes act as links between generations.

KEY QUESTIONS

1 What are chromosomes and where are they found?
2 How many matching sets of chromosomes are present in a normal human body cell?
3 By what other word are all types of sex cells also known?
4 Compare the number of sets of chromosomes present in a cell before and after gamete formation.
5 Name and briefly describe the process by which a zygote receives a double set of chromosomes.

Genes and genotype

Chromosomes are made of smaller units called **genes** (see figure 21.10) which control inherited characteristics. The set of genes that an organism possesses is called its **genotype**.

At least two forms of a particular gene normally exist amongst the members of a species. For example, the gene controlling wing type in fruit fly may be the form

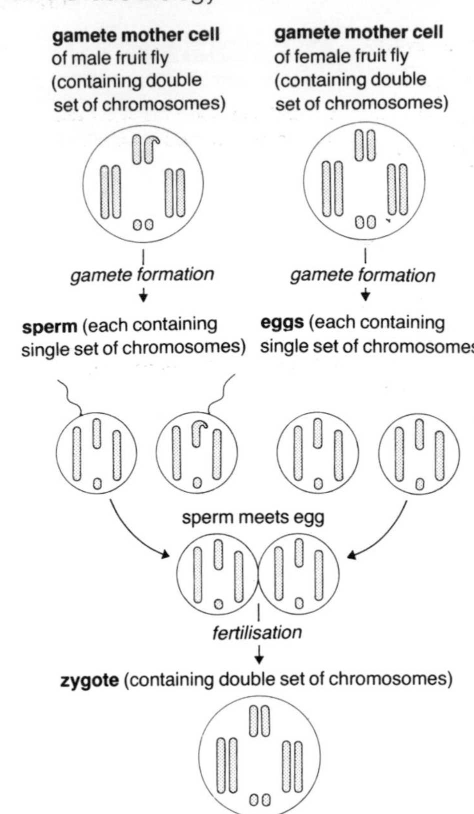

Figure 21.9 Fertilisation

that produces straight wing or the form that gives curved wing.

Every normal body cell in an organism carries two matching sets of chromosomes, one from each parent. Thus every body cell has two forms of each gene, one from each parent. These two gene forms may be the same or different depending on the organisms' parents as shown in the cross illustrated in figure 21.11.

Every gamete has one set of chromosomes and therefore carries only one of the two forms of a gene.

Symbols

For convenience the dominant and recessive forms of a gene are often represented by symbols. For example, in fruit fly, the two forms of the gene for wing type can be represented by **S** (straight) and **s** (curved).

Since every body cell has two forms of each gene, one from each parent, an organism's genotype is always represented by two letters per gene. A gamete only carries one form of each gene and is always represented by one letter per gene. The gene forms referred to in figure 21.11 can be represented using

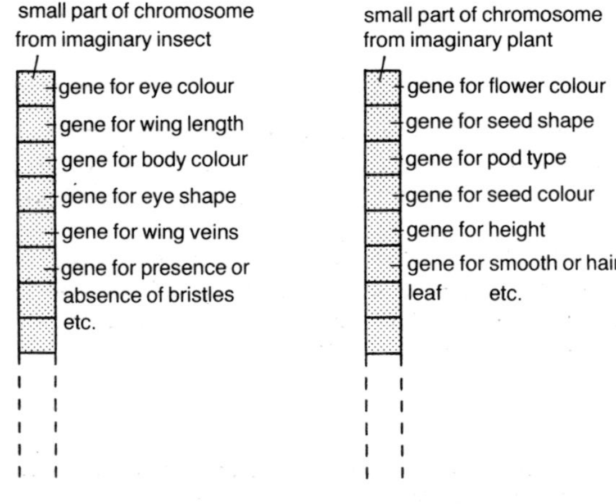

Figure 21.10 Genes

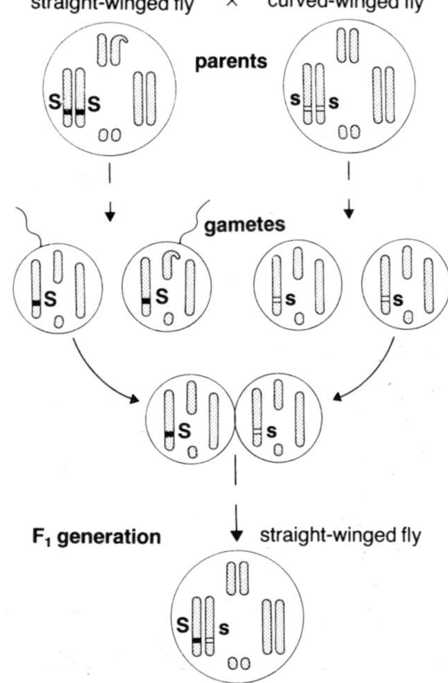

Figure 21.12 Symbols

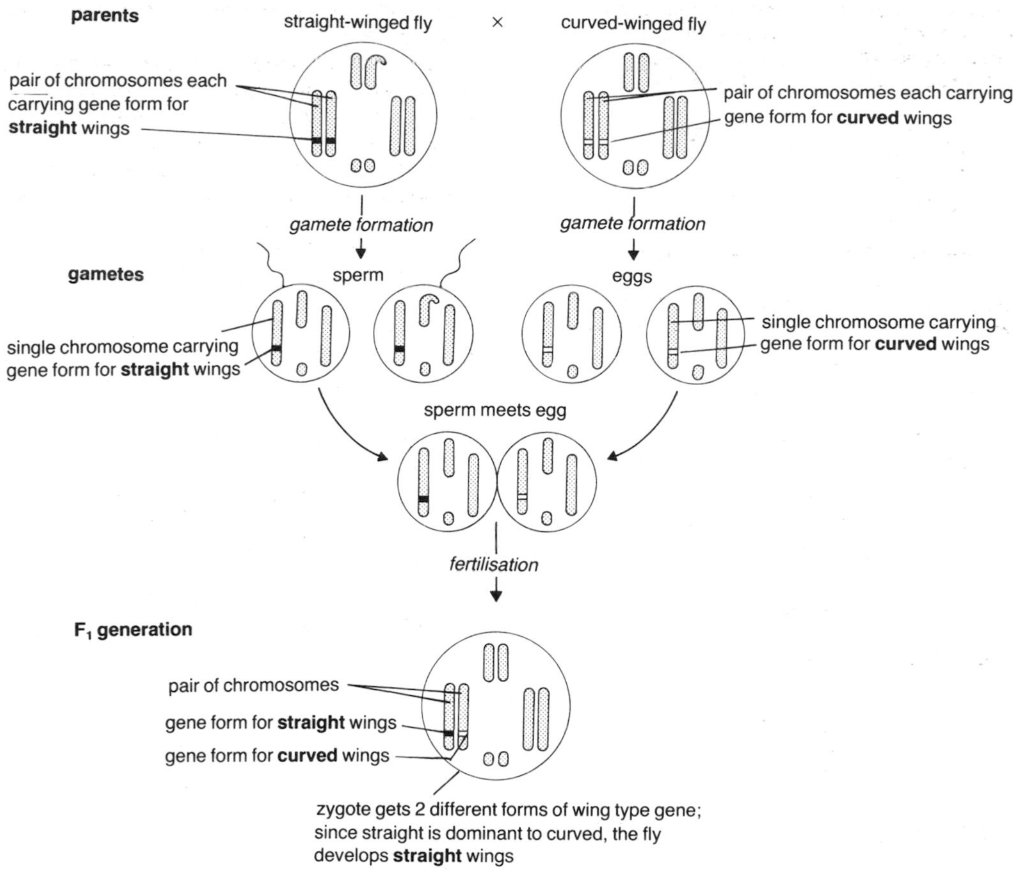

parents
straight-winged fly
×
curved-winged fly
pair of chromosomes each carrying gene form for **straight** wings
pair of chromosomes each carrying gene form for **curved** wings
gamete formation
gamete formation
gametes
sperm
eggs
single chromosome carrying gene form for **straight** wings
single chromosome carrying gene form for **curved** wings
sperm meets egg
fertilisation
F_1 generation
pair of chromosomes
gene form for **straight** wings
gene form for **curved** wings
zygote gets 2 different forms of wing type gene; since straight is dominant to curved, the fly develops **straight** wings

Figure 21.11 Fruit fly cross

symbols as shown in figure 21.12. This is often further simplified as follows:

parents (P)	SS (straight-winged)	×	ss (curved-winged)
gametes	all S	↓	all s
first filial generation (F_1)		all Ss (straight-winged)	

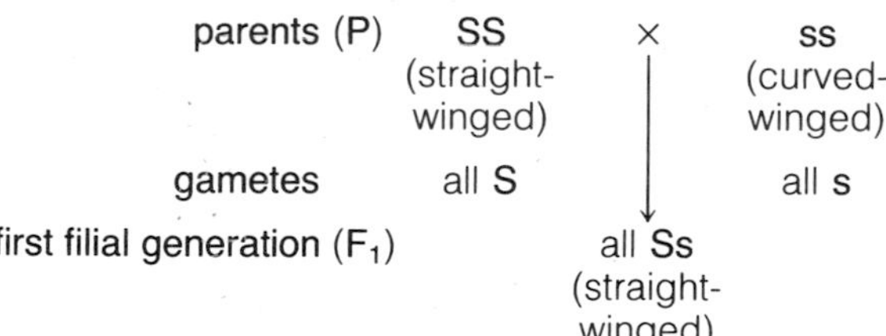

More to do

Homozygous and heterozygous

When an organism possesses two identical forms of a gene (e.g. S and S or s and s), its genotype is said to be **homozygous** and it is true-breeding.

When an organism has two different forms of a gene (e.g. S and s), its genotype is said to be **heterozygous** and it is not true-breeding.

In the cross shown above, each organism's phenotype (appearance) is given in brackets. Note that the two different genotypes SS and Ss have the same phenotype (i.e. straight wing).

Alleles

The different forms of a gene (e.g. straight and curved wing) are called **alleles**. An organism with a homozygous genotype (e.g. SS) makes gametes which all receive the same allele (S). An organism with a heterozygous genotype (e.g. Ss) makes two types of gametes. Half get one allele (S) and the other half get the other allele (s).

The monohybrid cross shown in figure 21.5 on page 156 can be represented using symbols as follows:

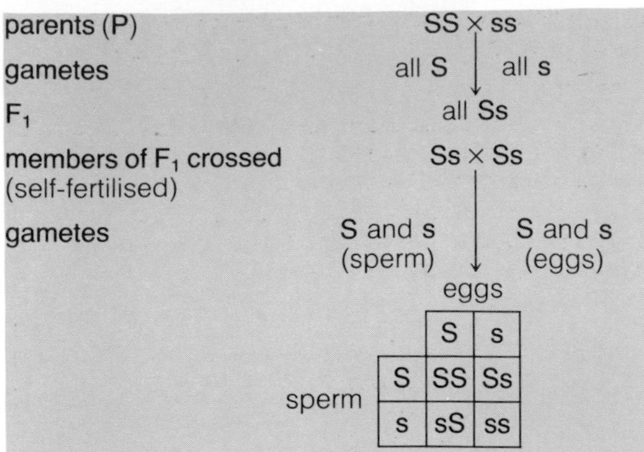

F_2 genotypic ratio = 1SS : 2Ss : 1ss

F_2 phenotypic ratio = 3 straight : 1 curved

Observed versus predicted figures

Monohybrid crosses of the type shown above always produce a 3:1 phenotypic ratio in the F_2 generation. However, there is often a difference between the observed and the predicted figures. Table 21.2 shows a set of actual (observed) results. The F_2 offspring do not show the predicted ratio of 3:1 exactly, although the ratio is very close to it.

	straight-winged	curved-winged
parents	6	6
F_1	198	
Members of F_1 crossed	6 males and 6 females	
F_2	147	53

Table 21.2 Observed results

An exact 3:1 phenotypic ratio would have occurred in the F_2 if during the **Ss** × **Ss** cross exactly half of the **S** eggs had been fertilised by **S** sperm and the other half of the **S** eggs by **s** sperm while at the same time exactly half of the **s** eggs had been fertilised by **S** sperm and the other half of the **s** eggs by **s** sperm. However, this rarely happens in nature because fertilisaton is a **random** process involving an element of **chance**.

Coin tossing

This principle can be illustrated by tossing two coins many times and keeping the score as shown in table 21.3. The same element of chance that affects fertilisation operates and an exact 3:1 ratio rarely occurs. (The greater the number of times the coins are tossed, the nearer the totals come to showing an exact 3:1 ratio.)

heads on both coins (equivalent to SS)	heads on one, tails on other (equivalent to Ss)	tails on both coins (equivalent to ss)
卌 卌 卌 卌 卌 III etc.	卌 卌 卌 卌 卌 卌 卌 卌 卌 卌 卌 IIII etc.	卌 卌 卌 卌 卌 卌 etc.
3		1 ratio (though rarely exact)

Table 21.3 Coin-tossing results

Figure 21.13 Backcrossing

B = dominant allele for brown eye
b = recessive allele for blue eye

■ = brown-eyed male ● = brown-eyed female
□ = blue-eyed male ○ = blue-eyed female

family tree	explanation
BB ■ — ○ bb; offspring: ■ Bb	Each offspring must get **B** from one parent and **b** from the other. All offpsring are **Bb** (brown-eyed) and there is no ratio.
Bb ■ — ● Bb; offspring: ■ BB, ● Bb, ■ bB, ○ bb	Each offspring gets either **B** or **b** from one parent and either **B** or **b** from the other as shown in the 'punnett' square: eggs: B, b; sperm: B, b B × B = BB; B × b = Bb; b × B = bB; b × b = bb Thus **BB**:**Bb**:**bb** are produced in a 1:2:1 ratio. Such a ratio normally results only if a large number of offpsring is produced. Regardless of number of offspring, each individual stands a 1 in 4 chance of being **BB** (brown-eyed), a 1 in 2 chance of being **Bb** (brown-eyed) and a 1 in 4 chance of being **bb** (blue-eyed).
Bb ■ — ○ bb; offspring: Bb ●, □ bb	Each offspring must get **b** from one parent and either **B** or **b** from the other. Thus **Bb**:**bb** are produced in a 1:1 ratio if a large number of offpsring are produced. Each individual stands a 1 in 2 chance of being **Bb** (brown-eyed) and a 1 in 2 chance of being **bb** (blue-eyed).

Table 21.4 Ratios of offspring

Human family trees – ratios of offspring

Table 21.4 shows three family trees involving the gene for eye colour and the ratios of offspring that result in each case.

Backcross

When an organism exhibits a dominant trait (characteristic), it is not obvious whether its genotype is homozygous or heterozygous for that trait. The identity of an unknown genotype can be found by **backcrossing** it with a homozygous recessive organism as shown in figure 21.13.

KEY QUESTIONS

1 a) What name is given to the basic unit of inheritance which controls a characteristic that is passed on from generation to generation?
b) Of what structures found in a cell's nucleus do such units form a part?
2 How many forms of a gene does a zygote receive from each parent?
3 Copy and complete the following paragraph.
One gene found in fruit flies controls wing type.
a) When a zygote receives the straight wing form of this gene from both parents, it develops ________ wings.
b) When a zygote receives the curved-wing form of the gene from both parents, it develops ________ wings.
c) When a zygote gets the straight-wing form of the gene from one parent and the curved form from the other, it develops ________ wings. This is because the straight-wing form of the gene is ________ and masks the presence of the curved-wing form which is said to be ________.
4 State the meaning of the term **genotype**.

Extra Questions

5 a) What is the difference between a homozygous and a heterozygous genotype?
b) Using the letters **S** and **s**, give the genotypes of the zygotes referred to in question 3.
c) Which of these have the same phenotype but different genotypes?
6 What are alleles?
7 Six true-breeding male long-winged fruit flies were crossed with six true-breeding short-winged flies. The F_1 generation all had long wings. Some F_1 males were crossed with F_1 females. The F_2

generation was found to contain long-winged and short-winged flies in the ratio 603:197.
a) Using symbols of your choice, show this cross in diagrammatic form.
b) With reference to the F_2 genotypes, explain why a phenotypic ratio of 3:1 is expected in the F_2.
c) Why do the actual observed results differ slightly from the predicted figures?

8 Return to the family trees on page 153 and rewrite them using letter symbols of your own choice, where non red hair is dominant to red hair and tongue rolling ability is dominant to non tongue rolling ability.

Sex chromosomes

Every normal human body cell contains 46 chromosomes (see page 157) as 23 pairs. Of these, one pair make up the *sex chromosomes*. In each body cell of a human female these are the two equal-sized **X** chromosomes. In each male body cell they are the **X** chromosome and the much smaller **Y** chromosome.

Gametes and sex chromosomes

Look at figure 21.14, which shows gamete formation in humans where only the sex chromosomes are shown. Each female gamete (egg) gets an X chromosome from its gamete mother cell. However, each male gamete (sperm) gets either an X or a Y chromosome from its gamete mother cell.

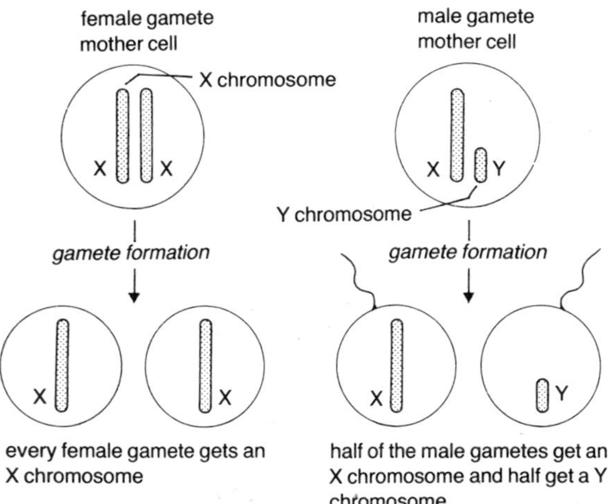

Figure 21.14 Sex chromosomes

Sex determination

The sex of a child is determined by these sex chromosomes as shown in figure 21.15. If an egg (which always contains an X chromosome) is fertilised by a sperm containing an X chromosome, a female child (XX) is formed. However, if an egg (X) is fertilised by a sperm containing a Y chromosome, a male child (XY) is the result.

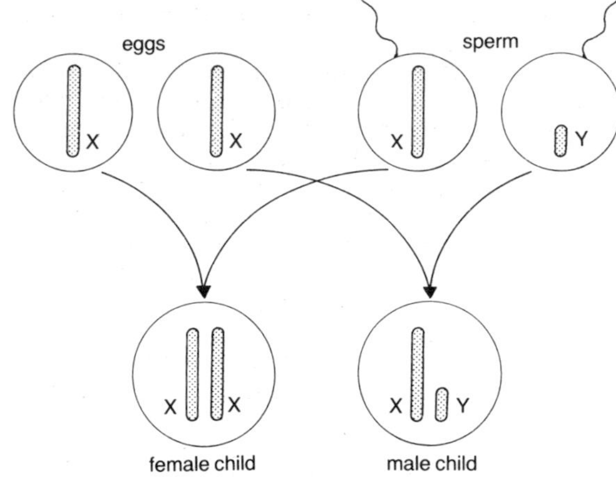

Figure 21.15 Sex determination

KEY QUESTIONS

1 How many pairs of chromosomes are present in a normal human body cell?
2 a) By what letters are the chromosomes that determine the sex of a male child represented?
b) Which of these would be found in a female gamete mother cell?
3 Which sex chromosomes could be found in (a) a sperm, (b) an egg?
4 With reference to sex chromosomes, explain how the sex of a child is determined.

22 Genetics and society

Selective breeding

Variation exists amongst the members of a species (see page 150). For centuries animal breeders have attempted to improve stocks by **selecting** for breeding purposes those males and females which have certain desirable characteristics.

Ayrshire cattle, for example, have been selectively bred over many generations for milk production, and Aberdeen Angus for meat production as shown in figure 22.1. As a result, these varieties differ greatly from their original ancestors which possessed little meat by comparison and produced much less milk.

Similarly, plant breeders have brought about revolutions in plant agriculture by selecting and breeding those varieties of plants with characteristics which are beneficial to humans. As a result of selective breeding, one original species of cabbage now provides us with four different vegetables as shown in figure 22.2.

By selective breeding man has produced many new varieties of crop plant which grow more quickly, produce more food and are more resistant to disease than their ancestors. Thus selective breeding of organisms is useful to people.

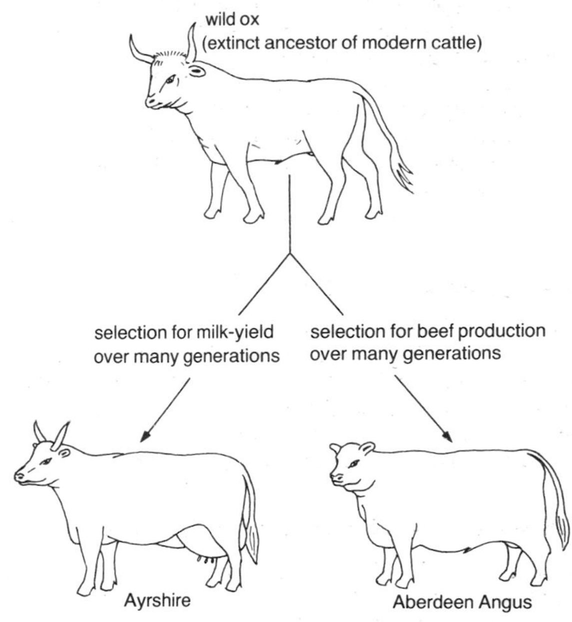

Figure 22.1 Selective breeding of cattle

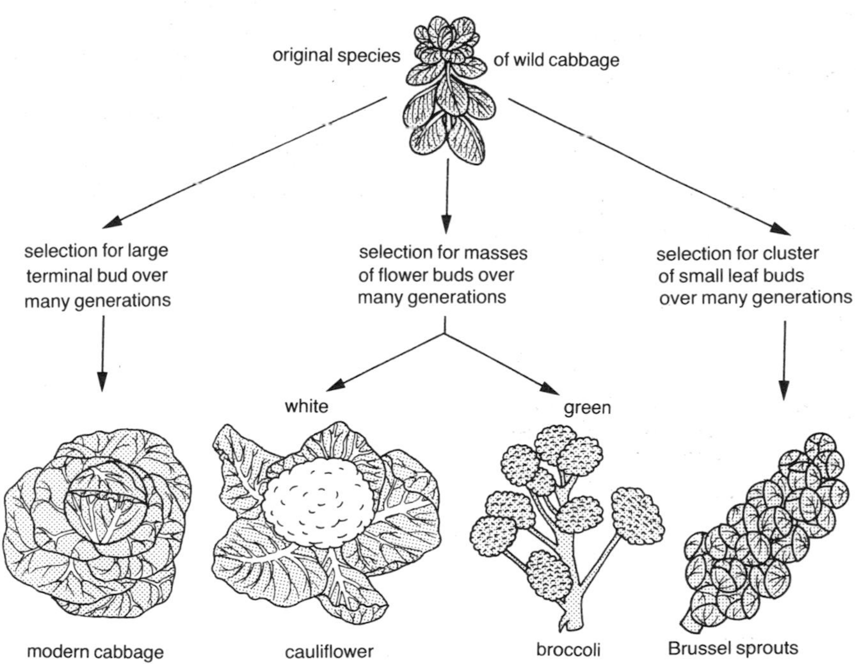

Figure 22.2 Selective breeding of cabbage plant

More to do

Enhancement of a characteristic through selective breeding

Plants

In 1895 a group of American scientists began a breeding experiment using a variety of maize (sweetcorn) whose seeds varied slightly in oil content. They selected only those plants producing seeds with the highest oil content for breeding strain X and only those with the lowest oil content for breeding strain Y. This procedure was repeated for 50 generations as shown in figure 22.3.

Figure 22.4 shows a graph of the results from the experiment. In strain X the oil producing characteristic has been enhanced by selection, whereas in strain Y the reverse has occurred.

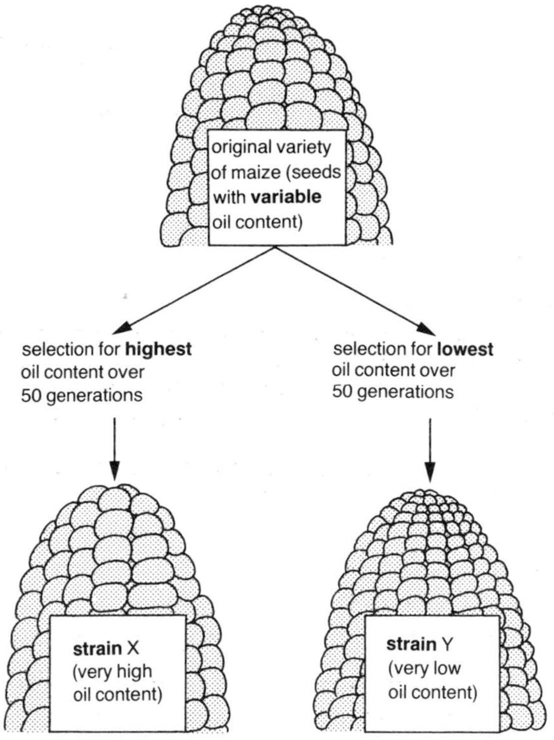

Figure 22.3 Selective breeding experiment

Animals

Certain characteristics of domesticated animals have also been enhanced through selective breeding. In addition to cattle that yield more milk or beef, the birds on a modern poultry farm are found to grow more rapidly and require only half the food per pound gain in weight compared with their ancestors of 50 years ago. Again this is the result of selective breeding.

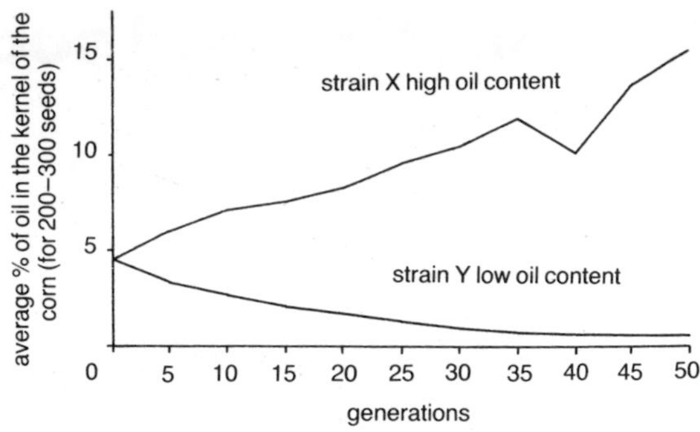

Figure 22.4 Graph of selective breeding results

KEY QUESTIONS

1 Within a population of animals or plants, which organisms are normally used for selective breeding?
2 Give TWO examples of improved characteristics that are present in modern plants as a result of selective breeding.
3 In what TWO ways do modern cattle differ from the wild ox, their extinct ancestor?

Extra Question

4 Describe an example of the enhancement of a named characteristic of (a) a plant, (b) an animal as a result of selective breeding.

Mutation

A **mutation** occurs when an organism's genes or chromosomes become altered in some way. If the mutation produces a change in phenotype, the affected organism is said to be a **mutant**.

Down's syndrome

One type of chromosome mutation that occurs in humans results in an egg receiving an extra chromosome at gamete formation. Following fertilisation by a normal sperm, the zygote contains 47 instead of 46 chromosomes. The presence of this one extra chromosome affects development and leads to a condition known as Down's syndrome. The sufferer is mentally deficient and has a Mongoloid appearance.

Amniocentesis

During all pregnancies, amniotic fluid contains cells which are genetically identical to the foetus since they have come from its skin. A sample of these cells can

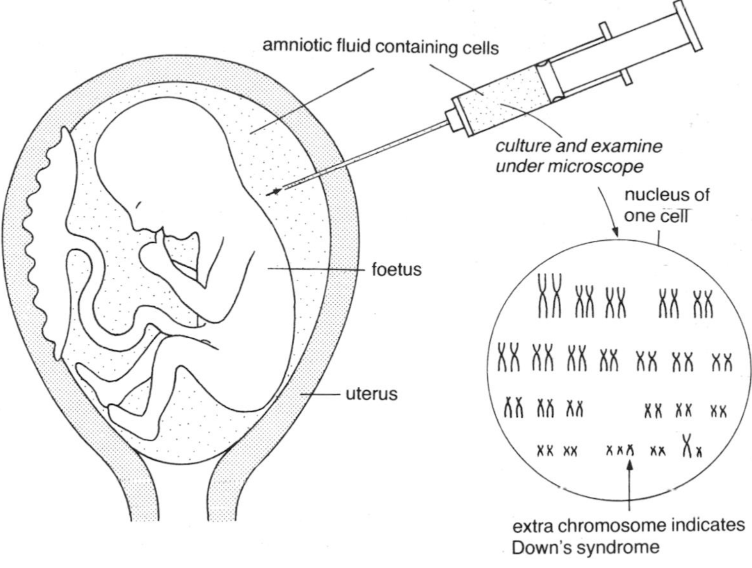

Figure 22.5 Amniocentesis

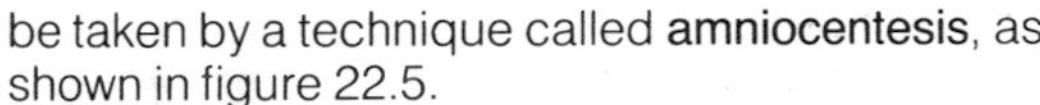

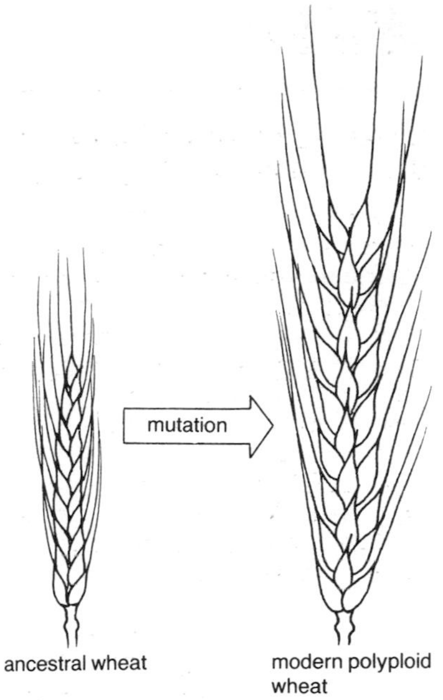

Figure 22.6 Wheat varieties

be taken by a technique called **amniocentesis**, as shown in figure 22.5.

Doctors can then study the chromosomes from these cells for certain tell-tale signs before birth. If the doctors detect an abnormality which indicates Down's syndrome or an inherited disease, then the parents may decide to have the pregnancy terminated.

Useful mutations

Most mutations lead to the formation of an inferior version of the phenotype which is often fatal. However, on very rare occasions there occurs by mutation a mutant organism which is better than the original. For example, unusual varieties of plants possessing several extra complete sets of chromosomes occur as a result of a type of chromosome mutation. These plants are called **polyploids**.

Compared to their normal relatives, polypoid tomatoes contain more vitamin C, polyploid apples are extra large and polyploid wheat (figure 22.6) produces more flour.

More to do

Mutagenic agents

Naturally occurring mutations are rare and the rate of mutation is low. However, mutations can be induced by **mutagenic agents**. These include various types of radiation (atomic, ultra-violet light and X-rays). Such factors speed up mutation rate in organisms.

KEY QUESTIONS

1 What is a mutation?
2 Describe a human condition caused by a mutation.
3 a) What is amniocentesis?
 b) What can be detected using this technique?
4 a) When are mutations in other organisms useful to man?
 b) Give TWO examples of mutant strains that are of economic importance to man.

5 Name TWO factors which can speed up mutation rate in an organism.

PROBLEM SOLVING

daisy number	number of ray florets present	daisy number	number of ray florets present	daisy number	number of ray florets present
1	27	16	44	31	50
2	31	17	44	32	51
3	33	18	44	33	51
4	34	19	45	34	52
5	36	20	45	35	53
6	37	21	45	36	53
7	38	22	45	37	54
8	38	23	46	38	55
9	39	24	46	39	55
10	39	25	47	40	57
11	40	26	47	41	58
12	41	27	48	42	58
13	43	28	49	43	59
14	43	29	49	44	61
15	44	30	50	45	62

1 The numbers of ray florets possessed by 45 daisies are listed in the above table in order of increasing number.
a) Make a copy of the axes given below and present the information in the table as a histogram. The first five daisies have been done for you.

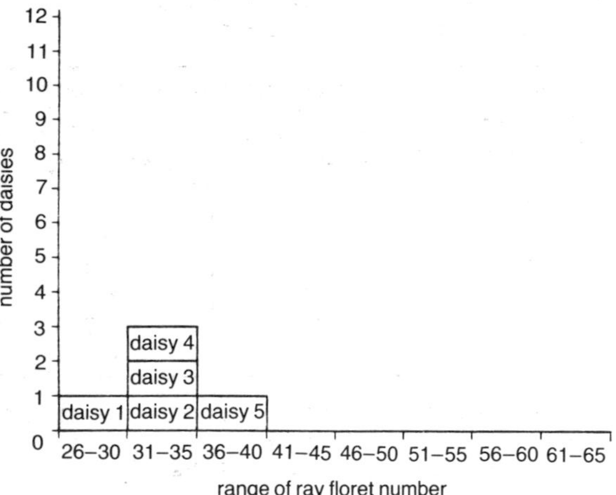

b) Is ray floret number an example of continuous or discontinuous variation?
c) How many daisies had between 51 and 55 ray florets in this sample?
d) What is the (i) most common, (ii) least common range of ray floret number from your histogram?

2 In humans there are two forms (alleles) of the gene which controls tongue rolling. Ability to roll the tongue (R) is dominant over inability to roll the tongue (r).
In the following family tree, some members can roll their tongues (rollers) and some cannot (non rollers).

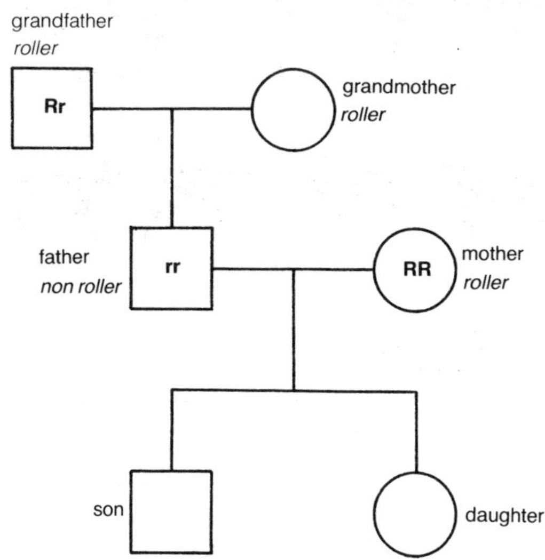

a) Copy and complete the diagram to include all of the missing genotypes and phenotypes.

Extra Questions

b) Name TWO individuals who have the same phenotype but different genotypes with respect to the tongue rolling gene.
c) Name the person who is homozygous for (i) the dominant trait, (ii) the recessive trait.
d) The son marries a woman who is a non roller. Add this information to your diagram. What chance is there of each of their children being a tongue roller?
e) The daughter marries a man with the same genotype as herself. Add this information to your diagram and work out what the chance is of each of their children being a non roller.

3 The cross shown below involves the gene for coat colour in cattle.
a) State the two phenotypes of coat colour in this cross.
b) Which form of the gene for coat colour is dominant?
c) Using symbols of your own choice, draw a keyed diagram to represent the cross.
d) When the calves reach sexual maturity, will they be true-breeding? Explain your answer.

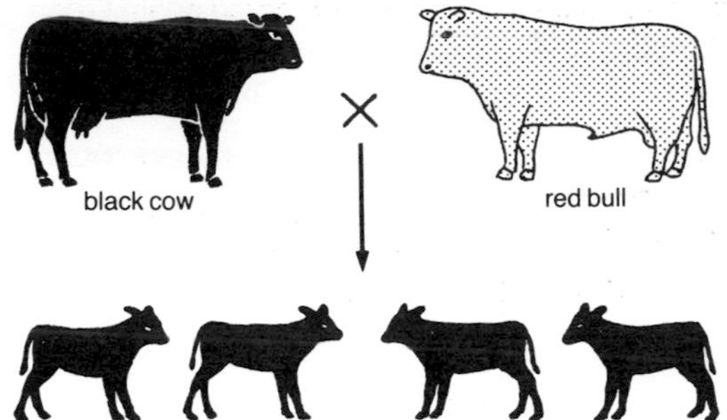

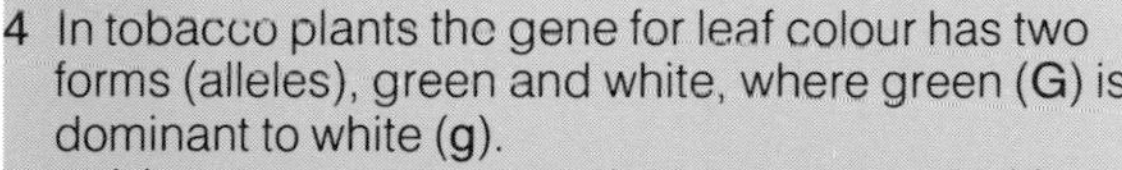

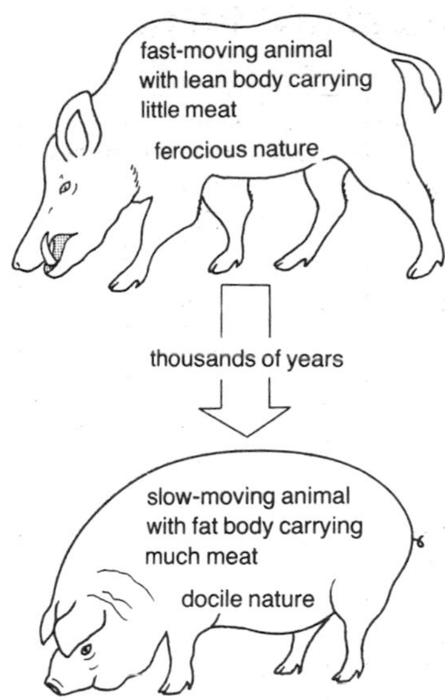

Extra Question

4 In tobacco plants the gene for leaf colour has two forms (alleles), green and white, where green (G) is dominant to white (g).

A heterozygous green plant was crossed with a white plant.

a) Give the genotypes of these two parent plants.

b) State the genotypes of the offspring that would be produced and the ratio in which they would be expected to occur.

c) What would be the genotype of a homozygous green plant?

5 a) Consider the table of results shown here and then state the overall effect that selective breeding has had on the annual milk yield of the British cow.

b) Calculate the average annual milk yield for the period of time referred to in the table.

year	annual milk yield of British cow (litres)
1957	3288
1958	3292
1959	3341
1960	3378
1961	3446

Extra Question

c) Calculate the percentage increase in milk yield for the period shown.

6 Man has brought about the changes shown in the diagram by selective breeding.

a) Describe the method he has used to do this.

b) Give two reasons why the modern pig is more useful to man than the wild boar.

Extra Question

7 The following table gives the results from an experiment to investigate the effect of artificial selection (selective breeding) on the protein content of a species of cereal grains.

number of generations of artificial selection	percentage protein present in grains	
	strain A (selected for most protein)	strain B (selected for least protein)
start	10.8	10.8
10	13.2	9.6
20	14.0	8.4
30	15.6	8.0
40	16.4	7.2
50	19.6	4.8

a) On the same sheet of graph paper, construct two line graphs of the results obtained.

b) From your graphs answer the following questions:

(i) What increase in percentage protein content occurred in strain A between 15 and 45 generations of artificial selection?

(ii) What difference in percentage protein content existed between strains A and B after 35 generations of selective breeding?

(iii) By how many times was the percentage protein content of strain A greater than that of strain B after 45 generations of artificial selection?

(iv) How many generations of selective breeding were required to produce a difference of 6.6 per cent protein content between strains A and B?

Section 7 Biotechnology

23 Living factories

Man has always made use of a wide variety of plant and animal products. Early man hunted animals for meat and skins and collected seeds and fruits for food. Eventually he discovered that by planting seeds and domesticating animals he could produce vast amounts of useful products. In recent years the process of **harvesting** a useful material made by a living organism has taken a gigantic step forward thanks to biotechnology.

In a biotechnological process, living cells are used to convert a raw material into a useful substance. Biotechnology began thousands of years ago when early man used yeast to make bread and wine. However it is only recently that man has begun to realise the enormous potential that biotechnology offers. By applying his vast scientific knowledge he is now developing exciting new biotechnological processes on a large scale. The cells of microbes (bacteria and fungi) are being employed to help solve some of mankind's problems by acting on raw materials to provide a wide range of **useful products** and **services** (see figure 23.1).

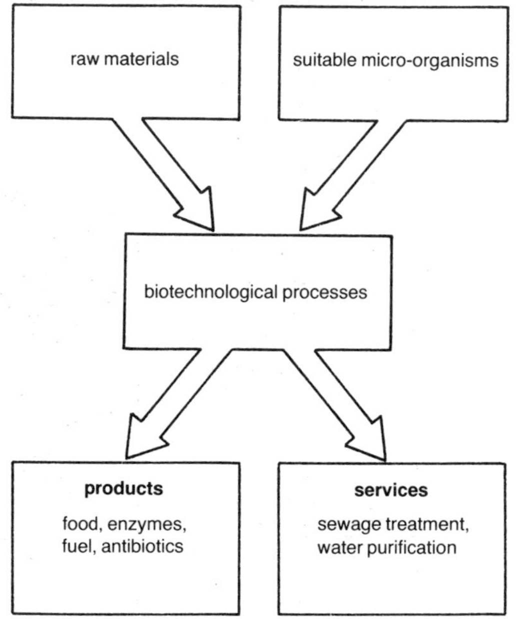

Figure 23.1 Biotechnology

Yeast

The raising of dough during bread-making and the manufacture of wine and beer depend on the activities of **yeast** as shown in figure 23.2. Yeast is a living organism. Under the microscope it is seen to be made up of single cells. It is a type of plant called a **fungus**. Unlike most other plants, fungi (see Appendix 2) do not contain green chlorophyll and cannot photosynthesise. Yeast cannot therefore make its own food.

Yeast is present as a '**bloom**' (grey dust) on the surface of many types of fruit such as grapes (figure 23.3) and plums. Yeast feeds on the fruit especially when it becomes over-ripe and bursts open.

Early man did not deliberately add yeast to grape juice during wine-making since he did not know of its existence. He simply crushed the grapes and allowed the mixture of juice, skins and 'bloom' to ferment.

Figure 23.3 Bloom on grapes

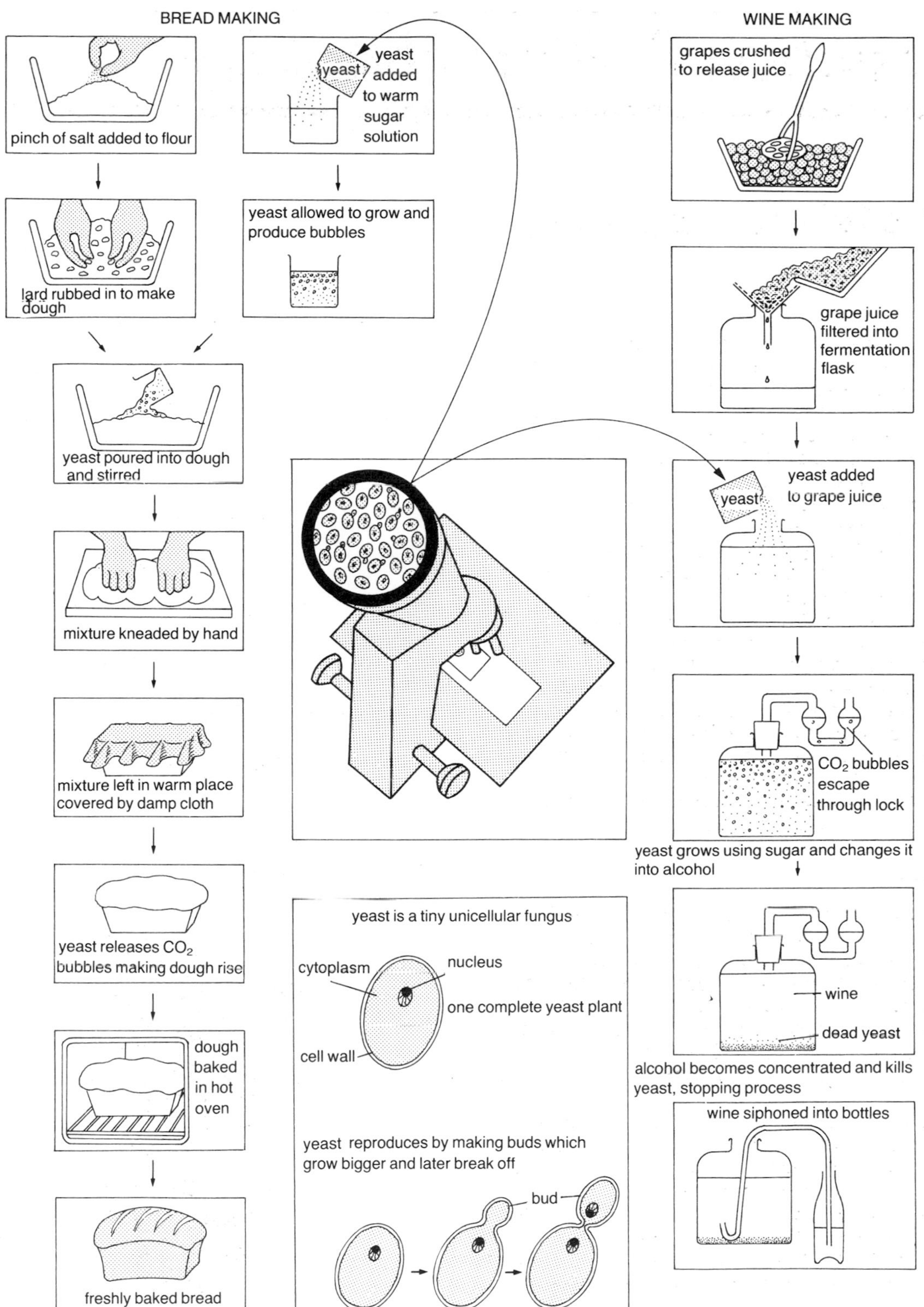

Figure 23.2 Yeast and its uses

Action of yeast cells on sugar

The experiment shown in figure 23.4. is set up to investigate the action of live yeast cells on sugar (glucose) in the absence of oxygen. The sugar solution is boiled before use to remove dissolved oxygen and to kill any other micro-organisms present. The oil layer keeps air, and therefore oxygen, out of the yeast and sugar mixture. A control is set up using dead yeast cells.

Table 23.1 shows a typical set of results. Since the control remains unchanged, it is concluded that living yeast cells (in the absence of oxygen) are able to use sugar as their source of energy. Some of this energy is lost as heat. Carbon dioxide and alcohol are released as waste products. This process is called **fermentation**. It is summarised in the following word equation.

sugar → alcohol + carbon dioxide + energy

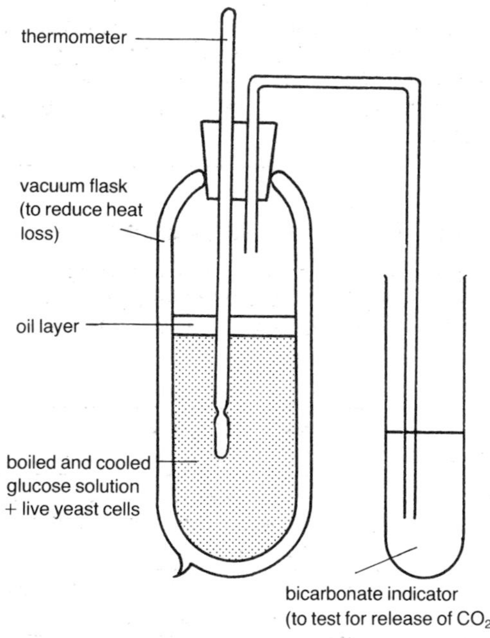

Figure 23.4 Fermentation experiment

More to do

Anaerobic respiration

Respiration is the process by which a living organism releases energy from its food. When respiration occurs in the absence of oxygen it is called **anaerobic** respiration. Fermentation of sugar by yeast cells is therefore an example of anaerobic respiration. Respiration which does involve the use of oxygen is called **aerobic** respiration (see page 60). Yeast is also able to respire aerobically. The two types of respiration are compared in table 23.2.

aerobic	anaerobic
oxygen always required	oxygen never required
efficient method of respiration, releasing much energy since sugar is completely broken down to CO_2 and water	inefficient method of respiration, releasing little energy, since much energy remains locked up in the molecules of the end product (e.g. alcohol)

Table 23.2 Comparison of aerobic and anaerobic respiration

Fermentation

The above process in yeast cells is called alcoholic fermentation to distinguish it from lactic acid fermentation (see page 174) which occurs in many bacteria and methane fermentation (see page 185) which occurs in other microbes.

Until very recently the term fermentation has been used by scientists to refer specifically to a process by which living cells, in the absence of oxygen, break down an organic substance to obtain energy. However, the term fermentation is now taking on a wider meaning. Many biotechnologists use the term to mean any process involving the activities of microbes where the product consists of microbial cells or their metabolites, regardless of whether the microbes respired aerobically or anaerobically during the process.

	results after 2 days	
	experiment	control
thermometer readings	temperature rises from 20 °C to 23 °C	temperature remains unchanged
bicarbonate indicator test for carbon dioxide (see also page 59)	colour changes from red to yellow	red colour remains unchanged
distillation at 80 °C of liquid remaining in flask	ethanol (ethyl alcohol) is collected	no ethanol is collected

Table 23.1 Fermentation results

KEY QUESTIONS

1 Briefly describe two ways in which man makes use of yeast.
2 Name the group of plants to which yeast belongs.
3 How many cells are present in a single yeast organism?
4 Which of the following does yeast use as food? alcohol, oxygen, sugar, water
5 Name TWO waste products made by yeast during fermentation.
6 Give the word equation of fermentation.

Extra Question

7 Give TWO differences between aerobic and anaerobic respiration.

Effect of germination on barley grains

Look at the experiment shown in figure 23.5. From the results it can be seen that dry barley grains contain starch but not simple sugar. However, barley grains that have been allowed to germinate for several days are found to contain simple sugar but only a little starch. It is therefore concluded that starch is broken down to simple sugar during germination. This degradation process is brought about by an enzyme. Under normal conditions of growth the sugar is used to feed the developing plant. In a brewery the sugar is used to make beer.

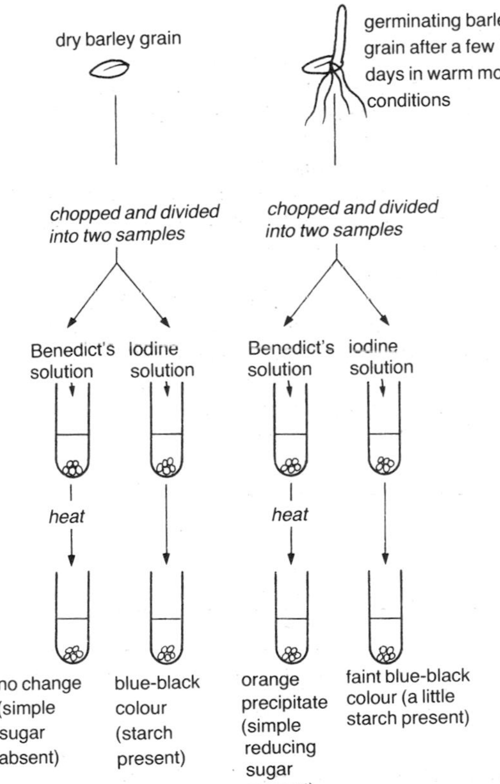

Figure 23.5 Effect of germination on barley grains

More to do

Malting

The brewing industry uses barley grains which are rich in starch for beer-making. However starch is a large molecule (see page 89) and yeast is unable to digest it. For fermentation to occur, yeast must be supplied with simple sugar.

During the early stages of beer-making, therefore, vast quantities of barley grains are spread out in moist, warm conditions on the floor of the malting house to promote germination. This process, which involves the conversion of starch to sugar inside the barley grains, is called **malting**. The malt formed is rich in simple sugar and provides the food needed by yeast during fermentation.

Commercial brewing of beer

Figure 23.6 shows the main practices involved in industrial brewing. Commercial brewers provide the best growing conditions for yeast as shown in table 23.3.

growing condition required by yeast	way in which growing condition is provided
food supply	starch converted to sugar for yeast by germinating barley grains on floor of malting house
suitable temperature	temperature of fermentation vessel controlled by thermostat
lack of competition	all other micro-organisms killed by boiling in wort kettle before yeast is added

Table 23.3 Best growing conditions for yeast

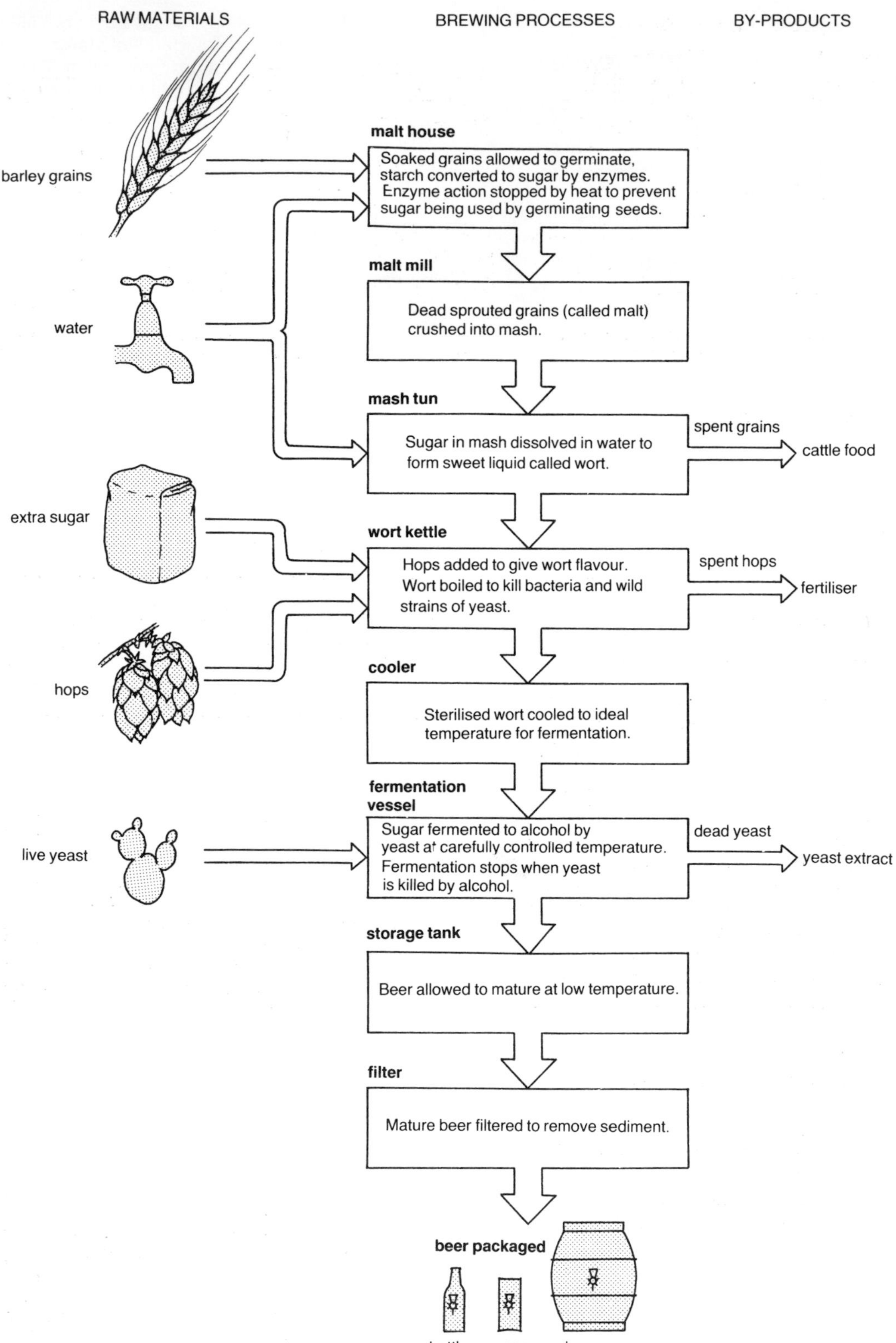

Figure 23.6 Commercial brewing of beer

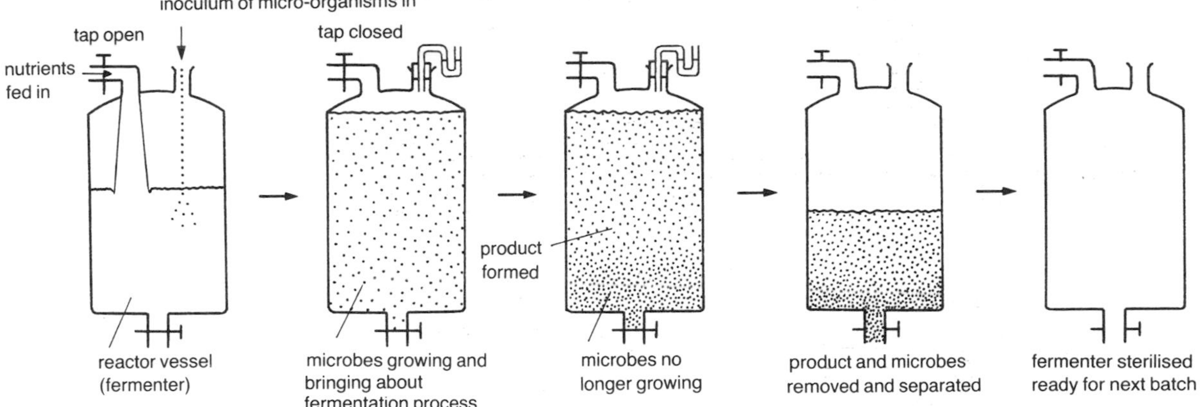

Figure 23.7 Batch processing

More to do

Batch processing

Figure 23.7 shows a simplified version of this type of fermentation operation. The reactor vessel (fermenter) is filled with a batch of nutrients and an inoculum of a certain micro-organism (e.g. yeast) is added. Optimum conditions (temperature, pH etc.) are maintained in the fermenter to promote growth of the micro-organisms which carry out the required chemical reaction.

After some time the process comes to an end when the micro-organisms stop growing. The fermenter is then emptied and the useful end product separated from the microbes (which are often discarded). The fermenter is cleaned out and sterilised and the process is repeated using a new batch of raw materials and a fresh inoculum of the microbe.

KEY QUESTIONS

1 a) Name the food found to be present in dry barley grains.
   b) What happens to this food during germination?
2 a) Name TWO growing conditions required by yeast during fermentation.
   b) For each of these describe how a commercial brewer attempts to provide the best possible growing conditions.

Extra Questions

4 a) What is meant by the term **malting**?
   b) Explain the need for the malting of barley grains before use in beer-making.
5 Explain what is meant by the term **batch processing**.

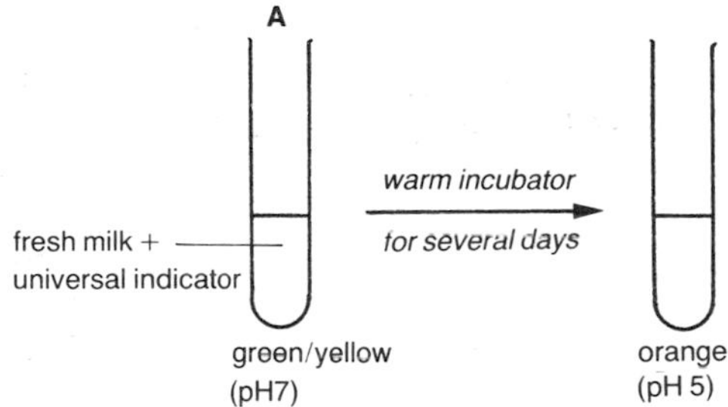

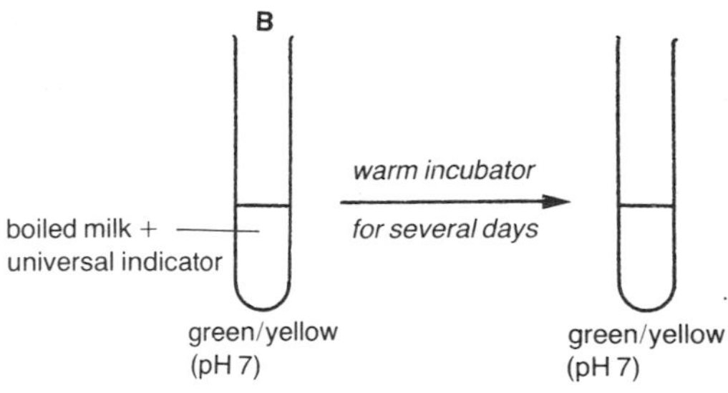

Figure 23.8 Effect of bacteria on pH of milk

Milk

Fresh milk from a cow normally contains many bacteria. Since milk is an excellent source of food, the bacteria will grow and multiply if the milk is left untreated.

Effect of bacterial growth on the pH of milk

In the experiment shown in figure 23.8 many drops of universal indicator solution are added to a small volume of fresh milk. After several days in a warm incubator the contents of test tube A are found to have changed in colour from greenish yellow (pH 7) to orange (pH 5). No change is observed in control tube B.

It is therefore concluded that the change in pH from neutral to acidic in tube A has been brought about by the activities of live bacteria which have fed on the milk and made an acid. This acid which turns milk sour is called **lactic acid**. The souring of milk is a further example of a fermentation process.

More to do

Fermentation of lactose

During the souring of milk, the bacteria growing in the milk respire anaerobically (without oxygen). They feed on a sugar in the milk called lactose and break it down into lactic acid as in the word equation:

lactose → lactic acid

This process is called **lactic acid fermentation**.

Cheese and yoghurt

The manufacture of cheese and yoghurt (see figure 23.9) depend on the **curdling** (coagulation) of milk. Lactic acid is needed to make the milk curdle. It is produced during fermentation by special strains of bacteria (see figure 23.10) added to the milk for this purpose. Although these lactic acid-forming bacteria are sometimes seen in groups or chains, they are single-celled organisms.

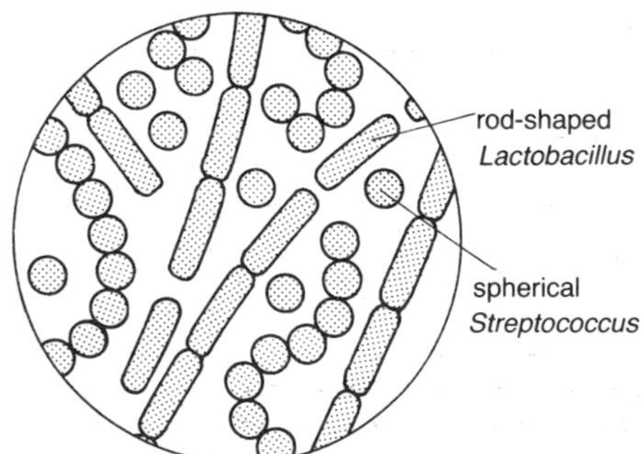

Figure 23.10 Lactic acid-forming bacteria

KEY QUESTIONS

1 a) What change in pH occurs in milk as it turns sour?
b) What substance produced by bacteria brings about this change?

2 What features do the manufacture of cheese and yoghurt have in common?

3 Rewrite the following sentence to include only the correct words.
Cheese and yoghurt both contain fungi/bacteria which are single-celled/multicellular organisms needed to convert sugar in milk to lactobacillus/lactic acid.

4 The following list gives five of the steps involved in the commercial production of a well known brand of yoghurt. Arrange them in the correct order.
A 'Sterilised' milk cooled to 44 °C.
B Milk kept at 44 °C for four hours to allow bacteria to convert it to yoghurt by making lactic acid.
C Milk heated to 73 °C for at least 30 seconds to make it virtually sterile.
D Yoghurt stored in cartons at 4 °C to slow down further bacterial action.
E Special 'yoghurt' bacteria added to 'sterilised' milk.

Extra Question

5 Using the terms fermentation, bacteria, lactic acid, anaerobic respiration and lactose, describe what happens when milk turns sour.

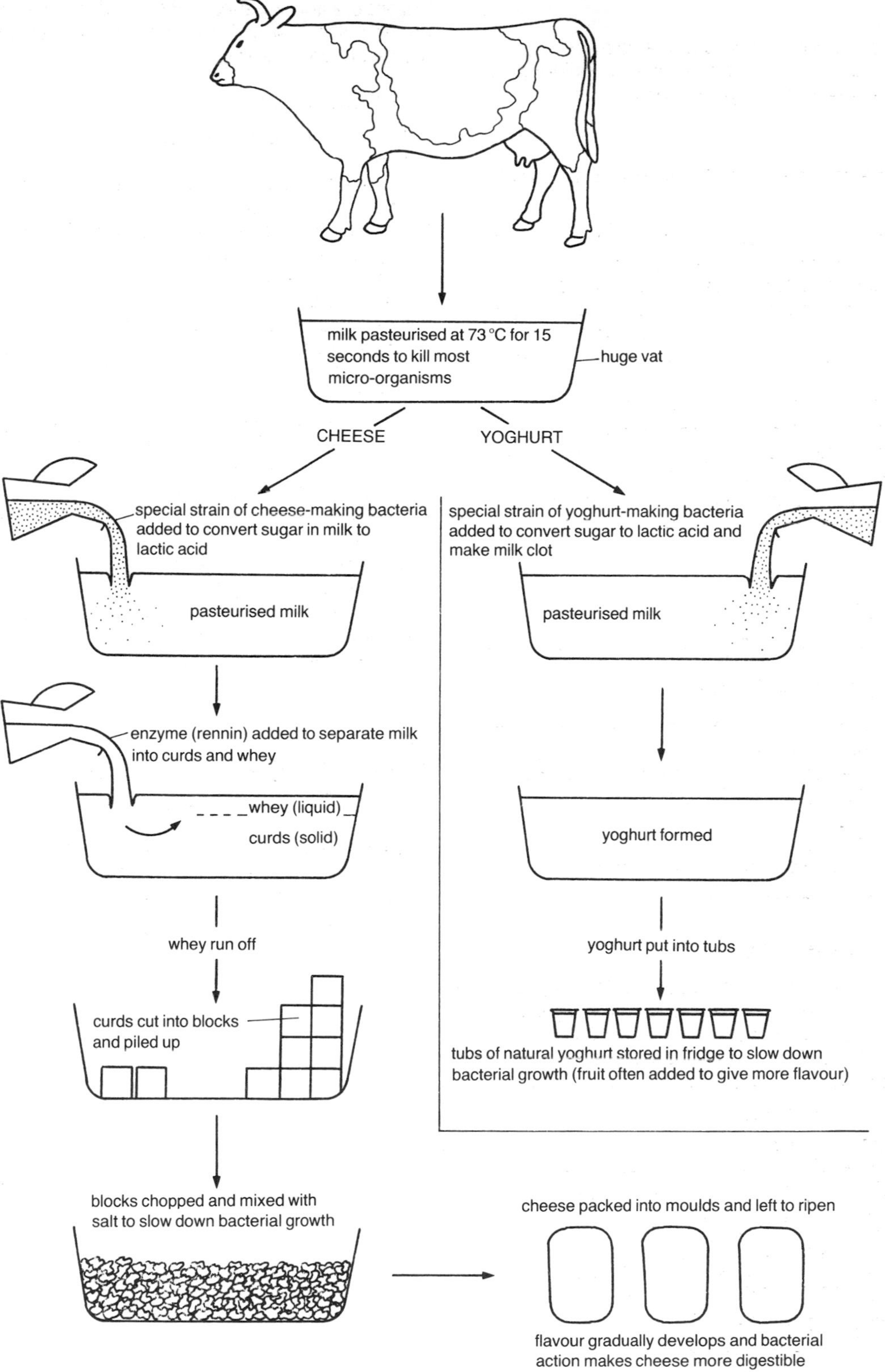

Figure 23.9 Manufacture of cheese and yoghurt

24 Problems and profit with waste

Effect of untreated sewage on a river

Untreated raw sewage contains organic material (faeces, food fragments, soap, etc.), mineral salts (e.g. nitrates and phosphates) and bacteria. Some of the bacteria can cause diseases.

The disposal of untreated sewage into a river can bring about many chemical and biological changes as shown in figure 24.1 (see also page 30). Since the river may suffer an algal bloom and the death of most of its animals and the release of foul-smelling gases, the local environment becomes badly damaged. These effects are not restricted to rivers. Even the great lakes of North America are at present suffering as a result of untreated sewage being released into them.

KEY QUESTIONS

1 Give THREE examples of the damage caused to the environment by disposal of untreated sewage.
2 Why do fish die in rivers where untreated sewage has been released?
3 Which type of plant thrives in a river polluted with raw sewage? Explain why.
4 What causes foul-smelling gases to be released from a polluted river?

Water-borne diseases

Untreated sewage often contains micro-organisms which cause diseases such as dysentery, typhoid and, in some parts of the world, cholera. The germs normally fail to infect the population provided that sewage is kept separate from drinking water. In Britain, sewage is converted to harmless materials at a sewage treatment works (see page 184) before being released into a waterway.

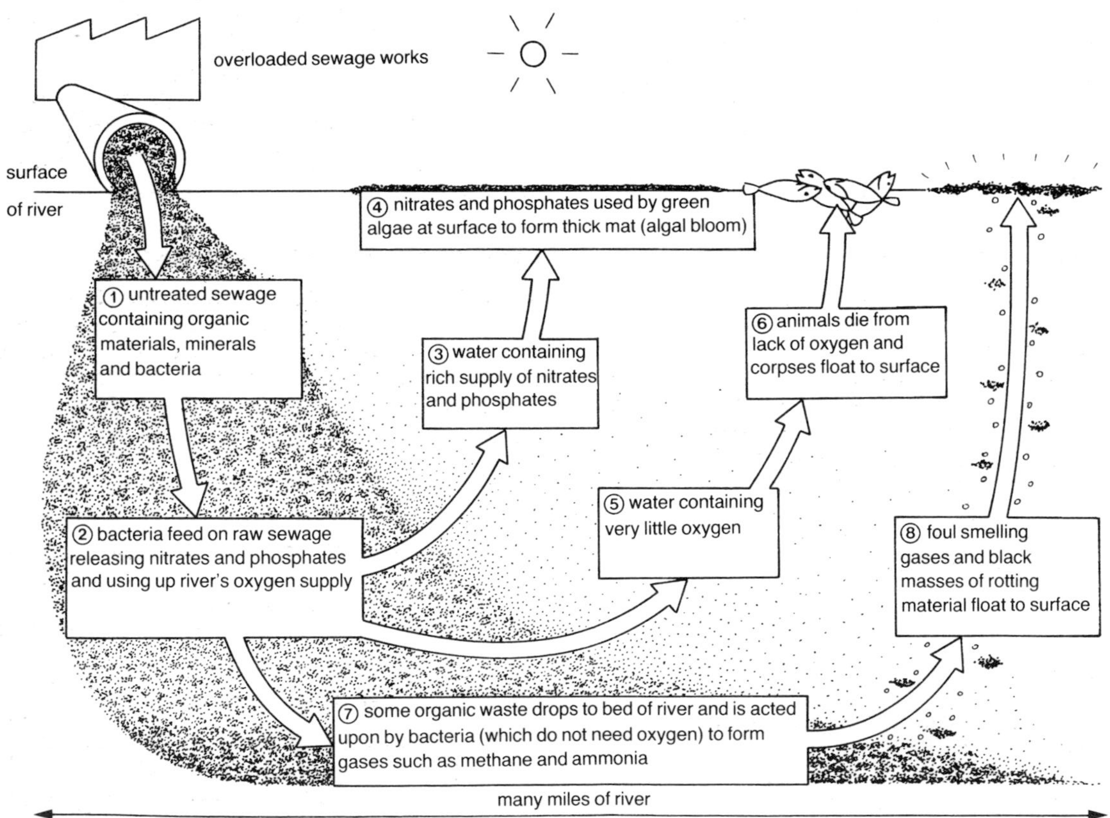

Figure 24.1 Effect of untreated sewage on a river

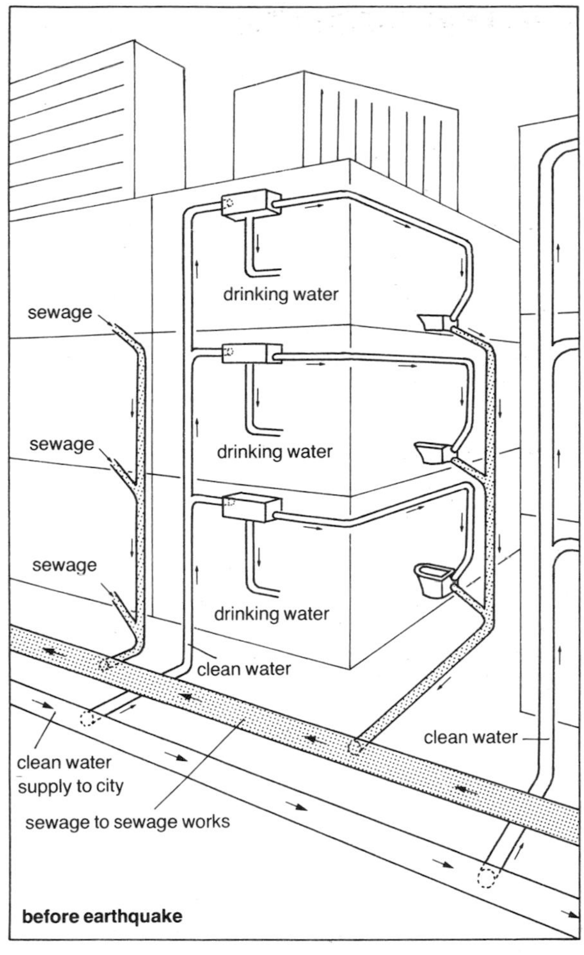

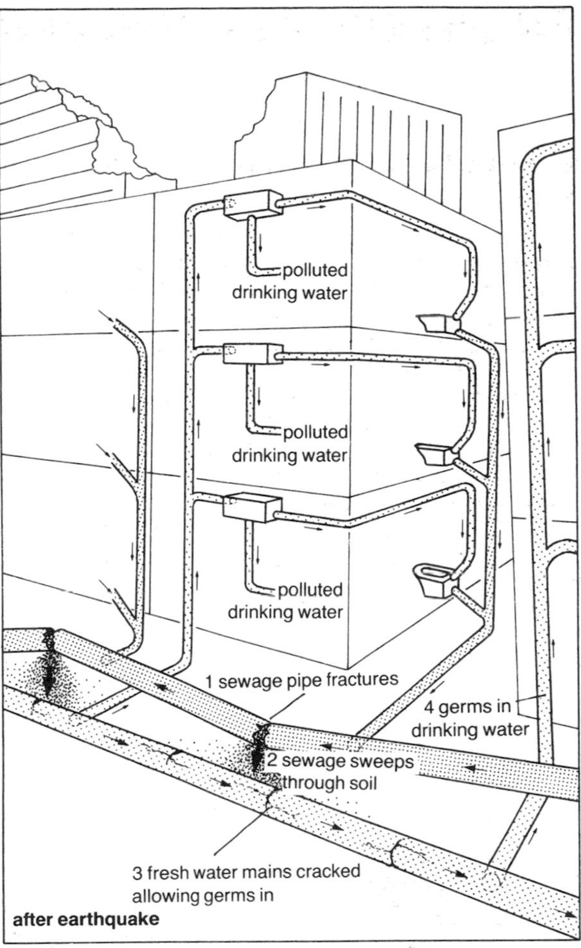

Figure 24.2 Effect of natural disaster

Natural disasters

If an earthquake occurs, underground pipes may become cracked or fractured as shown in figure 24.2. Untreated sewage may be able to seep through the soil from a sewer to the fresh water supply. Similarly, flooding may allow germs to pass from sewage to drinking water. Under these circumstances all drinking water must be boiled to ensure that disease-causing germs are not being consumed.

KEY QUESTIONS

1 Name TWO diseases that may be spread by untreated sewage.
2 Under what circumstances could a water-borne disease spread in a modern city?
3 Suggest a precaution that should be taken by people living in a city suffering a cholera epidemic.

Airborne micro-organisms

Look at the experiment shown in figure 24.3. The nutrient agar in plate A is exposed to air for 30 minutes and then the plate is incubated. After two days many micro-organisms are found to be growing on the nutrient agar.

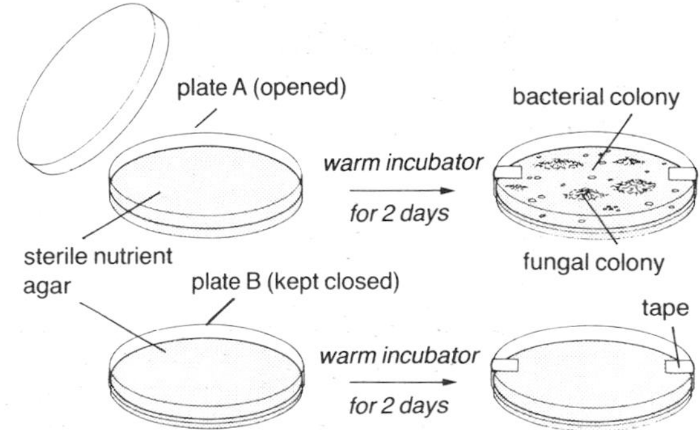

Figure 24.3 Micro-organisms from the air

Since plate B, the control, shows no microbial growth, it is concluded that air contains many types of micro-organisms. Many of these are present in dust as air-borne **spores** and are constantly landing on all surfaces.

During laboratory work with micro-organisms, it must be remembered, therefore, that stray microbes are always present in the air, on all surfaces and inside all open containers. Since a few of these are harmful to people, certain precautions must be taken to prevent unwanted microbes being given the chance to grow and perhaps cause diseases.

Precautions and sterile techniques

Work surface and microbiologist

In a microbiology laboratory all surfaces should be made of plastic and be cleaned with **disinfectant** before starting an experiment. Microbiologists must be careful not to touch their faces with their hands during an experiment and they should wear **lab coats** to prevent clothing from becoming contaminated.

Agar preparation

A few micro-organisms can even survive boiling water. All glassware must therefore be heated to 121 °C for twenty minutes in an **autoclave** or pressure cooker to ensure that all microbes are dead. Plates of agar are prepared by the method shown in figure 24.4. to try to make sure that the agar is sterile at the start of an experiment.

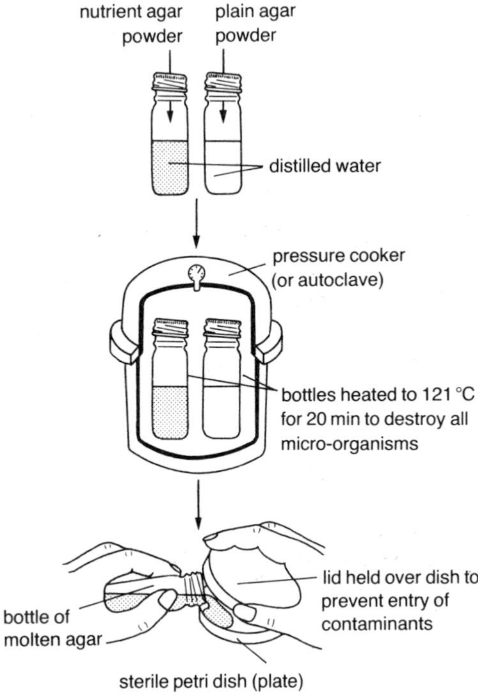

Figure 24.4 Preparation of sterile agar plates

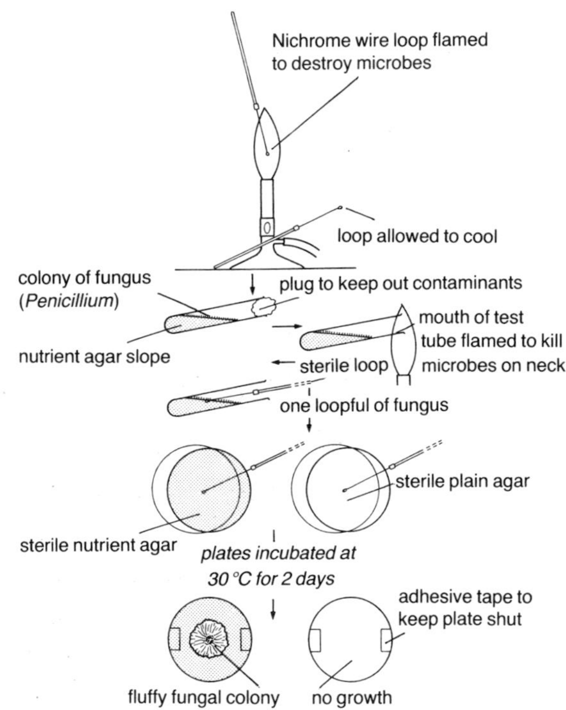

Figure 24.5 Growth of a fungal culture

Culturing the micro-organism
When the agar is inoculated with the particular micro-organism to be cultured, the further sterile techniques shown in figures 24.5 and 24.6 are carried out to prevent foreign microbes entering and contaminating the plate or culture bottle. Despite all of these precautions, unwanted **contaminants** do occasionally get into a culture.

Keeping stray microbes out is very important in a large-scale biotechnological process. A contaminant could slow down growth of the proper micro-organism by using up much of its food supply. This would prevent the useful microbe from carrying out the job it had been set to do. The whole operation would then have to be scrapped and restarted. This would involve extra time, resources and money being spent and would therefore increase the cost of the end product.

Summary
The principal precautions and sterile techniques employed during laboratory work with micro-organisms are summarised in figure 24.7. Once an investigation has been completed, the petri dishes should be sealed in plastic bags and autoclaved to kill all micro-organisms.

Resistant spores

Under unfavourable conditions some bacterial cells are able to protect themselves by becoming **endospores** (see figure 24.8). Some fungi can also

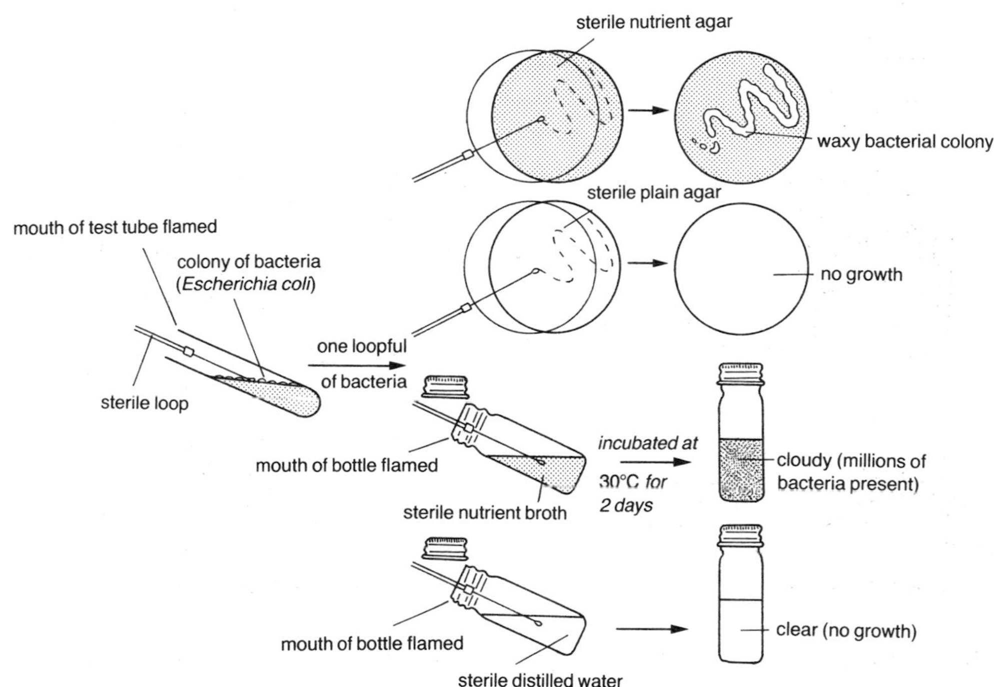

Figure 24.6 Growth of a bacterial culture

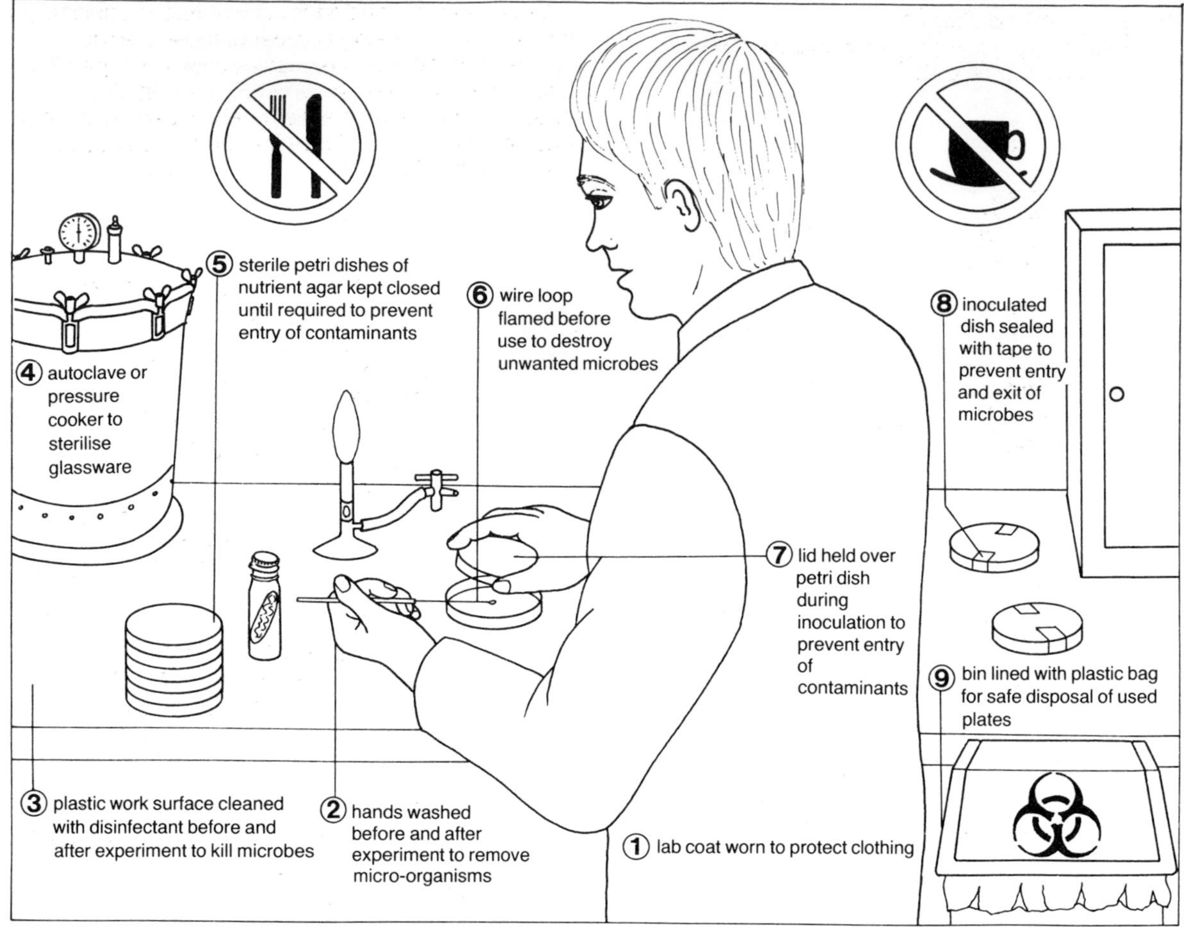

Figure 24.7 Summary of precautions

produce spores with thick coats. Such spores are resistant to factors such as extremes of temperature, drying out, pH changes, disinfectants, etc., that are fatal to normal cells. Only a temperature of 121 °C applied under pressure for several minutes guarantees the death of all resistant spores. This precaution must therefore be taken during a manufacturing process such as the canning of food to ensure that no resistant spores survive and end up in the product.

KEY QUESTIONS

1 State TWO principal precautions that should be taken in a microbiology laboratory before beginning an investigation. Explain the importance of each of these.

2 Copy and complete the following table:

sterile technique	reason
heating glassware in autoclave	
holding lid over open petri dish	
flaming wire loop	
flaming mouth of culture tube	
applying adhesive tape to plates	

3 Name ONE source from which a contaminant micro-organism may gain entry to a plate of nutrient agar.

4 How should petri dishes be disposed of at the end of an investigation?

5 Look again at figure 24.7 and explain the meaning of the notices displayed on the wall. Why are such precautions important?

6 What is the meaning of the sign on the bin in figure 24.7?

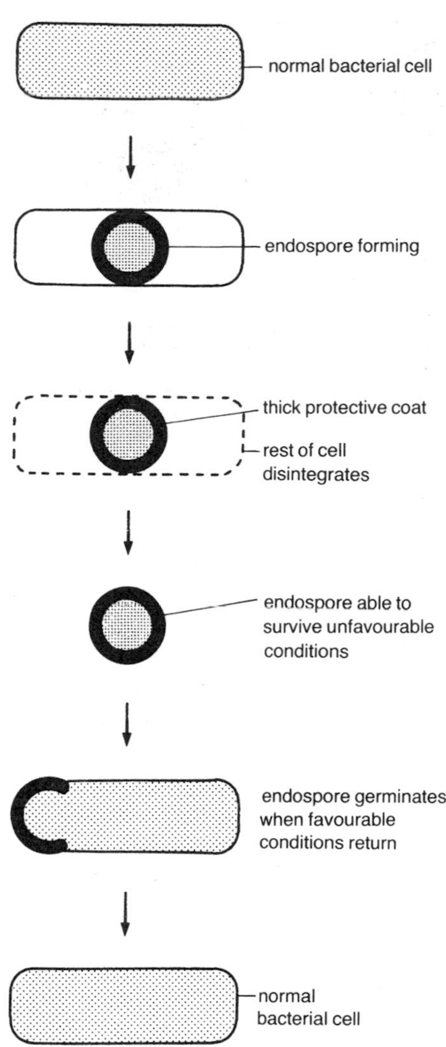

Figure 24.8 Endospore formation

Extra Questions

7 What is a bacterial endospore able to resist?

8 a) Why must precautions be taken against resistant spores during manufacturing processes?
b) By what means can all resistant spores be killed?

Process of decay

The experiments illustrated in figures 24.5 and 6 show that fungi and bacteria only grow when the culture medium contains food (nutrients).

Under natural conditions some micro-organisms get their food by attacking and decomposing a wide variety of organic substances as shown in figure 24.9. **Decay** is a natural process which ocurs in all ecosystems. It is the means by which certain microbes (the **decomposers**) obtain the energy and building materials that they need to stay alive and grow.

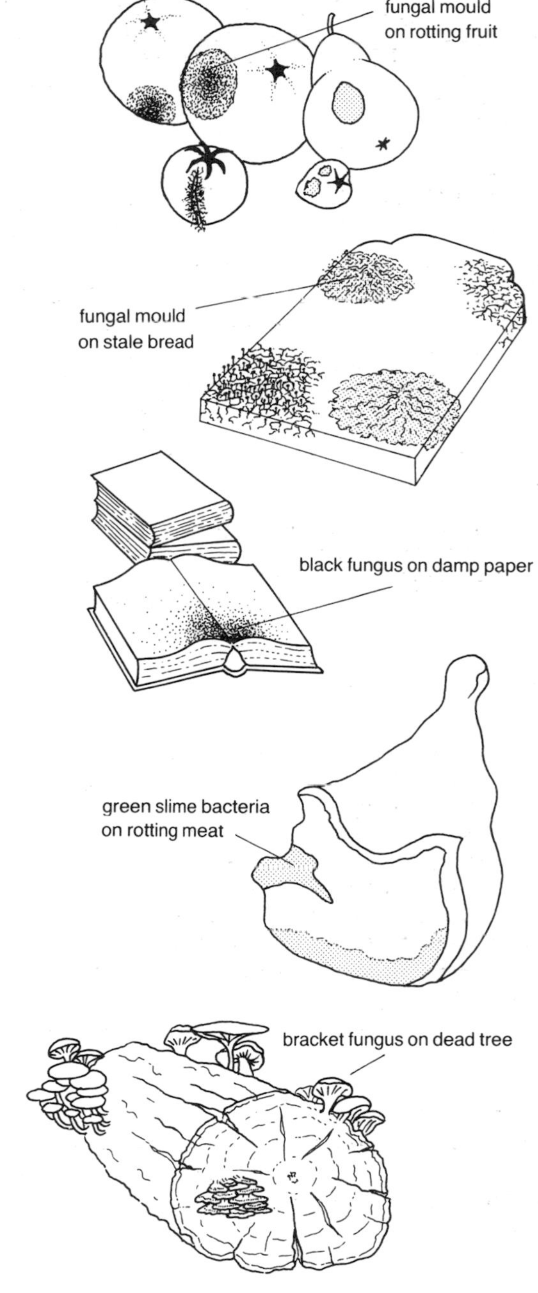

Figure 24.9 Decay

More to do

Recycling of mineral salts

A **saprophyte** is an organism that obtains its food from dead or decaying organic matter. It secretes enzymes onto the food to digest it externally before absorbing some of the products into its body for its own use.

Many bacteria (and fungi) are saprophytes. Under natural conditions in the soil they act as decomposers by digesting and decomposing dead organisms and waste materials such as faeces. **Mineral salts** which were present in the dead organic matter are now released into the soil solution and **recycled** as shown in figure 24.10. By bringing about both the process of decay and the release of mineral salts, bacteria play an essential role in this cycle. Without their activities nature's 'rubbish' would accumulate and mineral salts would remain locked up in dead bodies and be unavailable for use by new living organisms.

The maintenance of a balanced ecosystem depends on the repeated cycling of many chemical elements. Two of these are **carbon** and **nitrogen**. Each has its own particular cycle dependent upon the activities of bacteria.

Carbon cycle

The carbon cycle is shown in figure 24.11. This is similar to the recycling of mineral salts in that it depends on the activity of saprophytic bacteria. In the carbon cycle these bacteria release the gas **carbon dioxide** into the air. It can then be used by plants during photosynthesis. When plants are eaten by animals the carbon is passed on round the cycle as shown in the diagram.

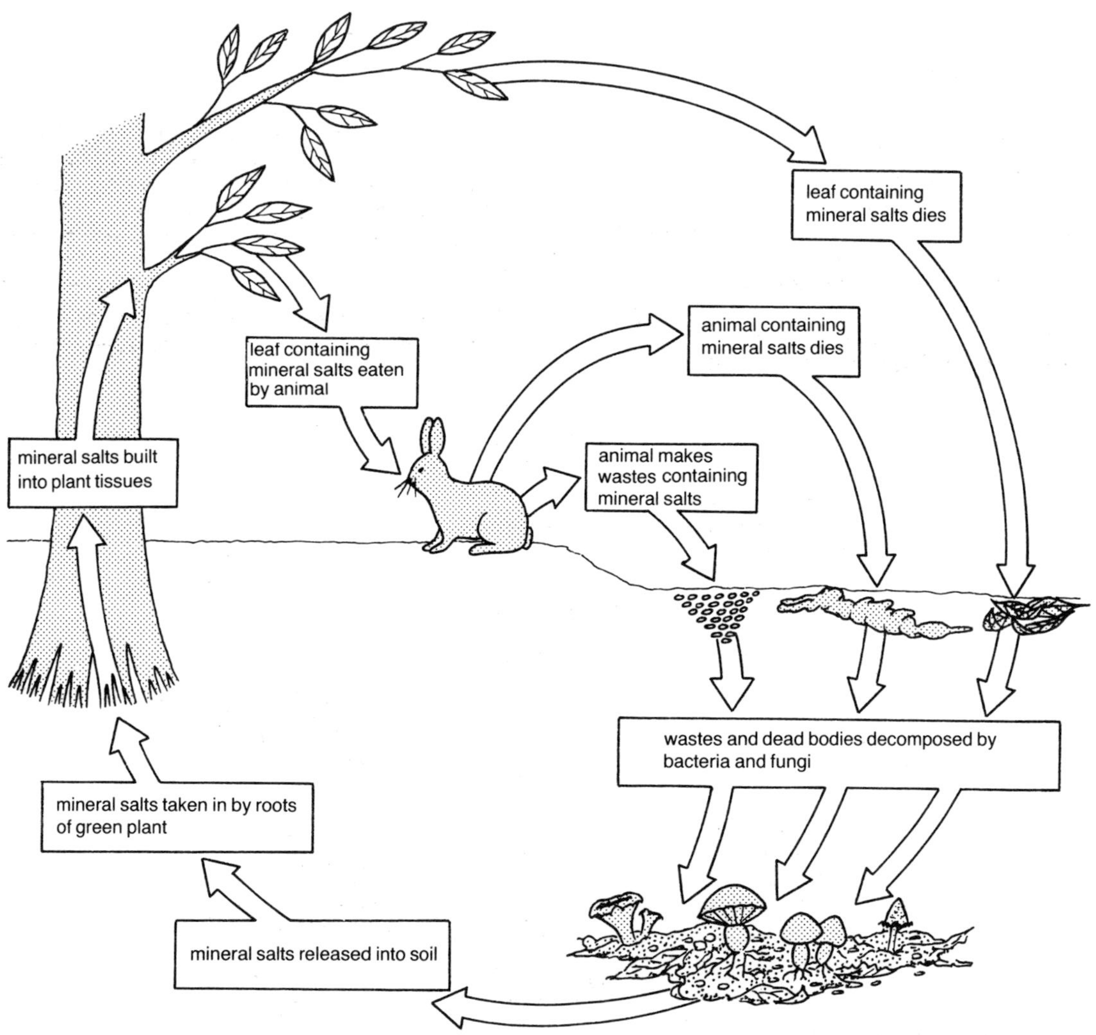

Figure 24.10 Recycling of mineral salts

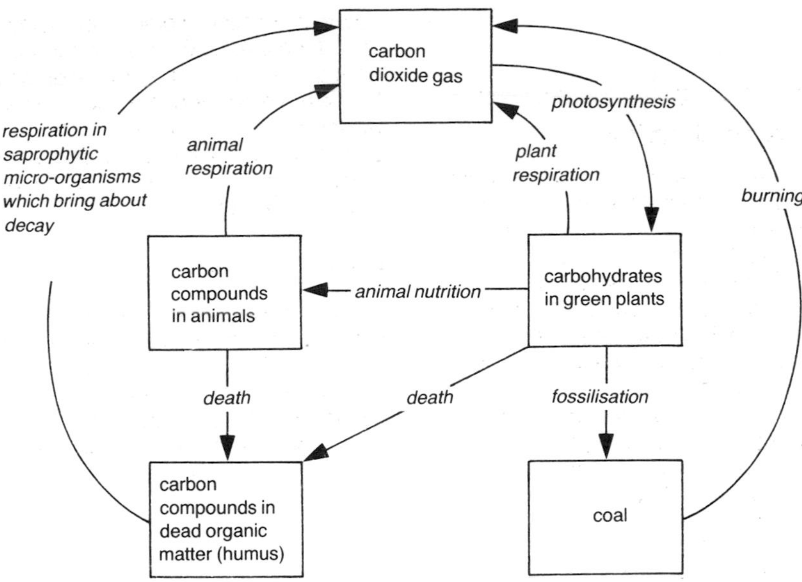

Figure 24.11 Carbon cycle

Nitrogen cycle

The sequence of processes involved in the nitrogen cycle has already been described on page 24. This cycle is dependent on the activities of several different types of bacteria, each playing a key role as indicated by the numbered arrows in figure 24.12.

At arrow 1, saprophytic bacteria bring about the **decomposition** of dead bodies and wastes releasing compounds containing **ammonia** into the soil.

The arrows marked 2 represent the process of **nitrification**. This is brought about by bacteria (called *Nitrosomonas*) converting ammonium compounds to **nitrites** and then bacteria (called *Nitrobacter*) converting the nitrites to **nitrates** which can be used by plants.

Arrow 3 represents the process of **denitrification**. Denitrifying bacteria (called *Pseudomonas denitrificans*) deprive the soil of nitrogen compounds

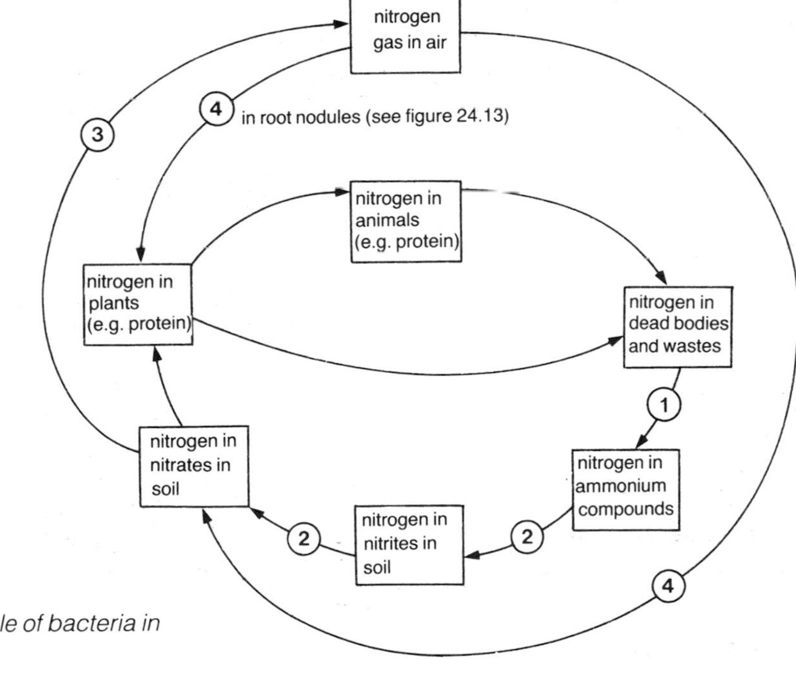

Figure 24.12 Role of bacteria in nitrogen cycle

by breaking down nitrates and releasing **nitrogen gas** into the air.

This loss is compensated by the activities of bacteria at the arrows marked 4 which bring about **nitrogen fixation**. Free-living soil bacteria (called *Azotobacter*) absorb atmospheric nitrogen and 'fix' it into **nitrate** (or even protein). Other bacteria (called *Rhizobium*) live inside swellings called **nodules** on the roots of leguminous plants such as pea, bean and clover (see figure 24.13). These bacteria also absorb nitrogen gas from the air and 'fix' it into a form which the plant can use to build protein. It is for this reason that leguminous plants are able to thrive in soil deficient in nitrogen.

Without the activities of all of these different types of bacteria, the nitrogen cycle would not turn and life on Earth would come to a halt.

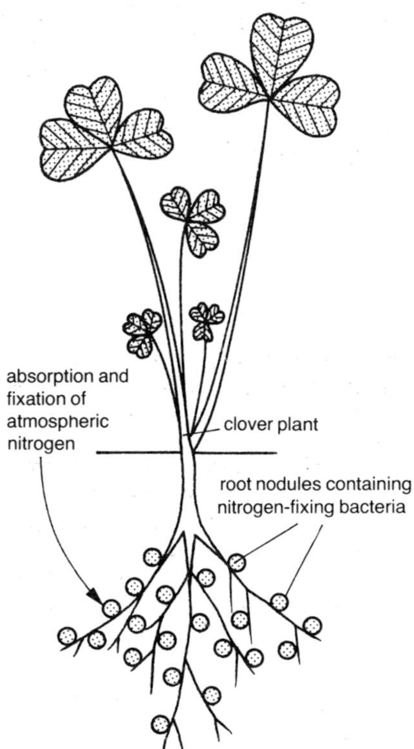

Figure 24.13 Root nodules

KEY QUESTIONS

1 a) Name FOUR non living organic substances that may be attacked by micro-organisms.
b) Give TWO reasons why the micro-organisms attack them.

Extra Questions

2 What is a saprophyte?
3 Describe the role played by a saprophyte in the cycle shown in figure 24.10.
4 What role is played by saprophytes in (a) the carbon cycle, (b) the nitrogen cycle?
5 Explain the difference between the processes of nitrification and denitrification.
6 Why are nitrogen-fixing bacteria beneficial to leguminous plants?
7 Starting with saprophytic bacterium, arrange the following bacteria into the correct order in which they would be met by an atom of nitrogen passing round the nitrogen cycle: *Nitrobacter*, *Rhizobium*, *Nitrosomonas*, saprophytic bacterium, *Pseudomonas denitrificans*.

Treatment of sewage

Raw sewage consists of water containing organic wastes such as faeces, soap, detergent and food fragments. It becomes mixed with plastic bags, rags, road-grit and many other materials that find their way into underground sewers.

On arrival at a **sewage treatment works** (see figure 24.14) this mixture undergoes preliminary treatment at stages 1, 2 and 3. The sludge formed at stage 3 passes to the sludge digester (stage 6) for further treatment.

The liquid passing out of a container is called an **effluent**. The effluent from stage 3 is taken to stage 4 for **secondary treatment**. This is the main process in the treatment of sewage. It is here that all the organic compounds are broken down by micro-organisms into products that are harmless to the environment. There are two main methods of secondary treatment. Each provides the rich supply of **oxygen** needed by the decay micro-organisms to do their work.

Biological filtration (stage 4A)

The effluent from stage 3 is pumped up a central pipe and out along arms which rotate, sprinkling the sewage onto a bed of stones. These stones are coated with a film of micro-organisms which feed on the organic material as it trickles through on its way down to the bottom of the filter. Between the stones there are many **air spaces** which supply the microbes with oxygen.

Activated sludge process (stage 4B)

The sewage effluent from stage 3 is mixed with sludge which is activated (i.e. rich in bacteria). To provide these decay micro-organisms with plenty of oxygen, the mixture is aerated by blowing **compressed air** through it and stirring the surface with paddles. After about six hours the bacteria have digested the organic compounds in the sewage.

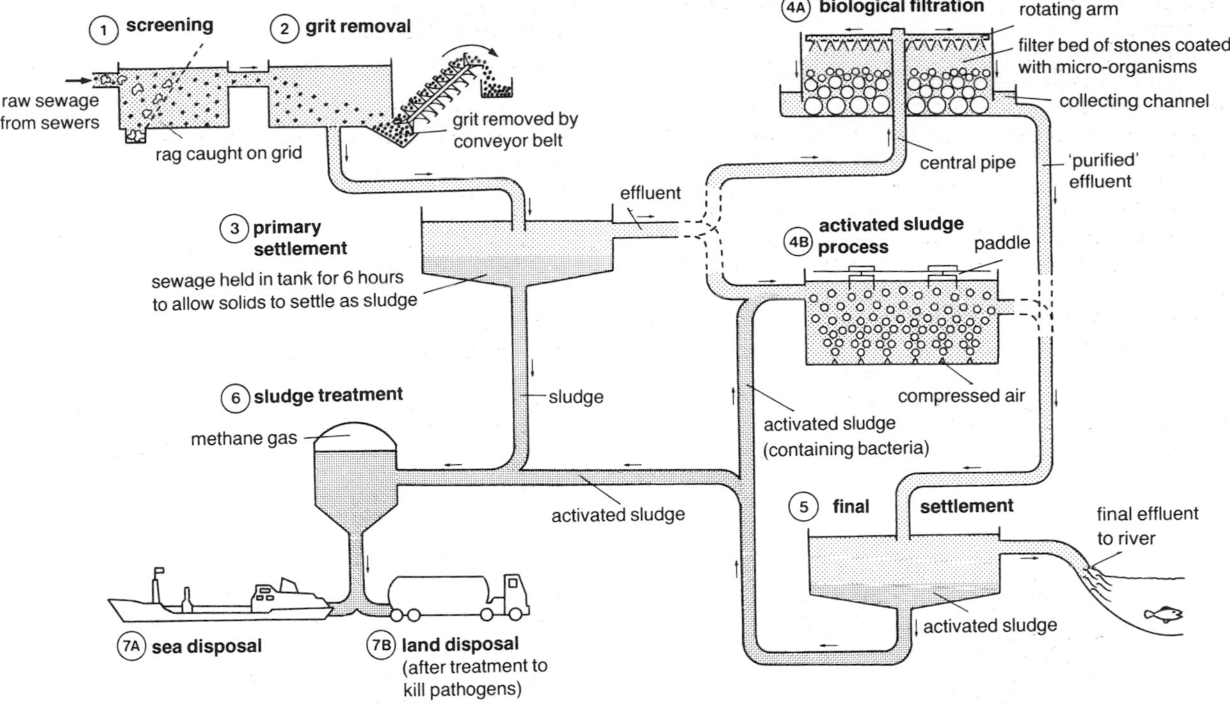

Figure 24.14 Sewage treatment plant

Following secondary treatment, the 'purified' sewage passes to the final settlement tank (stage 5) where remaining suspended matter settles out as **activated sludge**. Any substances now present in the liquid above the sludge are harmless to the environment and the effluent can therefore be safely released into the nearest waterway.

Stages 5–7 show what happens to the sludge. Many microbes can feed on sewage sludge and respire in the absence of oxygen to form **methane** gas. This is a further example of a fermentation process. A sewage treatment works may have a methane fermenter at stage 6 where sludge is treated. The methane (like natural gas) can be used as a fuel to drive some of the other processes at the sewage works. Methane from sewage is used in developing countries (e.g. India) for cooking and heating.

Need for a range of micro-organisms

Sewage is a mixture of many different complex organic chemicals such as fats, carbohydrates, proteins and vitamins. Each species of micro-organism is only able to break down a few substances. Many different species of bacteria are therefore needed at stages 4A and B in figure 24.14 to ensure **complete breakdown** of all sewage materials to carbon dioxide, water and simple inorganic substances.

Most of the inorganic substances are harmless, however in high concentrations some of them could later pollute the waterway by causing an algal bloom (see figure 24.1). Special bacteria may therefore be used to remove excess ammonium compounds, nitrates and phosphates as shown in figure 24.15.

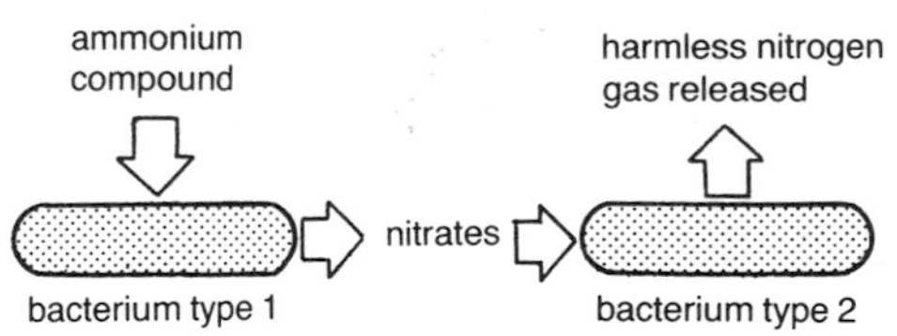

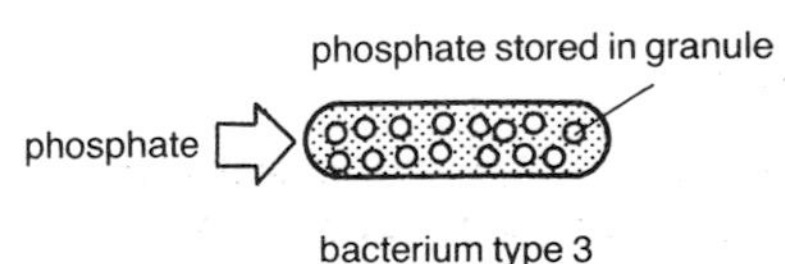

Figure 24.15 Range of micro-organisms

More to do

Complete breakdown of sewage

Anaerobic respiration takes place in the absence of oxygen. Compared with aerobic respiration, anaerobic respiration in less efficient since the organic compound being acted upon is only partly broken down. Much of its energy still remains in the product (e.g. alcohol, lactic acid or methane).

Complete breakdown of sewage is only possible in aerobic conditions since this form of respiration using oxygen completely breaks the organic material down to carbon dioxide and water. It is for this reason that the decay microbes acting at stage 4A and B in the diagram of the sewage treatment plant are provided with a rich oxygen supply.

Chemical oxygen demand

If the products of anaerobic respiration were released into a waterway, they would be used by further micro-organisms which would at the same time use up the river's oxygen supply. Such semi-treated sewage is said, therefore, to have a **chemical oxygen demand (COD)**. The aim of a sewage treatment plant is to reduce the COD of the effluent to the lowest possible level before releasing it into the waterway.

KEY QUESTIONS

1 Name the main process in the treatment of sewage and describe the role played in it by micro-organisms.
2 By what means is oxygen provided for use by microbes during stages 4A and B in the diagram of the sewage treatment plant?
3 Why are many different species of bacteria necessary for the breakdown of sewage?

Extra Questions

4 Explain why the complete breakdown of sewage is only possible under aerobic conditions.
5 Which has the higher COD, treated or untreated sewage?
6 Why is it desirable to reduce the COD of sewage to a minimum before releasing it into a waterway?

Upgrading waste

Many manufacturing processes release large amounts of organic material as waste products. In the past these were often thrown away. However, many of the wastes or excess materials are now fed to certain micro-organisms which convert them to useful products of a **higher economic value** than the original substances. This process involving biotechnological methods is called **upgrading**. Several examples are given in table 24.1.

Advantages of upgrading waste

It is much better to make use of a waste material than to dispose of it. Micro-organisms can be used to change the chemicals in waste into other chemical substances which contain a high level of readily available **energy** or a rich source of **protein** (see figure 24.16).

Upgrading of waste therefore provides a supply of two valuable commodities – fuel such as methane gas and protein such as mycoprotein.

Since the fermentation methods used to upgrade low-value waste into, for example, high-value protein-rich animal food neither involve vast financial

industry	waste product	micro-organism used in upgrading process	useful product
cane sugar refining	molasses	*Aspergillus* (a fungus) yeast	organic acids (e.g. vinegar) alcohol
cheese-making	whey	yeast	protein (for cattle feed) and vitamins (in dead yeast cells)
agriculture	straw	*Agaricus bisporus* (a fungus)	edible mushrooms
potato crisp manufacture	starch	*Fusarium* (a fungus)	mycoprotein (suitable for human consumption)
gas and oil	methanol	*Methylophilus* (a bacterium)	protein (suitable for animal feed)

Table 24.1 Upgrading waste

investment nor high running costs, they are not restricted to wealthy developed countries. In addition, a considerable economic saving on waste disposal is made and environmental pollution is prevented.

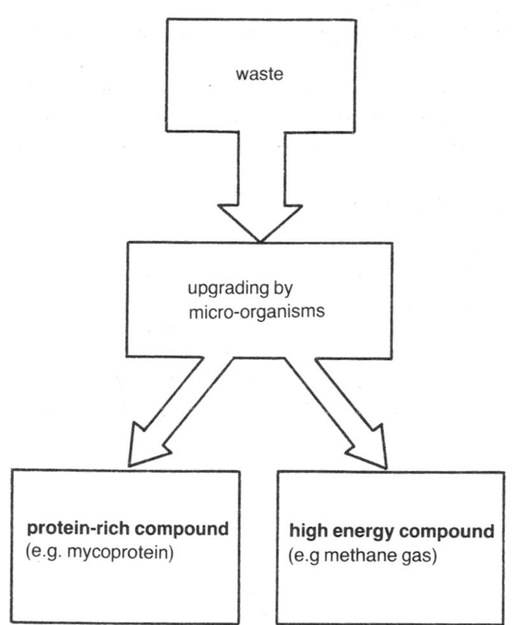

Figure 24.16 Upgrading waste

KEY QUESTIONS

1 a) Give TWO examples of useful products which can be obtained from waste materials.
b) For each of these identify the micro-organism responsible.

2 In terms of energy and protein content, explain the advantages of upgrading waste.

Fuels from micro-organisms

Methane

When fresh manure is kept in a flask (which is sealed to prevent the entry of oxygen) the micro-organisms in the manure carry out a fermentation process and produce **methane** gas (see figure 24.17). When a sample of this gas is collected in a syringe and sprayed into a Bunsen flame it burns. Methane can therefore be used as a fuel.

Methane production is a natural process which occurs when animal or plant material decays in the absence of oxgyen. Many sewage works and farms produce methane from wastes. The gas can be used to provide heat and electricity.

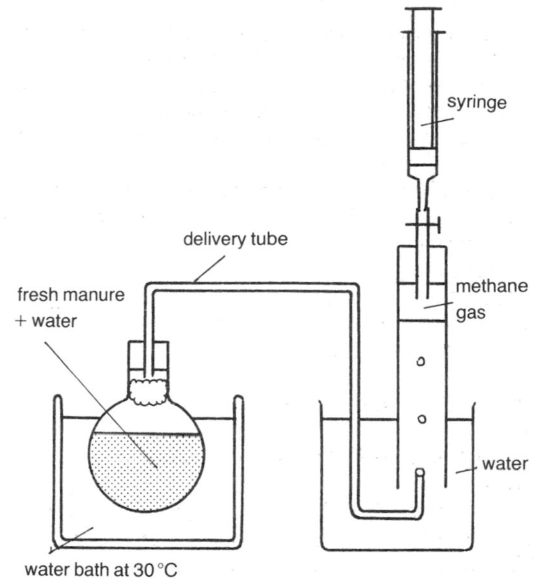

Figure 24.17 Methane fuel

Alcohol

When sugar solution is fermented by yeast, **alcohol** (ethanol) is formed (see page 170). The fermented solution contains only a small amount of alcohol because alcohol is a poison and most of the yeast cells die when the concentration reaches 12–15 per cent. The alcohol can be separated from the fermentation mixture by distillation. The pure alcohol is found to be flammable as shown in figure 24.18. Alcohol can therefore be used as a fuel.

Fuel from sugar

Brazil does not have rich supplies of fossil fuels (coal, oil and gas). However, it does have a warm climate and plenty of land. For these reasons sugar cane is grown on a large scale and fermented to alcohol. (Approximately one billion litres were being produced annually in the mid 1980s.) The alcohol is used instead of petrol to run cars.

Renewable and non-renewable resources

Fossil fuels are **non-renewable** resources. They have taken millions of years to form and are being used up at an ever-increasing rate. It is estimated that the world's oil reserves will be exhausted by the middle of the twenty-first century. Such a resource is therefore **limited** and finite.

The advantage of deriving fuel from fermentation of plant material is that it is a **renewable** resource.

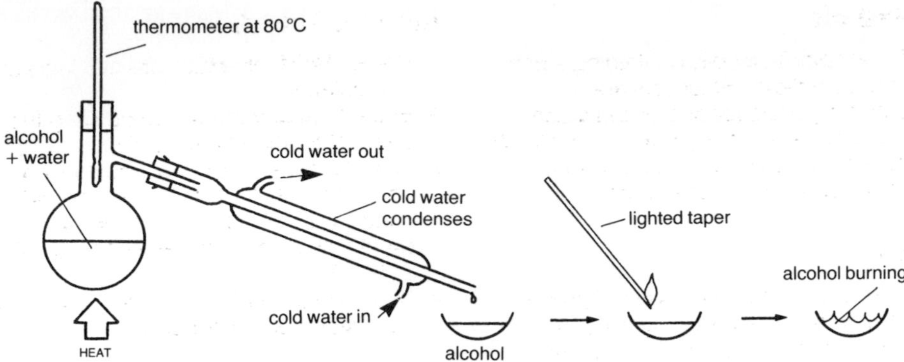

Figure 24.18 Alcohol fuel

Provided that the sun shines and photosynthesis occurs, light energy can be converted to chemical energy indefinitely. This resource is therefore **unlimited** and infinite.

Growth rate of micro-organisms

A bacterium reproduces asexually by dividing into two. Under suitable conditions (adequate supply of food and water, optimum temperature and pH, etc.) this process can occur as rapidly as once every 20–30 minutes in some types of bacteria. Consider the growth of one such bacterial cell over a period of five hours as shown in table 24.2 and graphed in figure 24.19. This pattern of unlimited growth only occurs under ideal conditions (see also page 20).

time on 24-hour clock	number of bacteria
09.00	1
09.30	2
10.00	4
10.30	8
11.00	16
11.30	32
12.00	64
12.30	128
13.00	256
13.30	512
14.00	1024

Table 24.2 Growth of bacteria

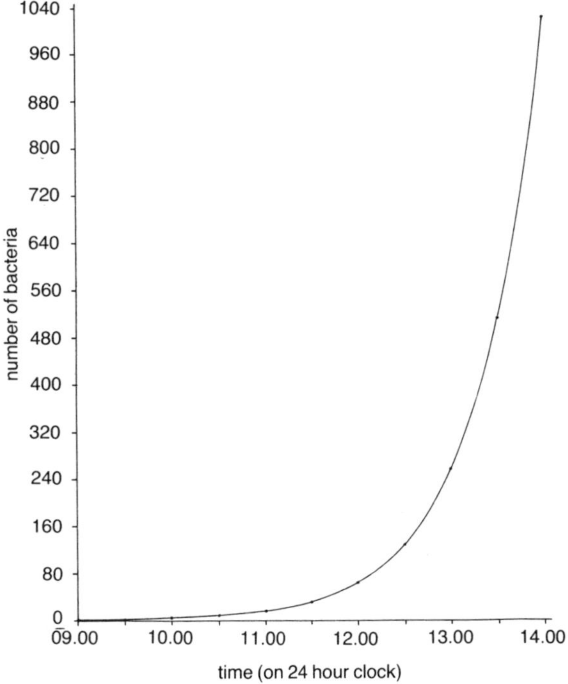

Figure 24.19 Graph of bacterial growth

organism	protein produced per day by 1000 kg of biomass (kg)
cattle	1
soya bean plants	100
yeast	100 000
bacteria	1 000 000 000 000 000

Table 24.3 Protein production

Single-celled protein

Since a high percentage of a bacterial cell's mass is composed of protein and bacteria grow very quickly, these tiny organisms are very efficient protein producers compared with other organisms (see table 24.3).

Certain types of bacterial cells can be grown, harvested and dried to form a protein-rich powder called **single-celled protein (SCP)**. SCP is used to feed animals such as chickens and calves.

Single-celled oil

Some strains of yeast produce cells containing a rich supply of **oil** in their cytoplasm. Scientists are at present working on the possibility of growing these yeasts on renewable waste carbohydrates (e.g. starch and molasses) to produce oil which would be cheap to manufacture and suitable for humans to eat.

Mycoprotein

Some fungi (e.g *Fusarium*) produce a form of protein called **mycoprotein** in their threadlike hyphae. This protein is suitable for human consumption and the threadlike strands can be spun into meat-like products.

KEY QUESTIONS

1 Name TWO fuels which are products of fermentation.
2 Explain the advantages of deriving fuel from sugar cane rather than from fossil fuels
3 Imagine that the time is 12.00 hours and you add one bacterium to a flask which provides it with ideal conditions. If this bacterium divides once every twenty minutes, how many bacteria will be present at 16.00 hours?
4 Name TWO protein-rich foods that can be harvested from micro-organisms.

25 Reprogramming microbes

Chromosomal material of a bacterium

A bacterium has one chromosome in the form of a complete **circle** and one or more small circular **plasmids** (see figure 25.1). The chromosome and plasmid(s) are made up of **genes**. Each gene carries the information necessary to make a certain **protein** (e.g. an enzyme). Each enzyme in turn controls a particular chemical reaction. The normal activity of a bacterial cell depends therefore on its chromosomal material.

The type of bacterium referred to in figure 25.2, for example, can grow on lactose sugar because it has the gene needed to make the enzyme lactase. This enzyme breaks lactose sugar down to simpler sugars which are then used by the cell as a source of energy. If this gene is absent or damaged, the enzyme is no longer made and the bacterium cannot make use of lactose.

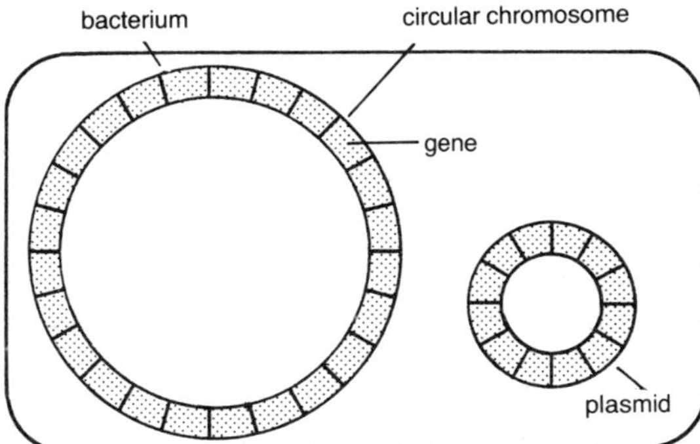

Figure 25.1 Chromosomal material of a bacterium

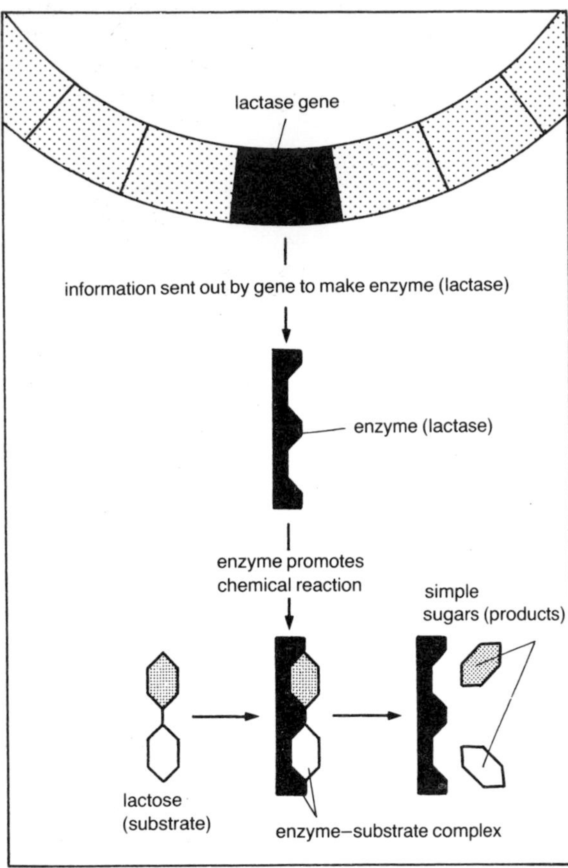

Figure 25.2 Action of lactase gene

Genetic engineering

Pieces of chromosome can be transferred from one organism (e.g. man) to another organism (e.g. bacterium). This process is called **genetic engineering**. The **reprogrammed** bacterium acts as a chemical factory and manufactures a new substance (useful to man).

Figure 25.3 shows a simplified version of genetic engineering. It normally involves the following stages:

1 identification of the required gene (e.g. gene for human insulin);
2 cutting of the chromosome using special enzymes (acting as chemical 'scissors') to release the gene;
3 extraction of a plasmid from a bacterial cell;
4 opening up of the plasmid using another enzyme;
5 insertion of the gene into the plasmid ring;
6 insertion of the plasmid into the bacterial 'host' cell;
7 growth and multiplication of the bacterial cell;
8 formation of duplicate plasmids which express the 'foreign' gene and make the desired useful product (e.g. insulin).

When given suitable conditions, reprogrammed bacteria grow very rapidly. Vast amounts of useful products can therefore be quickly produced as a result of genetic engineering.

More to do

Genetic engineering versus selective breeding

Genetic engineering and selective breeding are two methods by which people can **alter** the **genotype** of

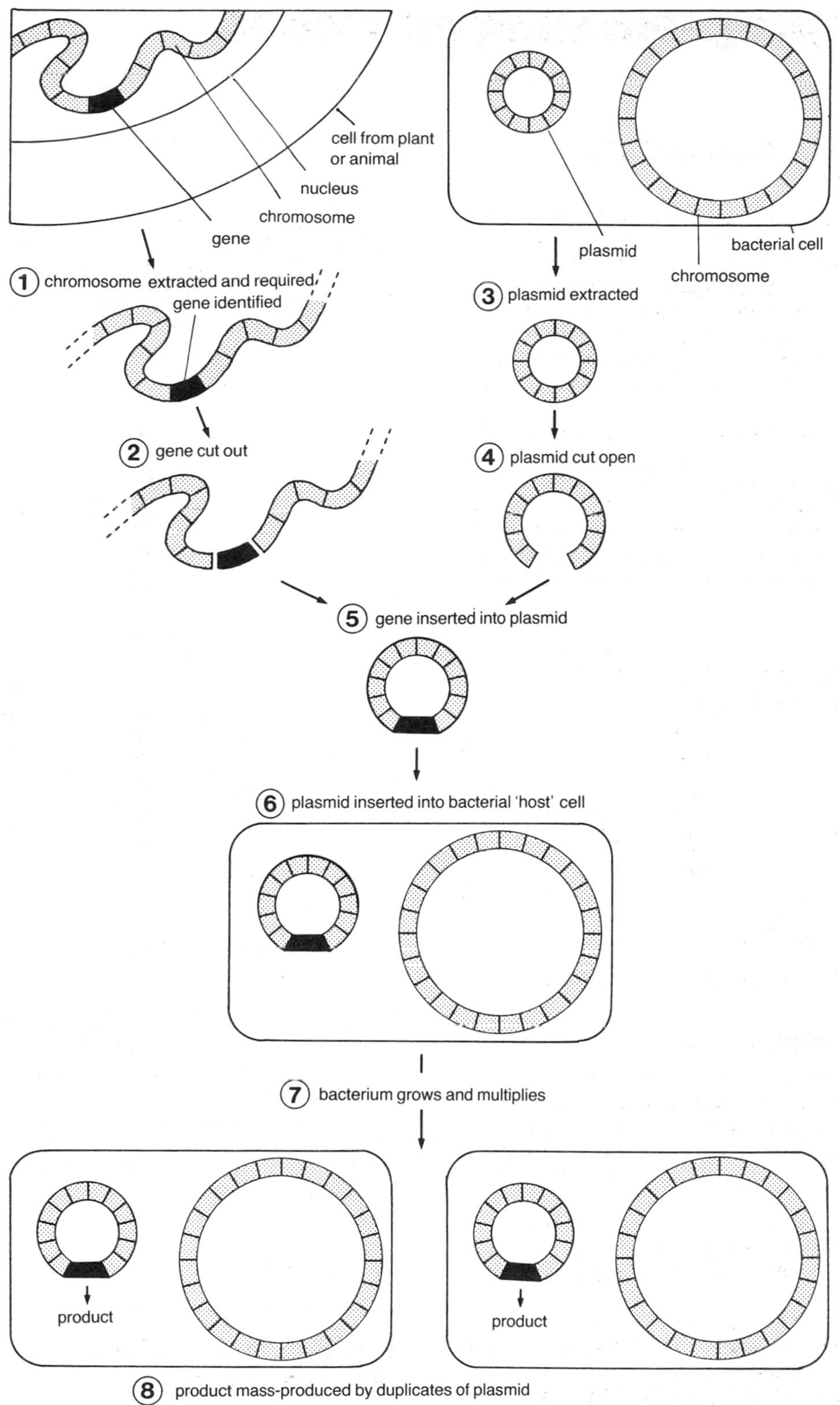

Figure 25.3 Genetic engineering

another species for their own benefit. Each involves producing new combinations of genetic material that would be unlikely to exist normally in nature.

Selective breeding
This traditional method of producing new improved varieties involves many years of careful selection and breeding over many generations. It is very time-consuming and is indirect in that it does not necessarily produce the 'ideal' organism to suit man's requirements. Selective breeding rarely allows people to do more than develop a variety of organism which produces an improved version or quantity of a product (e.g. milk) originally made by that species (e.g. cow). In addition, the product may take a long time to develop (e.g. only mature cattle can produce milk).

Genetic engineering
This modern method of producing new varieties offers several advantages over selective breeding. It allows man to alter the genotype of a species directly by manipulating its chromosomal material. Once a genetic engineer has identified the gene which controls the production of a useful substance (e.g. human insulin) he/she can immediately transfer that gene to a microbe and produce an organism with a new genotype ideally suited to mankind's needs. Genetic engineering therefore allows man to do something completely new – programme one species to make products previously only made by another species. In addition, vast quantities of the product can be produced rapidly by the 'new' microbe.

Products of genetic engineering

Medical value
Many gene products have important **medical applications** as shown in table 25.1. In each of these examples a particular gene has successfully been inserted into a bacterium which is then allowed to multiply and make the product of medical value.

Scientists have also managed to reprogramme yeast cells to produce human serum albumin (used in blood replacements), epidermal growth factor (which speeds up wound healing) and hepatitis B antigens (used in the manufacture of hepatitis B vaccine).

Commercial applications
Many reprogrammed microbes produce useful **enzymes** such as protease which is added to soap powders to digest difficult stains (see page 194). Other new varieties of bacteria are able to make ethylene glycol (antifreeze) and various other chemicals used in the plastics industry. Since these chemicals are at present derived from oil, production of them by bacteria in the future could be of great value when fossil fuels become exhausted.

The future
Genetic engineers are attempting to produce bacteria which will mop up oil slicks by digesting the oil. They are also trying to isolate the genes that enable certain bacteria to fix atmospheric nitrogen. If these genes could be transferred to crop plants, the plants would be able to fix their own nitrogen from the air and not need a rich supply (often as costly fertiliser) to be present in the soil.

Already scientists have produced a new strain of animal by genetic engineering, called 'mighty mouse', as shown in figure 25.4. It is possible that this form of genetic engineering may be used to change farm animals in the future.

Prevention of potential hazards
In the early days of genetic engineering, many people feared that new strains of micro-organisms would be produced which could turn out to be harmful to people

product of genetic engineering	normal source and function of substance	medical application of gene product made by bacteria
insulin	Made by pancreas cells. Controls the level of glucose in the blood.	Given to people who do not make enough insulin naturally and who would otherwise suffer from *diabetes mellitus*.
interferon	Made in tiny amounts by animal cells. Prevents the multiplication of a wide range of viruses. In the past it was extracted from white blood cells but 45 000 litres of blood yielded only 400 g of interferon.	May be of use in the future to combat viral diseases and some forms of cancer.
human growth hormone (somatostatin)	Made by cells in the pituitary gland. Essential during childhood and adolescence to control normal growth and development.	Given by regular injection to babies who do not make enough of their own. Prevents reduced growth and dwarfism.

Table 25.1 Medical applications of gene products

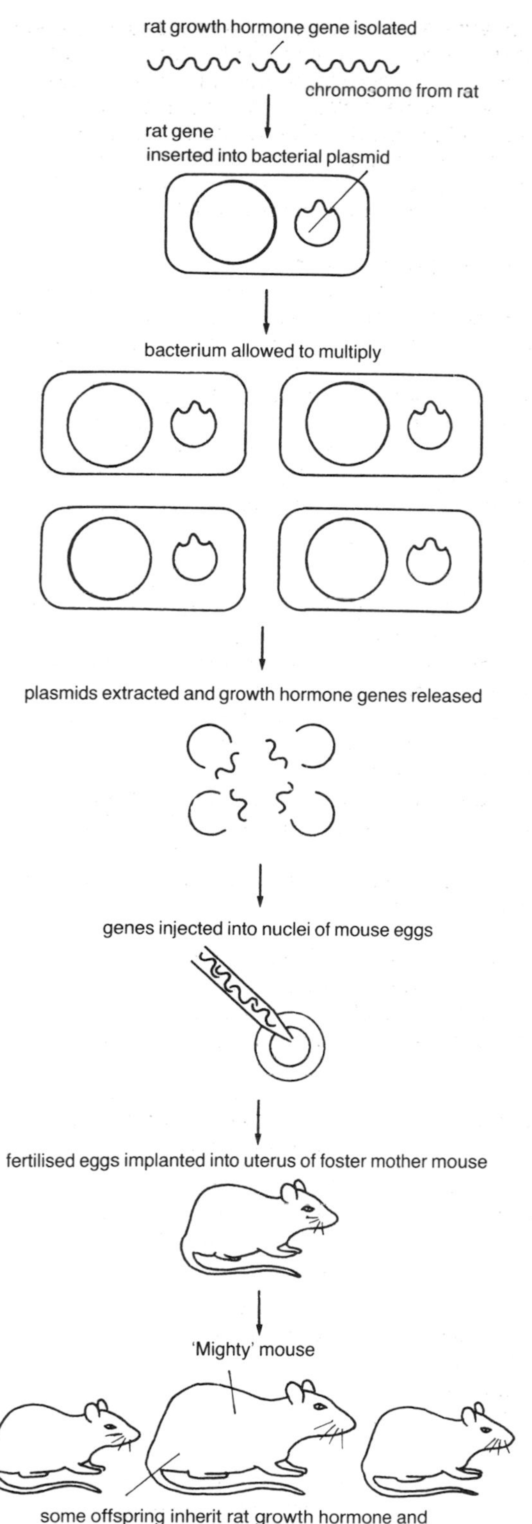

Figure 25.4 'Mighty mouse'

and/or be resistant to known antibiotics. The following strict precautions are therefore taken in genetic engineering laboratories:

1 Only highly trained personnel are allowed to work in the laboratories.
2 Only non disease-causing organisms are used.
3 The strains of these harmless organism used are specially produced by genetic engineering to require some nutrient that is unavailable in the ordinary environment.
4 Low air pressure is maintained inside the laboratories so that when a door is opened air can only pass in and not out.

More to do

Need for insulin

Insulin is a chemical messenger (hormone) which is made by the pancreas and passed into the bloodstream. It causes excess sugar to be removed from the blood and be stored in the liver as glycogen.

Diabetes mellitus is a fatal disease in which the level of sugar in the blood cannot be controlled due to the failure of the pancreas to make insulin. Until fairly recently, sufferers (diabetics) had to use insulin from cattle and pigs to stay alive. Since this insulin is not identical to the human type it sometimes caused side effects. This problem has been overcome by genetic engineering. Reprogrammed bacteria (see figure 25.3) now produce vast quantities of insulin which is pure and identical to human insulin.

Diabetes mellitus often develops in older people whose pancreatic cells produce inadequate amounts of insulin. As the expected life span of man extends, the demand for insulin is likely to increase in the future. Thanks to biotechnology there should be no problem in meeting this demand.

KEY QUESTIONS

1 a) Upon what cell material does the normal control of bacterial activity depend?
b) Explain why a bacterial cell with a damaged version of the gene shown in figure 25.2 would be unable to grow on lactose sugar.
c) An agent which acts as a carrier between two species is often called a vector. Which part of a bacterium's chromosomal material could this term be used to describe?

2 Arrange the following steps involved in the process of genetic engineering in the correct order:
a) insertion of gene into plasmid and plasmid into bacterial cell
b) identification and removal of required gene from organism (e.g. man)

c) growth and multiplication of bacterial cell forming product
d) extraction and opening up of bacterial plasmid

3 Why is genetic engineering of benefit to man?
4 Give THREE examples of products of genetic engineering and state why each is useful to man.

Extra Questions

5 What name is given to the traditional method by which man produces new varieties of useful organisms?
6 Give a brief definition of the term genetic engineering.
7 Describe TWO advantages of genetic engineering over selective breeding.
8 a) What is insulin and why is it essential?
b) When the body makes insufficient insulin what disease results?
c) Give ONE advantage of using insulin made by bacteria rather than the type obtained from cattle.
9 a) Why is it likely that there will be an ever-increasing demand for insulin in the future?
b) Explain why this is not a problem.

'Biological' detergents

'Biological' detergents contain **enzymes** produced by bacteria. In each of the investigations that follow, the action of a 'biological' detergent (soap powder) is compared with that of a similar but non-biological one.

(a) Effect of type of powder on grass stains at 40 °C

Grass stains contain a green substance called chlorophyll. The procedure shown in figure 25.5 is carried out and pieces of stained cloth X and Y are left to soak. After twenty-four hours the grass stain is found to have completely disappeared from X but not from Y.

The only difference between the two soap powders is the presence of an enzyme in the 'biological' one. It is therefore concluded that the enzyme has brought about the complete removal of the grass stain by digesting the green chlorophyll to soluble end products which pass into the washing water.

Effect of type of powder on mud stains at 40 °C

Some of the dirt particles in a mud stain are held to the fabric by 'protein glue'. When the procedure shown in figure 25.5 is repeated using cloth stained with mud, the results shown in figure 25.6 are obtained after twenty-four hours.

It is concluded that the enzyme present in the 'biological' powder has brought about the complete removal of the stain by digesting the 'protein glue' and releasing the dirt particles from the fabric into the water.

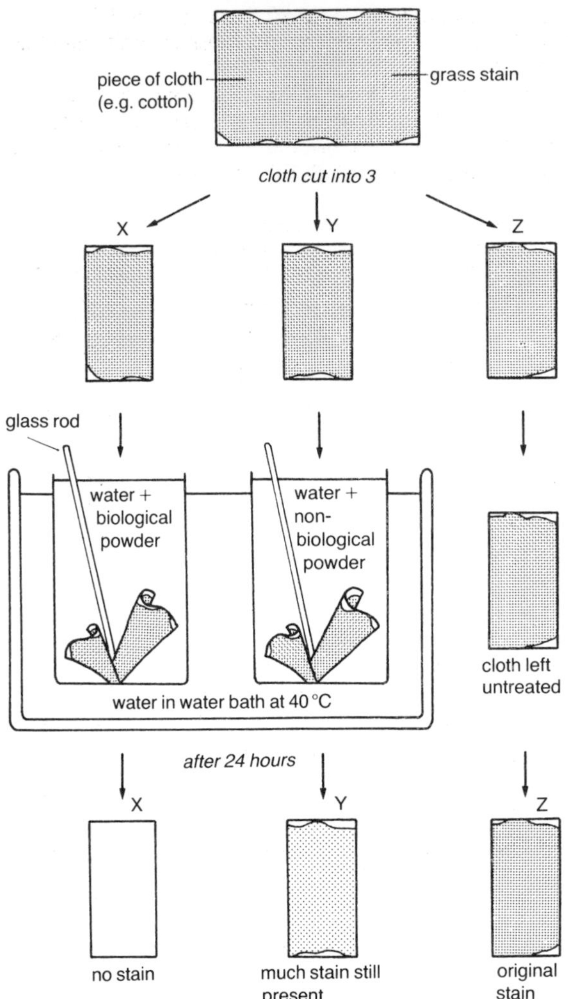

Figure 25.5 Grass stains at 40 °C

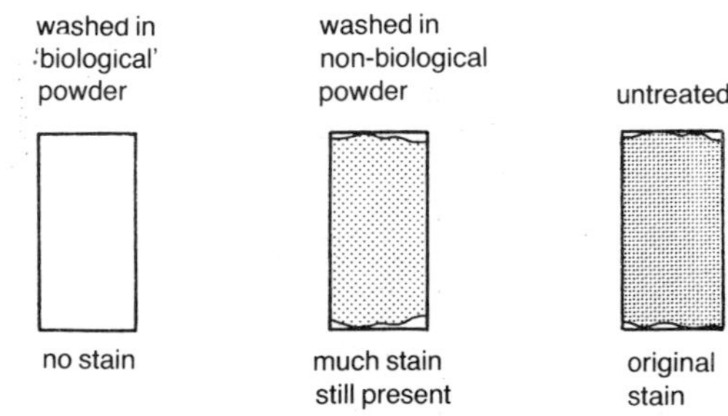

Figure 25.6 Mud stains at 40 °C

Effect of type of powder on grass stains at 100 °C
When the procedure is repeated using cloth stained with grass and kept at 100 °C for only thirty minutes, the results shown in figure 25.7 are obtained. It is concluded that neither powder is able to shift the stain at 100 °C.

The 'biological' powder is ineffective because its enzyme has been destroyed by the high temperature. The non-biological powder is ineffective against grass stains whatever the temperature of the wash.

Figure 25.7 Grass stains at 100 °C

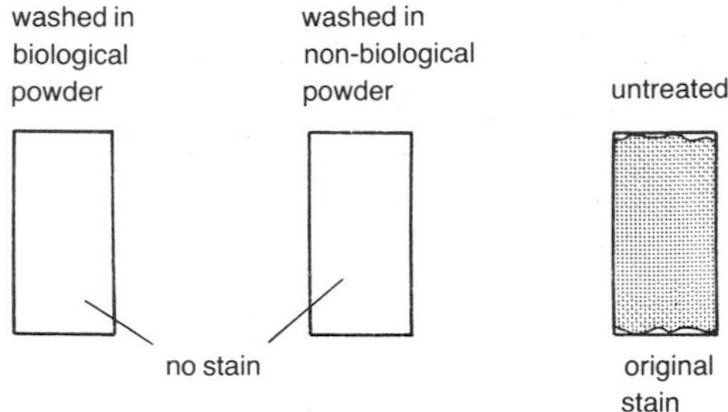

Figure 25.8 Mud stains at 100 °C

Effect of type of powder on mud stains at 100 °C
When the experiment is repeated using cloth stained with mud and kept at 100 °C for only thirty minutes, the results shown in figure 25.8 are obtained. Both powders have removed the mud stain at 100 °C.

Since the enzyme in the 'biological' powder has been destroyed by boiling it has not played a part in the process. It is concluded therefore that both powders have been effective because other chemicals which do work best at high temperatures have played their role in the cleaning process and have removed the dirt.

Table 25.2 gives a summary of the results from the four experiments.

More to do

Advantages of using 'biological' detergents

1 'Biological' detergents completely remove difficult biological stains such as grass, blood and egg yolk whereas much of the stain still remains after washing in a non-biological powder.
2 'Biological' powders are effective at low temperatures (e.g. 40 °C). This saves on fuel costs and prevents damage being done to delicate fabrics.

KEY QUESTIONS

1 a) What chemical substance is always present in a 'biological' detergent but absent from a non-biological one?
b) From what living things are supplies of this chemical substance obtained?

2 What advantages are gained by using a 'biological' rather than a non-biological detergent?

√ = stain completely removed
× = much of stain remains after washing

		biological detergent	non-biological detergent	
40 °C for 24 hours	grass	√	×	experiment (a)
	mud	√	×	experiment (b)
100 °C for 24 hours	grass	×	×	experiment (c)
	mud	√	√	experiment (d)

Table 25.2 Summary of results

Antibiotics and the discovery of penicillin

In 1928 Alexander Fleming found a fungal contaminant growing on one of his plates of bacteria (see figure 25.9). He noticed that the area around the fungal colony, instead of being cloudy with bacteria, was clear. He therefore concluded that some substance made by the fungus (*Penicillium*) was inhibiting the growth of the bacteria. This substance, an **antibiotic**, was later isolated and called **penicillin**. Tests showed that it was non-toxic to humans. Penicillin has now been used to cure many diseases such as pneumonia.

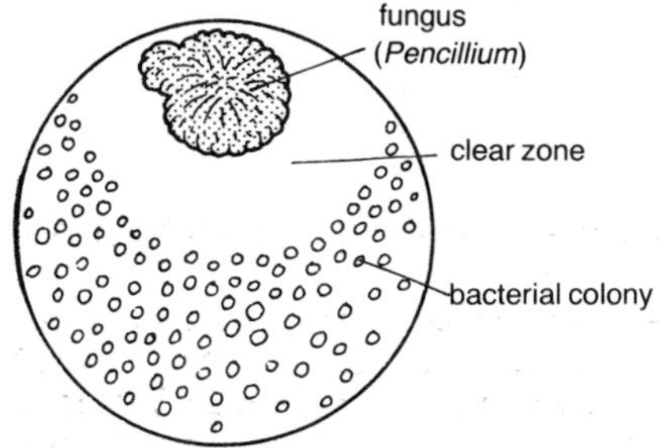

Figure 25.9 Fleming's famous plate

Other antibiotics

Many other antibiotics have been discovered. Each is a chemical produced by one micro-organism which prevents the growth, and may cause the death, of other micro-organisms.

If a micro-organism's growth is prevented by an antibiotic, the microbe is said to be **sensitive** to the antibiotic. If the antibiotic has no effect, the microbe is said to be **resistant**.

The experiment in figure 25.10 shows that each antibiotic is **specific** in that it only inhibits the growth of certain types of bacteria. There is no one antibiotic that is effective against all types of bacteria (see table 25.3).

Multidisc

The multidisc shown in figure 25.11 is made of sterile filter paper. The tip of each of its arms contains a different antibiotic. When a patient is suffering from an unknown bacterial infection it is necessary to quickly identify one or more antibiotics that will be effective against the germs. A multidisc makes it possible to test the sensitivity of the disease-causing bacterium to a range of different antibiotics at the one time. A sample of the germs is rubbed onto nutrient agar and then a multidisc is added.

Figure 25.12 shows a possible set of results after 24 hours. The bacterium is sensitive to any antibiotic that has a clear zone around it. In this case therefore,

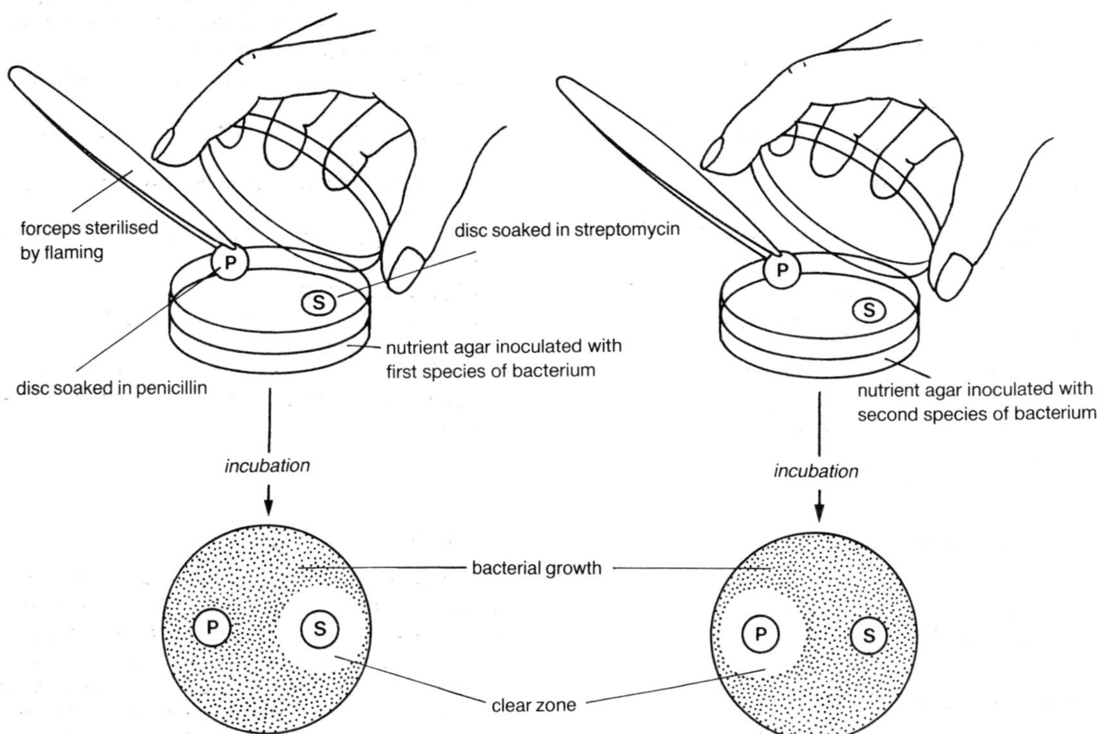

Figure 25.10 Action of antibiotic discs

+++ = very effective
+ = slightly effective
− = ineffective

disease-causing bacterium	disease caused	antibiotic			
		penicillin	streptomycin	tetracycline	chloramphenicol
Corynebacterium diphtheriae	diphtheria	+++	−	++	++
Mycobacterium tuberculosis	tuberculosis	−	+++	−	−
Salmonella typhii	typhoid	−	−	+	+++
Streptococcus pneumoniae	pneumonia	+++	−	+++	+++

Table 25.3 Effectiveness of different antibiotics

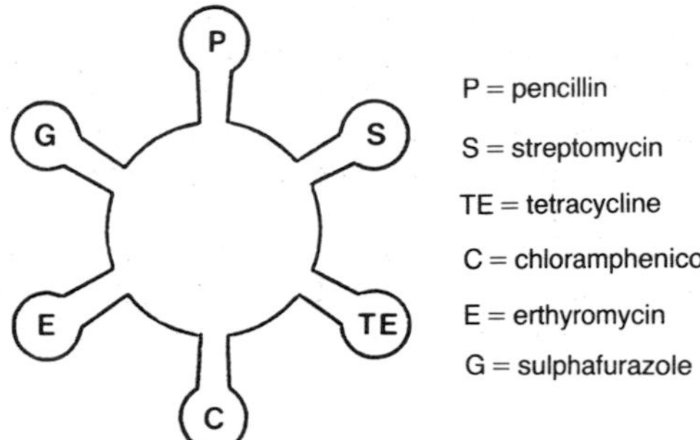

Figure 25.11 Multidisc

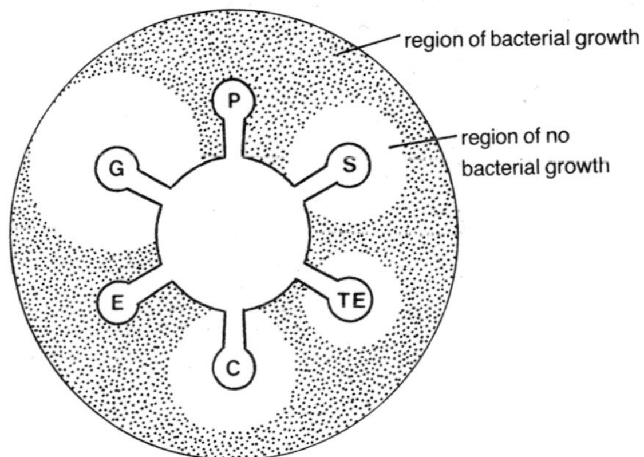

Figure 25.12 Use of multidisc

antibiotics S, TE, C and G could all be considered for use in treating the infection.

Range of antibiotics

A range of antibiotics is needed in the treatment of bacterial diseases for the following reasons:

1 If a patient is allergic to one antibiotic then another can be used.
2 New stains of disease-causing bacteria are appearing that are resistant to some antibiotics. If several different antibiotics are available then there is a good chance that the new bacterial strain will be sensitive to at least one of them.

KEY QUESTIONS

1 What is an antibiotic?
2 With reference to the experiment shown in figure 25.10, rewrite the following sentences choosing the one correct word in each case.
a) *Escherichia coli* is sensitive/resistant to penicillin.
b) *Escherichia coli* is sensitive/resistant to streptomycin.
c) *Staphylococcus albus* is sensitive/resistant to penicillin.
d) *Staphylococcus albus* is sensitive/resistant to streptomycin.
3 What is a multidisc?
4 a) Why is it not possible to treat all bacterial diseases with penicillin?
b) Give TWO advantages of having a choice of several antibiotics with which to treat a particular disease.

Immobilisation

An **immobilised** cell or enzyme is one which cannot move freely in its nutrient medium or substrate because it has been deliberately attached by scientists to another substance. Enzymes, for example, can be fixed onto a solid substance such as glass beads. Whole cells such as yeast can be entrapped inside pellets of gel as shown in figure 25.13.

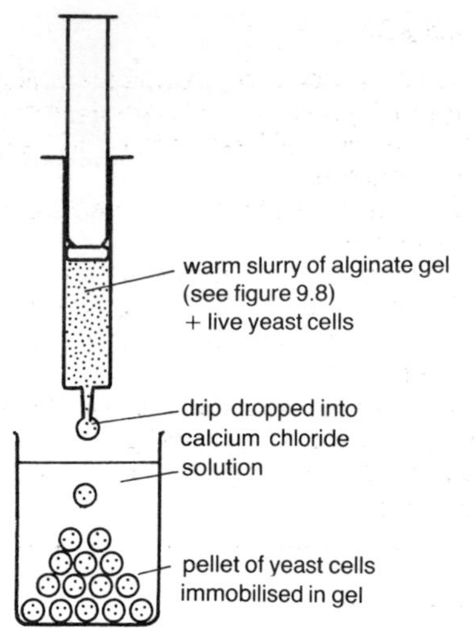

Figure 25.13 Pellets of immobilised yeast

Using immobilised yeast cells
Look at the experiment shown in figure 25.14. Once the immobilised cells have brought about the required chemical reaction, they can easily be separated from the product and used again.

Advantages of using immobilisation techniques
1 The end product does not have the cell or enzyme mixed with it.
2 Some cells and enzymes can be used many times.
3 The problem of waste disposal is reduced.

More to do

Batch processing

Figure 23.7 on page 173 gives a simplified version of **batch** processing. In this traditional method the microbe (or enzyme) is freely dispersed throughout the nutrient medium (or substrate).

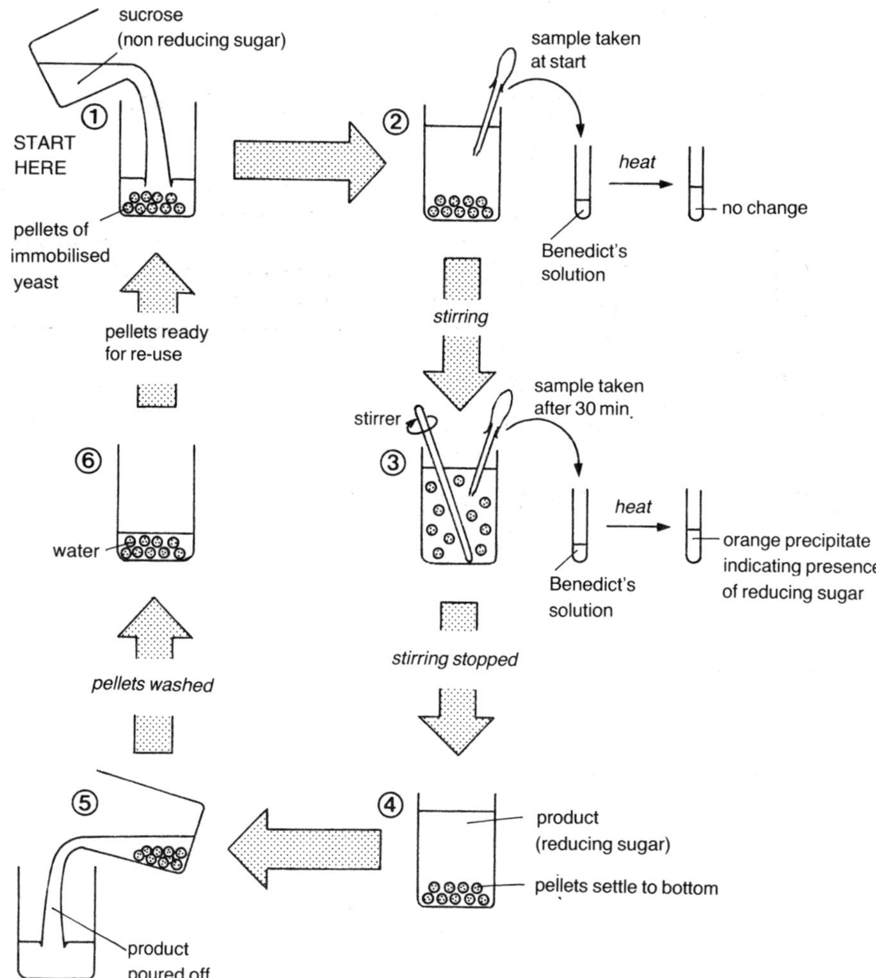

Figure 25.14 Use of immobilised yeast

At the end of a batch, the microbe is thoroughly intermixed with the product and has to be separated from it. The microbe is often in a useless state and is therefore discarded. Valuable time and labour are needed during the 'turn-around' between batches to empty, clean and sterilise the fermenter prior to the next batch being set up.

Continuous-flow processing using immobilised enzymes

The enzymes (or cells) that bring about the required chemical reaction are often expensive to produce and need to be used as efficiently as possible. If an enzyme (or cell) can be successfully immobilised, it can be used over and over again during **continuous-flow** processing.

Figure 25.15 shows a simplified version of such a process. The enzyme has been immobilised by attaching it to the surface of glass beads. Fresh nutrients are continuously fed into the reactor vessel while at the same time a similar quantity of product is removed. Although complex to develop and set up, this process can be run non-stop for long periods.

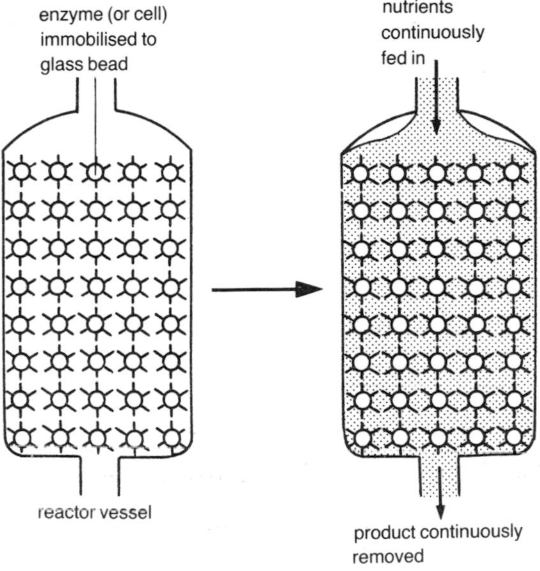

Figure 25.15 Continuous-flow processing

Advantages

Continuous-flow processing using immobilised enzymes (or cells) has many advantages over batch processing. It is more efficient and cost effective for the following reasons:

1 The product does not have to be separated from the enzyme (or cell).
2 The enzyme (or cell), which was probably expensive to produce, is used repeatedly.
3 Valuable time is not lost during 'turn-around' as in batch processing.

Continuous-flow processing also reduces the problem of waste disposal and possible pollution of the environment, since the enzymes (or cells) are not discarded.

KEY QUESTIONS

1 With reference to a cell or enzyme, explain what is meant by the term **immobilised**.
2 Give TWO methods of immobilisation.
3 Describe TWO advantages of using immobilisation techniques.

4 Compare batch processing with continuous-flow processing.
5 Give THREE reasons why continuous-flow processing is more advantageous than batch processing.

PROBLEM SOLVING

1 The following table refers to four different types of bacteria.

type of bacterium	ability of bacterium to use this food			motility
	sucrose	lactose	citrate	
Escherichia	+	+	+	–
Salmonella	–	–	–	+
Proteus	+	–	+	+
Klebsiella	+	+	+	+

(+ = yes, – = no)

a) Which type of bacterium is motile and able to grow on sucrose and citrate but not on lactose?
b) Using only the information given in the table, describe *Escherichia*.

2 In an experiment to investigate the effect of temperature on yeast activity, 50 cm^3 of glucose solution was added to live yeast cells in each of six measuring cylinders. Each cylinder was kept at a different temperature.

Yeast activity is indicated by the height of the froth of carbon dioxide bubbles formed in each cylinder. The following diagram shows the cylinders in random order after one hour.

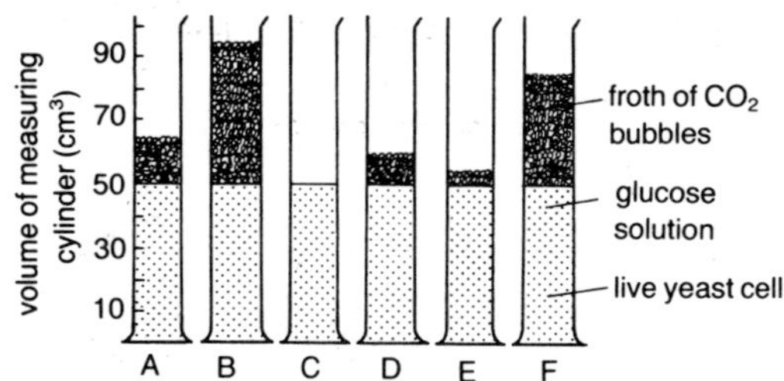

a) What was the one variable factor studied in this experiment?
b) How many conditions of the one variable factor were investigated?
c) State two factors shown in the diagram that were kept constant to make the experiment fair.
d) State one factor not referred to in the diagram which would also have to be kept constant to make this a fair test.

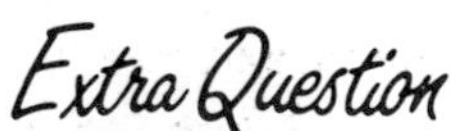

The accompanying diagram shows a graph of the results of the experiment.

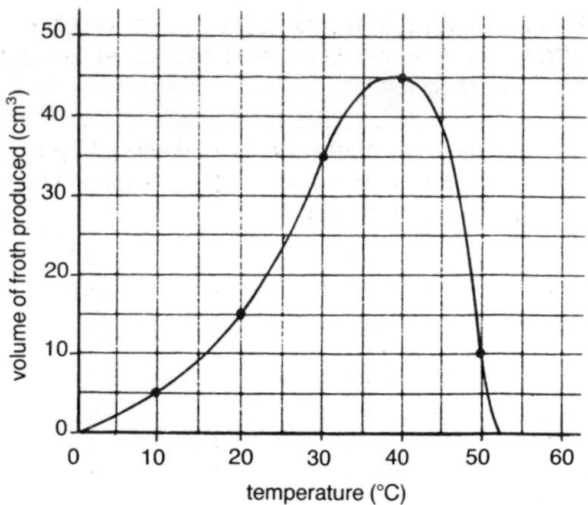

e) The six temperatures used were 10, 20, 30, 40, 50 and 60 °C. Match these with measuring cylinders A–F.

3 A thorough examination of the water at sample points 1–5 in the river shown in the following diagram gave the results tabulated below it.
a) Which letter in the diagram indicates the pipe from which untreated sewage was being added to the river?
b) With reference to the information in the table, explain your choice of answer.

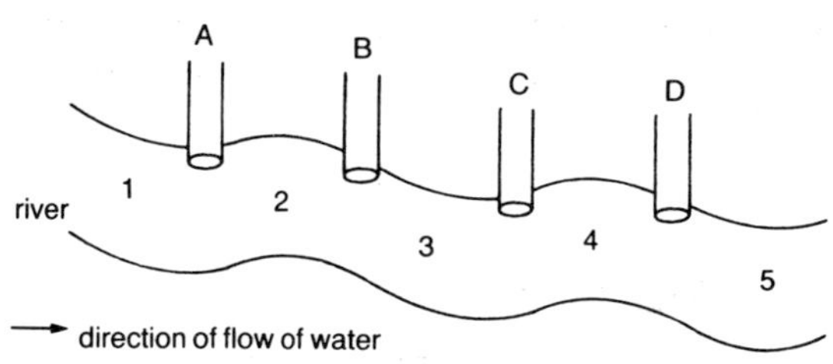

sample point	bacterial count	oxygen concentration	number of fish present
1	very low	very high	many
2	very high	very low	none
3	high	low	none
4	low	high	a few
5	very low	very high	many

4 Compare and contrast the two methods of biological treatment of sewage used in sewage treatment works as described on page 184.

5 The following table refers to the incidence of a certain water-borne disease in one of the world's developing countries.

period	annual number of cases per 1000 of population	death rate per 100 cases notified
1931–35	2.44	9.8
1936–40	2.39	9.7
1941–45	2.41	9.8
1946–50	2.42	9.6
1951–55	2.40	9.7
1956–60	1.27	6.5
1961–65	1.00	6.0
1966–70	0.96	4.7
1971–75	0.76	4.4
1976–80	0.61	4.1
1981–85	1.89	7.3

a) How many cases of the disease per 1000 of population were reported during period 1941–45?
b) During which period did a death rate of 9.6 per 100 cases occur?
c) The first significant change in the annual number of cases of the disease resulted from a programme of vaccination funded by the World Health Organisation. During which period of time did this change occur? Explain your choice of answer.
d) During the period 1981–85, the country's capital city (the main population centre) was affected by a series of earthquakes. What effect did this have on the annual number of cases of the disease per 1000 of population?
e) Suggest how the series of earthquakes can be linked to the change in disease incidence that resulted.

Extra Questions

f) (i) State the annual number of cases per 1000 of population during the period 1976–80.
(ii) Express this as a percentage of the number of cases per 1000 found to occur during 1931–35.
g) During the period 1961–65 the country's total population was 11 million.
(i) How many people suffered from the disease?
(ii) How many people died of the disease?

6 Some 'biological' soap powders contain an unusual enzyme (made by the bacterium *Bacillus subtilis*) which digests protein in conditions of 60 °C and pH 10.

Three different brands of soap powder were tested using pieces of fabric stained with blood. The results are shown in the following table.

	conditions in washing machine			
brand of soap powder	pH 7 30 °C	pH 7 60 °C	pH 10 30 °C	pH 10 60 °C
X	✓	×	×	×
Y	×	×	×	✓
Z	×	×	×	×

✓ = blood stain removed
× = blood stain not removed

a) Which brand of soap powder contains the enzyme made by *Bacillus subtilis*?
b) One of the other brands contains a different biotechnological product. Identify the brand and explain your choice of answer.
c) Which brand is the non-biological soap powder? Explain your answer.

7 The diagram below shows an experiment set up to investigate growth of bacteria. Single germinating spores of three different bacterial species (X, Y and Z) were placed on each of three nutrient agar plates. The plates were incubated at different temperatures for one week with the following results:

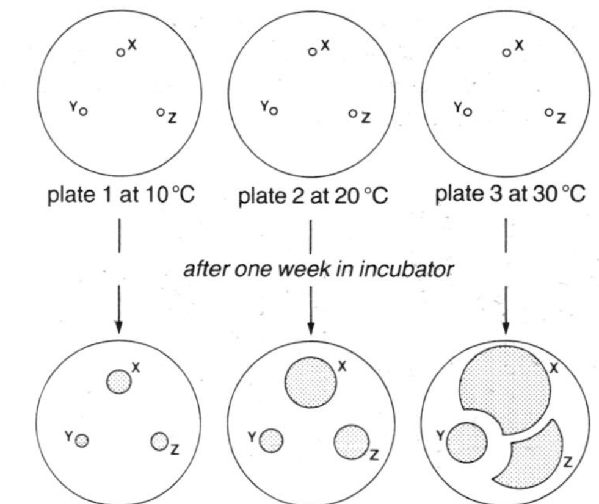

a) In which species was the rate of growth most stimulated by increasing the temperature?
b) Identify the species that makes an antibiotic which slows down growth of both of the other species.

Extra Questions

c) Which species fails to make an antibiotic able to slow down growth of either of the other two species?
d) Which species is sensitive to one of its neighbours and resistant to the other? Explain your choice of answer.

Appendix 1 Classification of animals

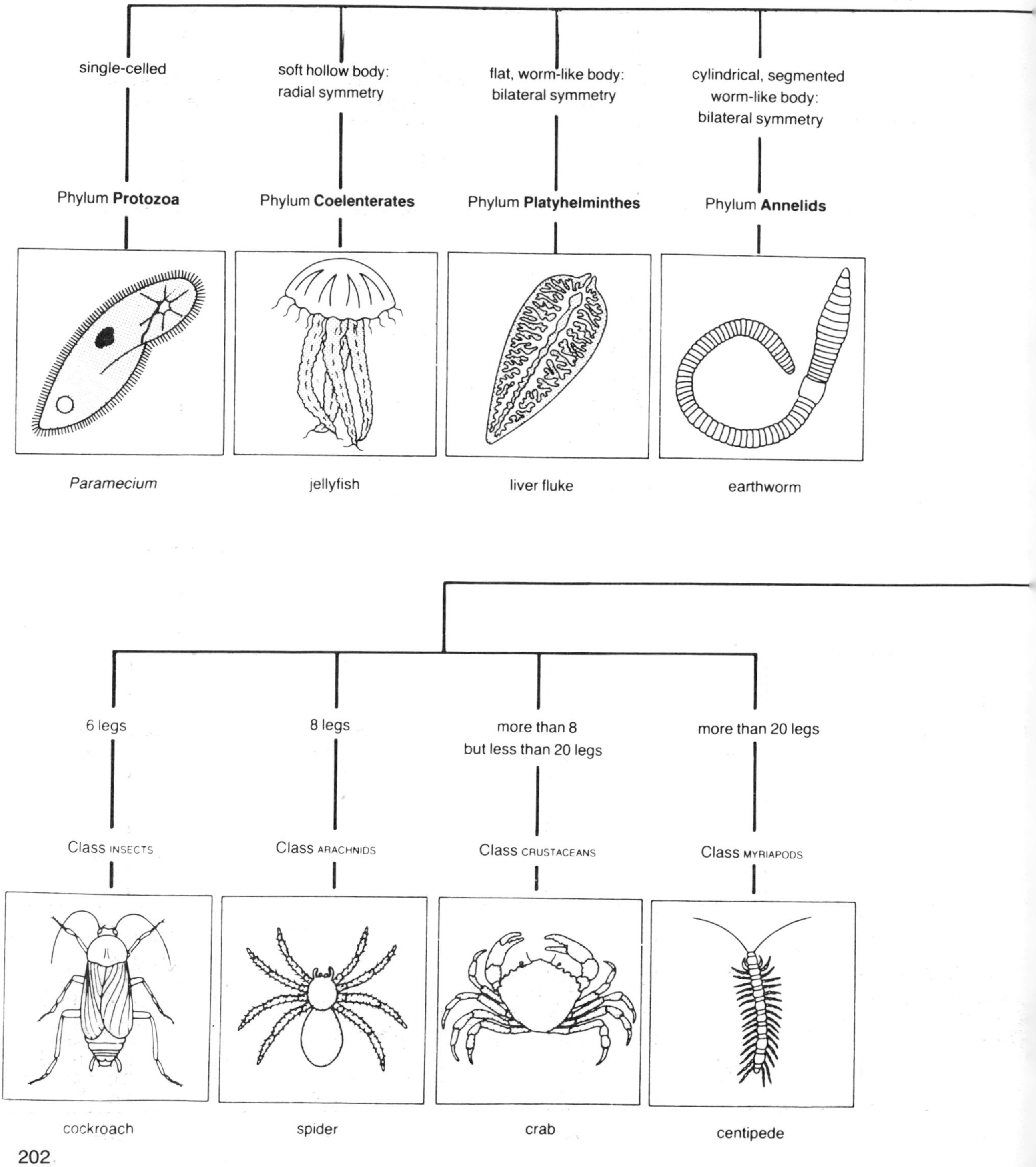

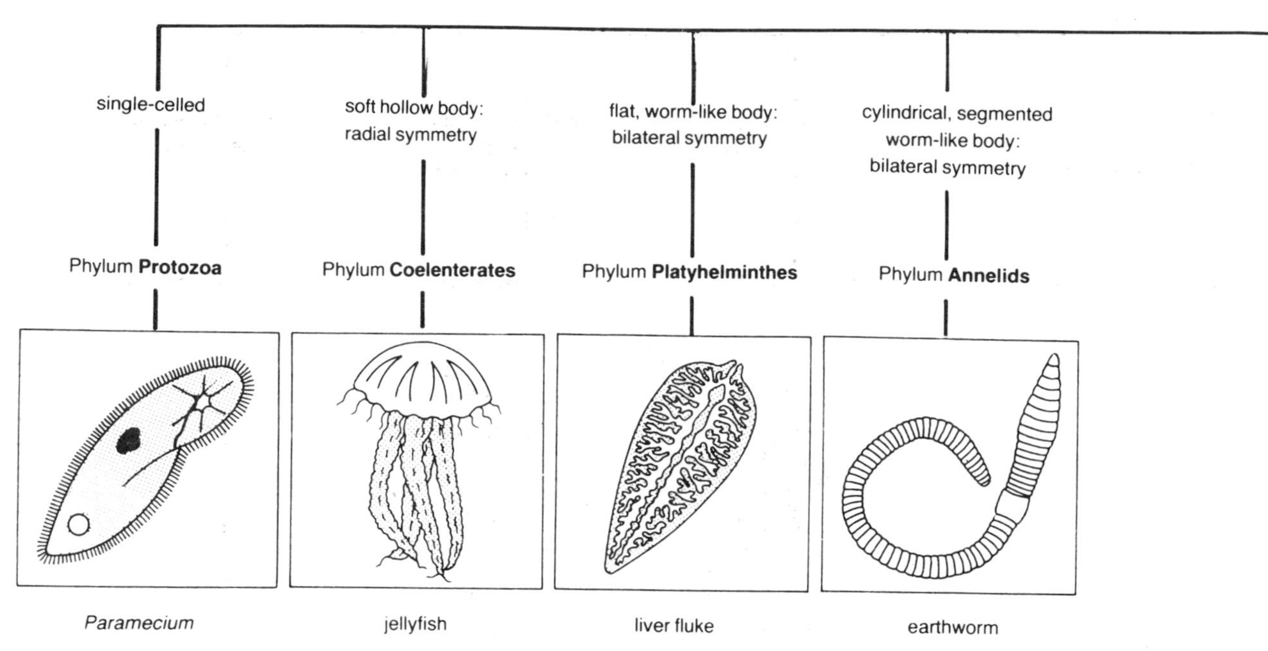

6 legs

8 legs

more than 8 but less than 20 legs

more than 20 legs

Class INSECTS

Class ARACHNIDS

Class CRUSTACEANS

Class MYRIAPODS

cockroach

spider

crab

centipede

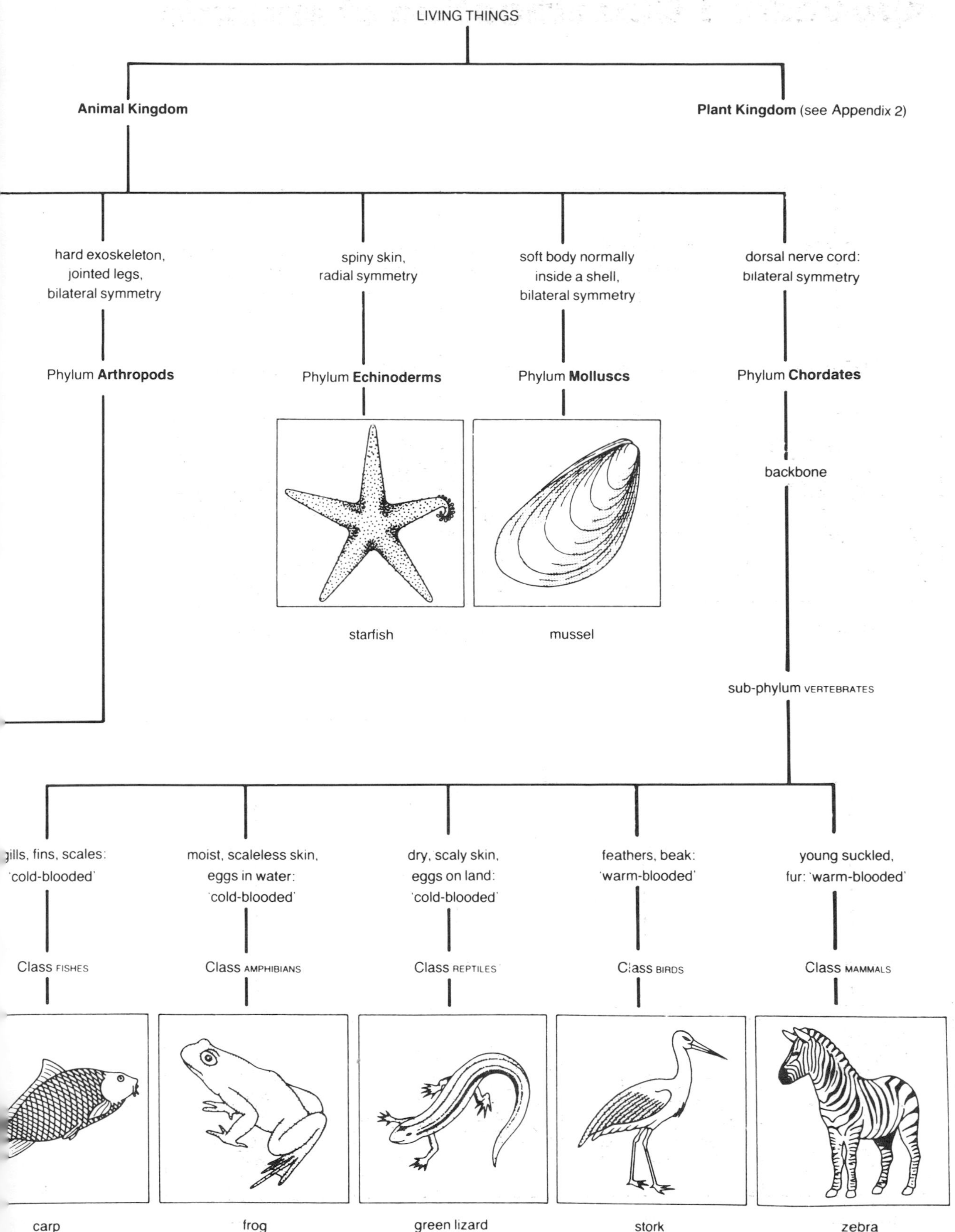
LIVING THINGS
Animal Kingdom
Plant Kingdom (see Appendix 2)
hard exoskeleton, jointed legs, bilateral symmetry
Phylum **Arthropods**
spiny skin, radial symmetry
Phylum **Echinoderms**
soft body normally inside a shell, bilateral symmetry
Phylum **Molluscs**
dorsal nerve cord: bilateral symmetry
Phylum **Chordates**
starfish
mussel
backbone
sub-phylum VERTEBRATES
gills, fins, scales: 'cold-blooded'
Class FISHES
moist, scaleless skin, eggs in water: 'cold-blooded'
Class AMPHIBIANS
dry, scaly skin, eggs on land: 'cold-blooded'
Class REPTILES
feathers, beak: 'warm-blooded'
Class BIRDS
young suckled, fur: 'warm-blooded'
Class MAMMALS
carp
frog
green lizard
stork
zebra

Appendix 2 Classification of plants

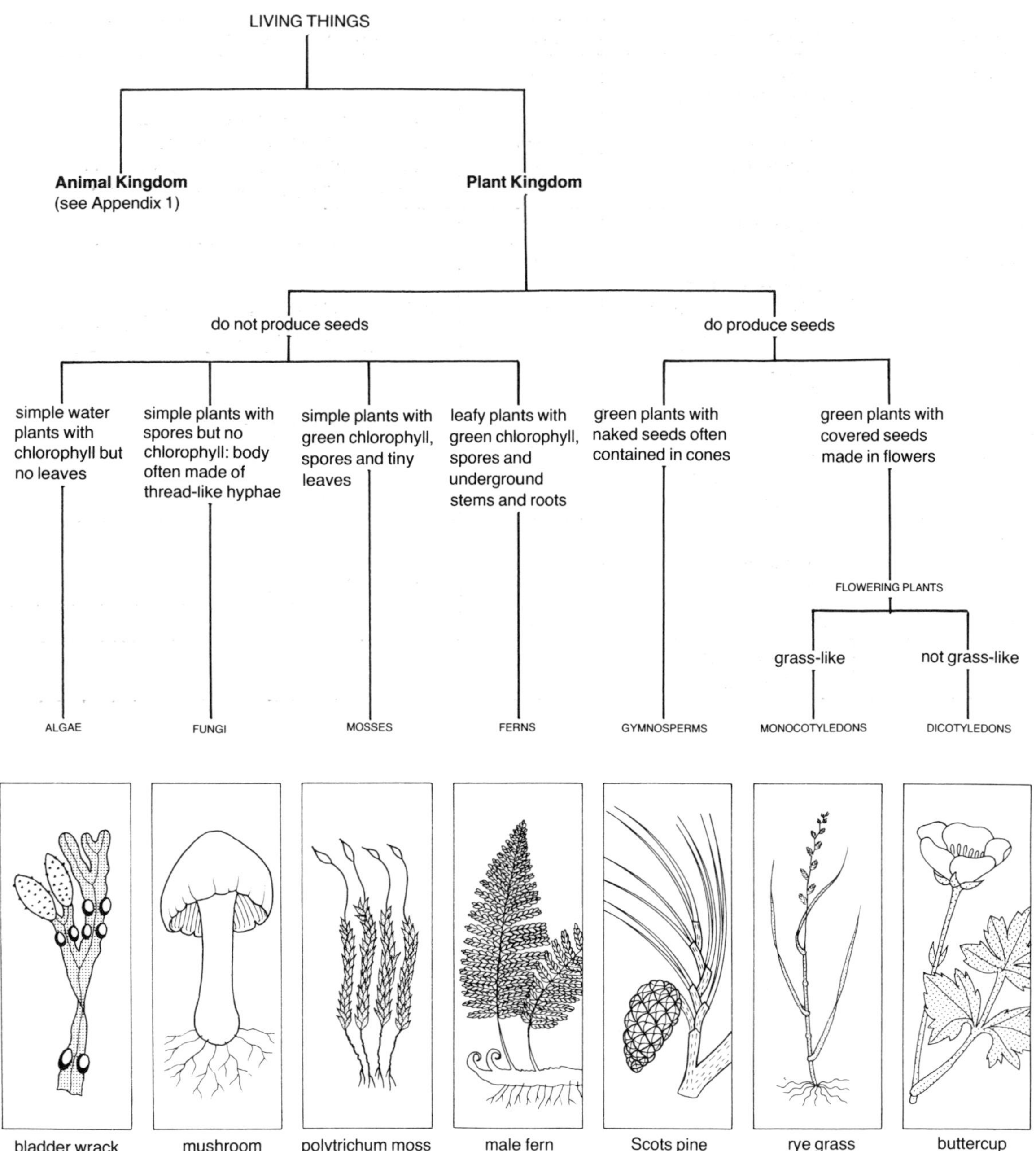

Appendix 3 Units of measurement

quantity	standard unit	symbol	relationships between units
length	**metre** other useful units: centimetre millimetre micrometre	m cm mm μm	 1 cm = 1/100th of a metre = 10^{-2} m 1 mm = 1/1000th of a metre = 10^{-3} m 1 μm = 1/1000th of a millimetre = 10^{-3} mm 1 μm = 1/1 000 000th of a metre = 10^{-6} m
area	**square metre**	m^2	
volume	**cubic metre** other useful units: cubic decimetre (litre) cubic centimetre	m^3 dm^3 cm^3	 1 dm^3 = 1/1000th of a cubic metre 1 cm^3 = 1/1000th of a cubic decimetre (litre)
mass	**kilogram** other useful unit: gram	kg g	 1 g = 1/1000th of a kilogram
energy	**joule** other useful unit: kilojoule	J kJ	 1 kJ = 1000 J
time	**second** other useful unit: minute	s min	 1 min = 60 seconds
force	**newton**	N	
temperature	**degree Celsius** (centigrade)	°C	

Appendix 4 Size scale of plants and micro-organisms

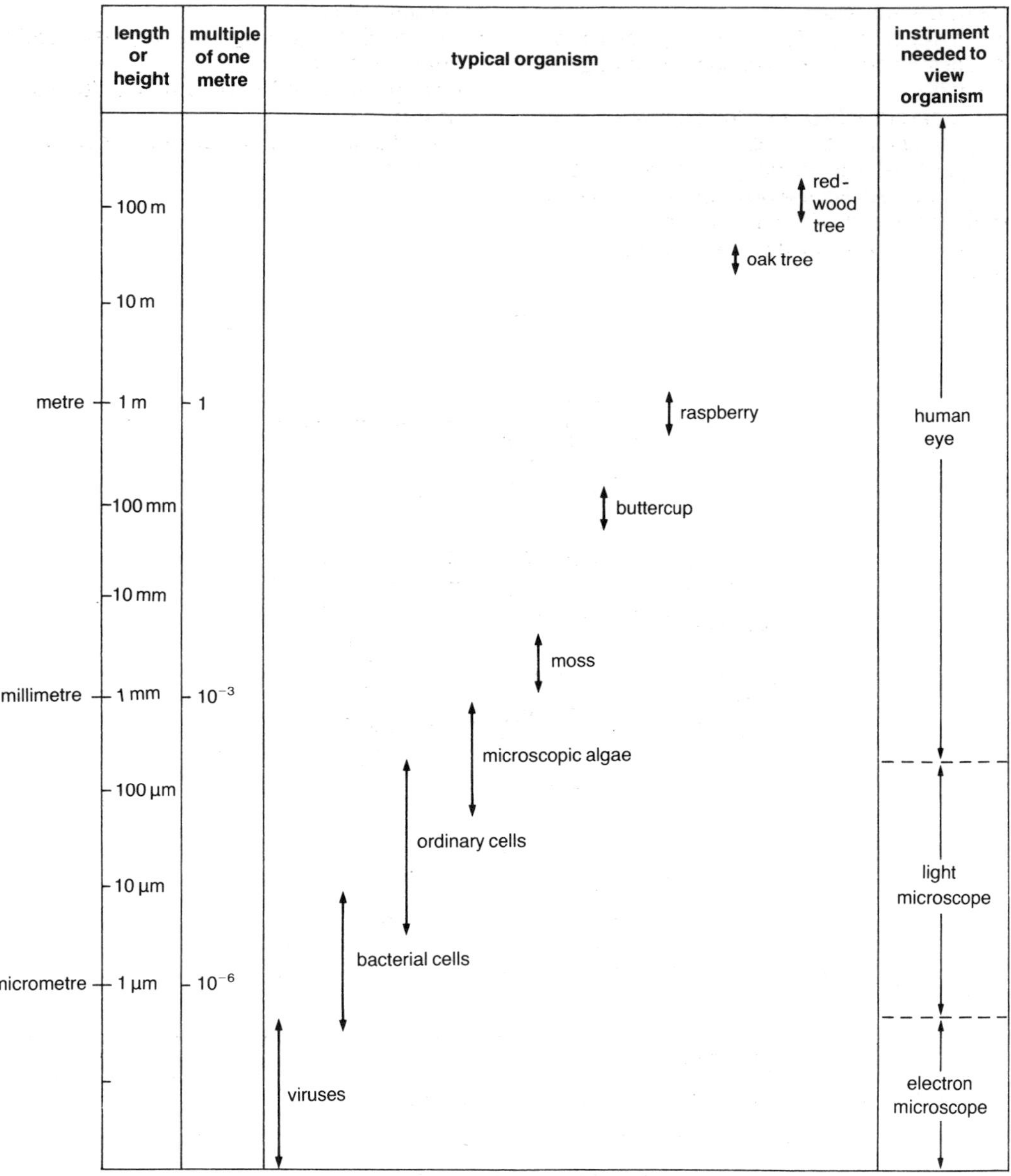

Appendix 5 Use of microscope

1 Turn the microscope's nosepiece (see figure A5.1) until the low power objective clicks into position above the hole in the stage.
2 Adjust the mirror until light (e.g. from a bench lamp but **not** direct sunlight) is seen to pass up through the microscope.
3 Place the prepared slide so that the specimen is in the centre of the hole in the stage.
4 With your eyes level with the stage, use the coarse adjustment knob to lower the objective to a position about 5 mm from the slide. (Note: in some microscopes the stage moves instead of the objective.)
5 Look down the microscope through the eyepiece and slowly raise the objective until the specimen comes into focus.
6 Use the fine adjustment knob to make slight changes in focus.
7 Change from low to high power by turning the nosepiece.
8 If in difficulty, consult table A5.1 or ask your teacher for help.

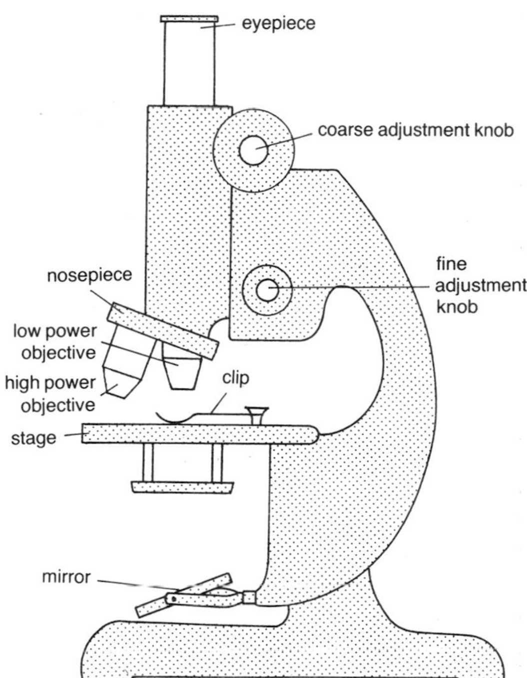

Figure A5.1 Microscope

difficulty	possible remedy
specimen cannot be found	check that specimen is in centre of hole in stage
image is very dark	adjust mirror to improve lighting
image is half light and half dark	rotate nosepiece to click objective into place
image is blurred	polish eyepiece lens and/or objective lens with lens tissue; remove, dry and clean coverslip then remount
many dark circles (air bubbles) are present	remove coverslip and remount by lowering it more slowly onto specimen

Table A5.1 Solving microscope difficulties

Appendix 6 Food tests

Testing for starch

1 Place a small sample of the food material to be tested on a white tile.
2 Using a dropper, add a few drops of **iodine solution** to the food.
3 Observe the result. If a **blue-black** colour is produced, then starch is present in the food.

Testing for reducing (simple) sugar

1 Pour a small sample of the food into a test tube.
2 Add a few drops of **Benedict's solution**.
3 Place the test tube in a water bath of boiling water for a few minutes as shown in figure A6.1.
4 Observe the result. If an **orange** (brick-red) precipitate is formed then reducing sugar (e.g. glucose or maltose) is present in the food.

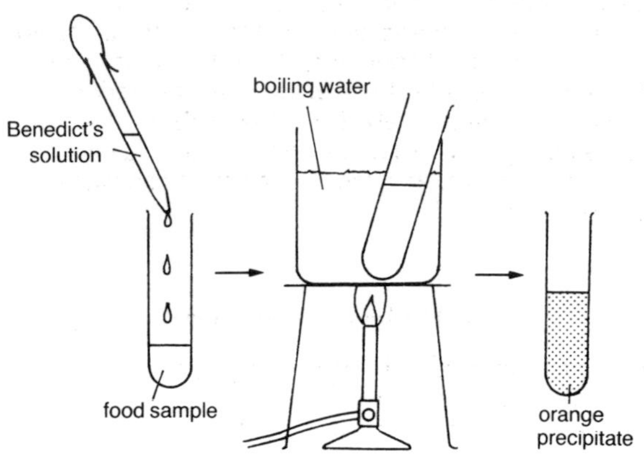

Figure A6.1 Testing for reducing sugar

Summary table

food being tested for	reagent used for test	original colour of testing reagent	heat or no heat required	colour resulting when food is present (positive result)
starch	iodine solution	brown	no heat	blue-black
reducing sugar	Benedict's solution	blue	heat	orange (brick-red)

Appendix 7 pH scale

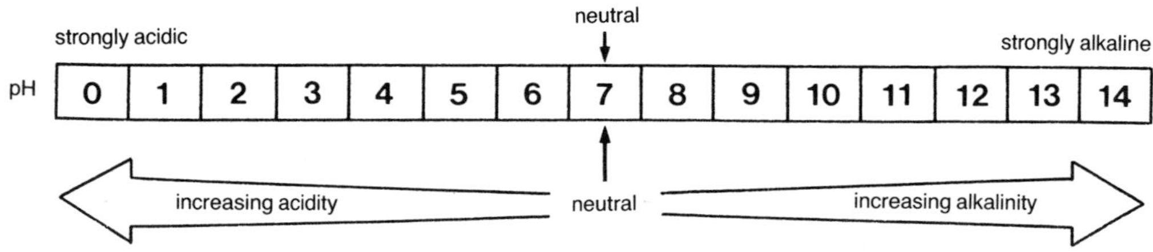

Appendix 8 Variable factors

Each of the pots shown in figure A8.1 contains the **same number** of seeds planted at the same **soil depth**. These words in bold print refer to factors which are constant in figure A8.1 but which could be varied. These are therefore called **variable factors**.

In a scientific investigation, a test is only fair if it deals with one variable factor at a time as shown in figures A8.2 and 3.

However if the parts of an experiment differ from each other by more than one variable factor the test is unfair. In figure A8.4 the pots differ from one another in two ways – seed number and depth of planting. This is an invalid test because it will be impossible to say whether the seeds that grow best have done so due to their number or their depth of planting.

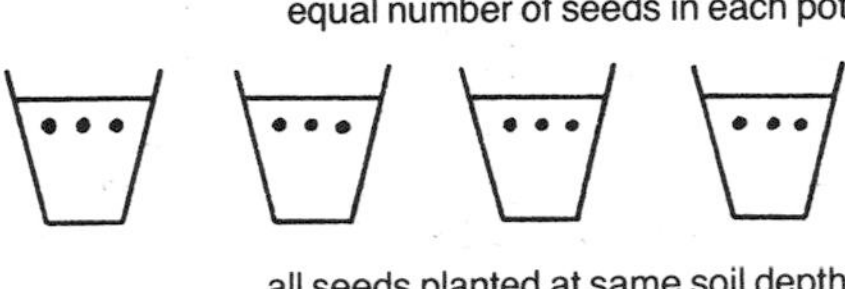

Figure A8.1 Variable factors kept constant

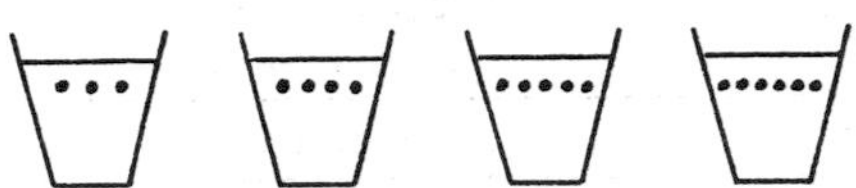

Figure A8.2 Seed number as variable factor

Figure A8.3 Depth of planting as variable factor

Figure A8.4 An unfair test

Appendix 9 Drawing a line graph

The axes

The axes are two straight lines – the horizontal (x) axis and the vertical (y) axis. They meet at the origin as shown in figure A9.1.

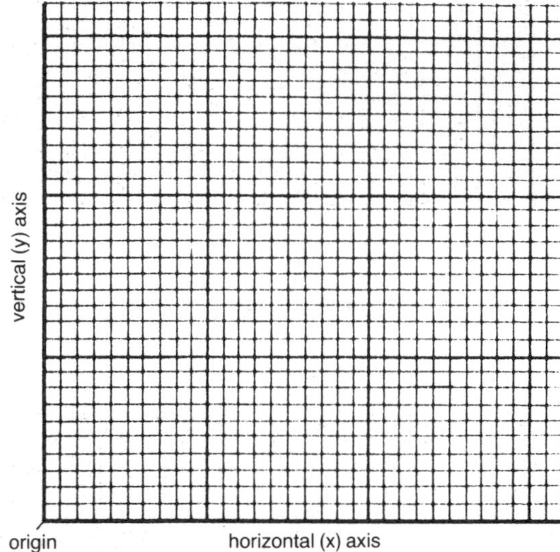

Figure A9.1 The two axes

temperature (°C)	number of oxygen bubbles released by water plant per minute
0	0
10	3
20	11
30	23
40	29
50	4

Table A9.1 Results to be graphed

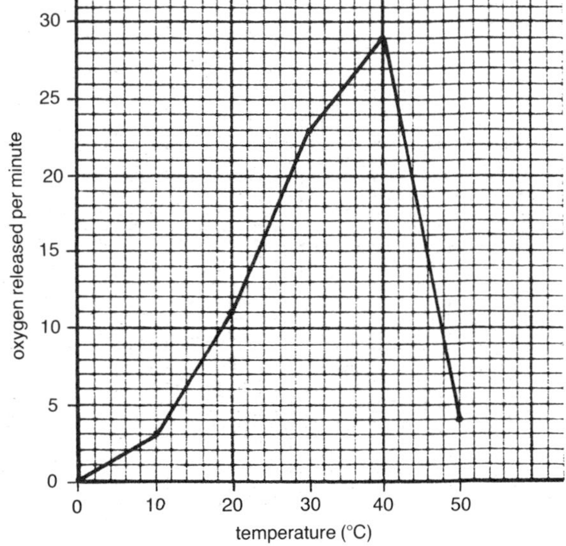

Figure A9.2 A line graph

Deciding which information to put on which axis

When presented with a set of results, decide which figures refer to the variable factor which was **under the control** of the investigator. In table A9.1, this was temperature. Other variable factors could be pH, CO_2 concentration, light intensity, length of time, distance, etc. In each case the investigator decides in advance which conditions of the one variable factor to include in the experiment. Put this variable factor on the x-axis in a line graph.

The other set of figures refers to the **actual results** of the experiment, i.e. the readings that were taken during the experiment. In table A9.1, this information was the number of bubbles of oxygen released per minute. Other examples could be volume of CO_2 released, number of organisms produced, decrease in amount of substrate, etc. Put this information on the y-axis in a line graph.

Figure A9.2 shows the information given in table A9.1 plotted as a line graph.

Choosing suitable scales

The information given in table A9.2 can be presented as the line graph shown in figure A9.3. However, this does not make the best use of the available space on the graph paper. It would be better to spread out the variable factor values (in this case time intervals) along the x-axis as shown in figure A9.4.

This line graph can be further improved. The lowest pulse reading is 70, therefore there is little point in having a large part of the y-axis used up by the scale 0–69. It would be much better to make the origin of the y-axis begin at 70 and spread out the readings as shown in figure A9.5.

time (min)	pulse rate (beats/min)
0	70
1	150
2	160
3	165
4	170
5	110
6	100
7	85
8	80
9	75
10	70

Table A9.2 Results to be graphed

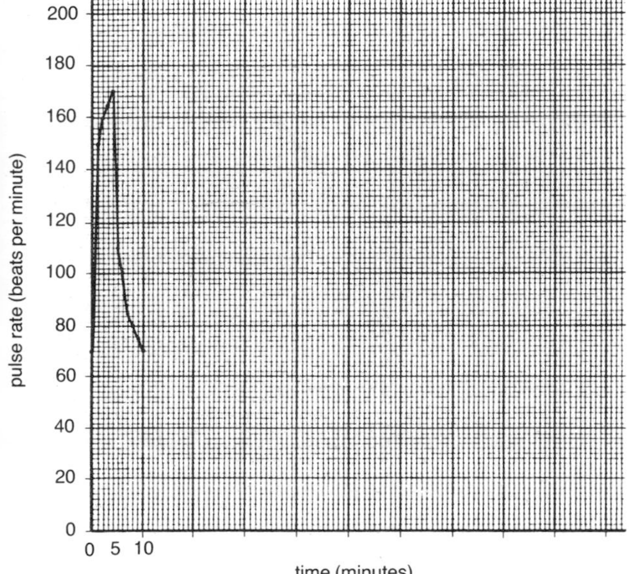

Figure A9.3 Poorly planned line graph

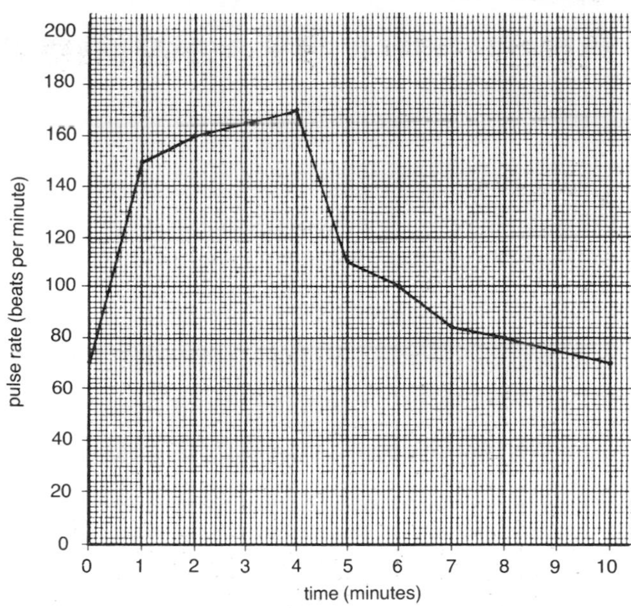

Figure A9.4 Improving the x-axis

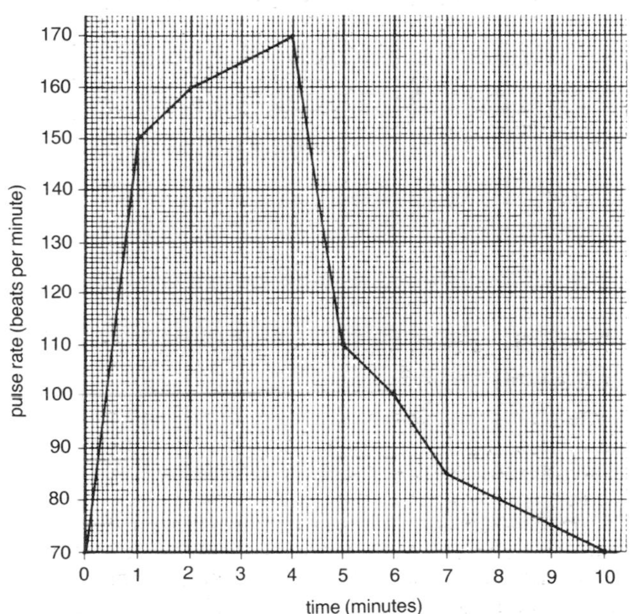

Figure A9.5 Improving the y-axis

Appendix 10 Constructing and testing a hypothesis

A scientist begins an investigation by making observations. He thinks carefully about each observation and then constructs a **hypothesis**. (A hypothesis is an untested idea that attempts to explain an observation using information based on known facts.) The scientist then tests the hypothesis by setting up an experiment. Consider the following example.

The plant shoot in figure A10.1 has been observed to bend towards light coming from one side only. To try to explain this observation, the scientist constructs a simple hypothesis: 'It is possible that the shoot bends because the region of shoot below the tip is sensitive to light.'

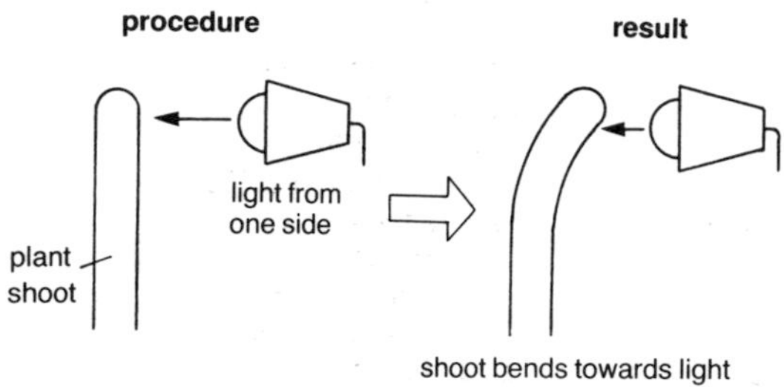

Figure A10.1 Making an observation

Having put forward this hypothesis, he makes a prediction: 'A shoot receiving light from one side only **at the region below the tip** will bend.' This prediction is then tested by an experiment as shown in figure A10.2, where the tip but not the region of shoot below is covered.

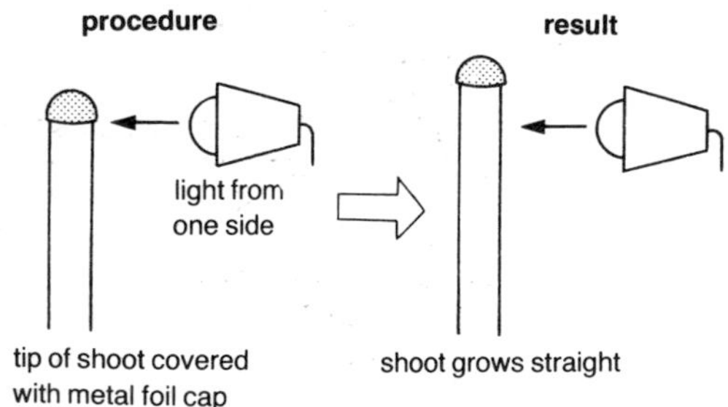

Figure A10.2 Test fails to support hypothesis

If the result does not confirm the prediction, as is the case here, then the hypothesis remains unsupported. It must be altered, or abandoned and replaced by a second hypothesis: 'It is possible that the shoot bends because the shoot tip is sensitive to light.'

The scientist now makes a new prediction: 'A shoot receiving light from one side only **at its tip** will bend.' He then puts this to the test as shown in figure A10.3, by leaving the tip uncovered but covering the region of shoot below it.

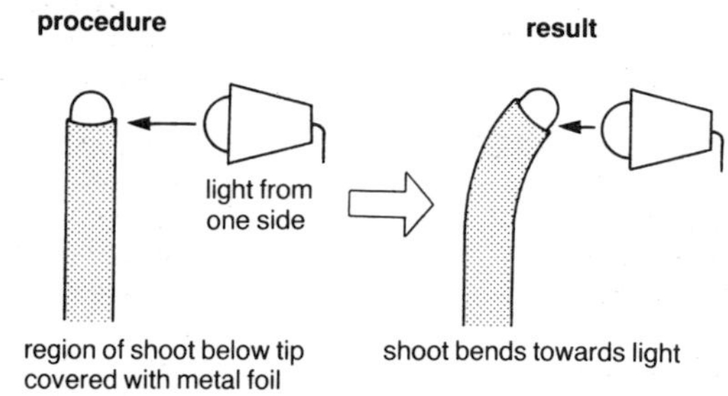

Figure A10.3 Test supports hypothesis

If the result does confirm the prediction, as is the case here, then the hypothesis is said to be supported and the scientist continues along this avenue of thought. The more experimental evidence that he can find to support the hypothesis, the more likely it is that the hypothesis is valid. This is the means by which all branches of science make progress.

Index